FIAS Interdisciplinary Science Series

The Frankfurt Institute for Advanced Studies (FIAS) is an independent research institute pursuing cutting-edge theoretical research in the areas of physics, life-science and chemistry, neuroscience, and computer science. A central aim of FIAS is to foster interdisciplinary co-operation and to provide a common platform for the study of the structure and dynamics of complex systems, both animate and inanimate. FIAS closely cooperates with the science faculties of Goethe University (Frankfurt) and with various experimentally oriented research centers in the vicinity. The series is meant to highlight the work of researchers at FIAS and its partner institutions, illustrating current progress and also reflecting on the historical development of science. The series comprises monographs on specialized current research topics, reviews summarizing the state of research in more broadly-framed areas, and volumes of conferences organized by FIAS.

David Vasak · Jürgen Struckmeier ·
Johannes Kirsch

Covariant Canonical Gauge Gravity

Springer

David Vasak
Frankfurt Institute for Advanced Studies
Frankfurt am Main, Hessen, Germany

Jürgen Struckmeier
Frankfurt Institute for Advanced Studies
Frankfurt am Main, Hessen, Germany

Johannes Kirsch
Frankfurt Institute for Advanced Studies
Frankfurt am Main, Hessen, Germany

ISSN 2522-8900 ISSN 2522-8919 (electronic)
FIAS Interdisciplinary Science Series
ISBN 978-3-031-43716-8 ISBN 978-3-031-43717-5 (eBook)
https://doi.org/10.1007/978-3-031-43717-5

This Springer imprint is published by the registered company Springer Nature Switzerland AG
The registered company address is: Gewerbestrasse 11, 6330 Cham, Switzerland

Preface

The idea for this book was born a few years ago, after we, the authors, had already published several articles on the canonical covariant gauge theory of gravity, or "CCGG" for short. Then, during the drafting phase, frequent adjustments in the content became necessary to incorporate new insights from our own research and from recently issued literature on the subject. Fortunately, Springer Publishers generously tolerated the resulting delays.

The goal was to find a formulation that is based on as few and as plausible assumptions as possible, but embedded in a resilient mathematical framework, and thus, unlike Einstein's derivation of his field equations, be less ambiguous and less dependent on speculation or random trial-and-error approaches. With the CCGG theory, this has been achieved and has led to an extension of Einstein's equations and thus to a generalized description of spacetime dynamics. However, the value of new theories only becomes apparent when compared with experimental data. To this end, the predictions of CCGG theory must, of course, be tested against selected cosmological observations. It turns out that torsion, a dynamical quantity that does not appear in Einstein's General Relativity and thus neither in conventional cosmology, has the potential to explain the previously not understood phenomenon of the expansion of the universe without the aid of a cosmological constant. This sounds promising, but the CCGG theory must pass many more tests, which are the subject of current research.

Our book is intended for readers familiar with the main concepts and methods of classical physics, special and general relativity, relativistic field theory, and quantum mechanics. For example, the calculus of variations, which plays a central role in the theory of canonical transformations, is used without explicit explanation of the procedure. We assume that senior students have internalized these topics. The same applies to mathematical knowledge of calculus, linear algebra, and topology. Basic knowledge of differential geometry would be helpful, but, we provide an introduction to this topic in the second chapter to allow readers with different prior knowledge to access it.

Many of the equations in this book are not readily understandable. We have derived some of them with the help of the mathematics software "Maple", a tool developed by

Maplesoft^{TM}, which has proved helpful and has also been used for verification, i.e., to check longer handwritten calculations. This reflects our experience that complex equations, such as those involving tensors of higher order, more often cannot be solved without unreasonable effort. Therefore, important longer calculations are listed explicitly, but in order not to interrupt the reading flow, these are placed within a gray box titled "Explicit calculation". Of course, it is up to the reader how to deal with this; for the overall understanding, those individual steps are not mandatory. Also, up to the reader is to study all chapters in consecutive order as some provide alternative formulations for specialists that might be skipped. Those chapters are marked as such.

In the course of the years during which the thoughts, concepts, and calculations reproduced here were developed, we had numerous discussions with colleagues and guests of our institute, to whom we express our sincere thanks for the valuable suggestions, in particular (in alphabetical order), David Benisty, Vladimir Denk, Eduardo Guendelman, Peter Otto Hess, Dirk Kehm, Adrian Königstein, Johannes Münch, Andreas Redelbach, Horst Stöcker, and Armin van de Venn. Furthermore, we would like to thank the Springer Publishers for their patience and good cooperation during the development of this book, the "Walter Greiner-Gesellschaft zur Förderung der physikalischen Grundlagenforschung e.V." (WGG) and the Fueck Stiftung for their financial support. A final word of thanks goes to our teacher and mentor Walter Greiner, who inspired and encouraged us over many years. We will always remember him as a friend and brilliant physicist.

Frankfurt am Main, Germany
July 2023

David Vasak
Jürgen Struckmeier
Johannes Kirsch

Contents

Chapter 1
Introduction

The theory of gravitation and matter as laid down by Einstein and Hilbert more than hundred years ago has since been facing a dilemma. On the one hand, it is a fairly accurate description of the macroscopic world, be it star formation, black holes and gravitational waves, or practical daily applications like satellite navigation and GPS. On the other hand, the mathematics of the Einstein-Hilbert theory fails to provide a pathway to the microscopic—quantum—world, nor does it satisfactorily account for many cosmological observations. Even though many alternative approaches have been investigated in the meantime, they all suffer from ambiguities and conceptual inconsistencies, and none has so far delivered any convincing and consistent theory.

The early accounts of gauge theories of classical (c-number) fields describing spacetime and matter have been carried out in the Lagrangian picture [1–12]. In contrast, our approach is based on the framework of covariant canonical transformation theory in the De Donder-Weyl-Hamiltonian picture, pioneered by Struckmeier and Redelbach [13]. Its generalization to a dynamic spacetime background is referred to as Canonical Covariant Gauge theory of Gravitation (CCGG). It is based just on four postulates:

- *Hamilton's Principle* or the Principle of Least Action holds, hence, the dynamics, i.e. the equations of motion of a system of classical physical fields, must be derived by variation of an action integral.
- *Regularity* (aka non-degeneracy) of the Lagrangian is mandatory for the Legendre transform from the Hamiltonian to the Lagrangian picture (and vice versa) to exist, i.e. the determinant of the associated Hesse matrix must not vanish. This ensures the applicability of the Hamiltonian canonical transformation theory.
- Einstein's *Principle of General Relativity* must hold, i.e. physics must be independent of the observer's frame. In the language of differential geometry that means the equations of motion must be independent of the atlas that maps the base manifold on $\mathbb{R}^4$. This is achieved if the Lagrangian is invariant under arbitrary coordinate transformations represented by the group of passive diffeomorphisms, $\mathrm{Diff}(M)$.

D. Vasak et al., *Covariant Canonical Gauge Gravity*, FIAS Interdisciplinary Science Series, https://doi.org/10.1007/978-3-031-43717-5_1

- The *Equivalence Principle* must apply, which means that *locally* the system's description in a spacetime frame must be equivalent to its description in an inertial frame and be invariant under the group of Lorentz transformations, $SO(1, 3)$.

The restriction to just four fundamental underlying assumptions is possible as the Hamiltonian framework enforces by construction the form-invariance of the action integral under canonical transformations. It thus provides a strong formal guidance to ensuring consistency of the emerging equations of motion. The validity of this approach was proven on conventional gauge theories, and shown to deliver from first principles the correct Hamiltonian for any $SU(N)$ gauge theory [14, 15]. Its generalization to scalar and vector fields in a dynamic spacetime background was published in Ref. [16].

This book is an extension of that previous work, aiming at the inclusion of spin-$1/2$ fields. In Chap. 2 we provide upfront a concise review of the underlying mathematics starting with the definition of spacetime as a Lorentzian manifold and the postulates of kinematics, and then listing the key physical quantities and implications arising with the formulation of the coupled system of dynamical matter in curved geometry. Throughout this book we retain the tensor calculus which is more complex to handle than the formulas of differential geometry, but easier to understand for physicists. We also present a useful but yet quite unknown theorem for absolute and relative tensors in Chap. 3 that constitutes the tensor analogue of Euler's theorem on homogeneous functions of calculus.

Chapter 4 is the key part of the book in which the framework of canonical transformations in the De Donder-Weyl Hamiltonian formalism is introduced and applied to scalar, vector and spinor fields in curved spacetimes. Crucial property of the matter fields that drives the formalism is their transformation behaviour w.r.t. the local symmetry group, $SO(1, 3) \times \text{Diff}(M)$. Requiring form invariance of the action integral upon those field transformations leads to the identification of the spin connection as a Yang-Mills type gauge field. The theory is then shown to unambiguously fix the spacetime-matter interaction for any matter field and any theory of gravity. Only in the next step do we specify the Lagrangians and Hamiltonians of matter (Klein-Gordon, Maxwell-Proca and Dirac fields) and gravity. A gravity Lagrangian compatible with the CCGG postulates must contain a term at least quadratic in the Riemann tensor. Based on analogy with other classical field theories, Einstein himself already proposed such a term [17], which will be discussed here as a "deformation" of to the conventional Hilbert Lagrangian which, with its linear, Ricci scalar term, is compatible with solar-scale observations. By Occam's razor we thus choose an extended, quadratic-linear ansatz for gravity, and derive a set of field equations for coupled matter and dynamic spacetime represented by the curvature and torsion tensors. Remarkably, the field equation emerging from this Lagrangian is equally satisfied by the Schwarzschild metric in the case of classical vacuum [18]. That generalizes the "affine space" of the Poincaré gauge theory formulated in the Lagrangian picture [5, 6, 19]. However, in the Hamiltonian formulation presented here, many ambiguities inherent to that Lagrangian formulation [7] are resolved [20].

In Chap. 5 we include an alternative, so called spinorial representation of the gauge formalism, that is aimed at specialists interested in that particular approach. For simplicity we restrict in the derivation the variety of matter fields to Dirac spinors only, but show in a straightforward analogy how to generalize the result to spin-1/3 fields. This chapter can be skipped without compromising the understanding of the following.

In Chap. 6 a derivation of the field equations along the lines of the Noether theorem is presented following Ref. [21]. Here we alternatively drop the spinor field and show how the vierbein formalism can be converted into the coordinate one, and how after variation the modification of the independent spacetime fields modifies their physical content.

In Chap. 7 we prove that the pure quadratic gravity approach admits, in addition to the Schwarzschild metric, further spherically symmetric, radially oscillating or time-dependent solutions, that restrict the general validity of Birkhoff's theorem, regardless of their physical relevance. The existence of the Schwarzschild metric as solution of the quadratic-linear ansatz of CCGG is proven, though, with the oscillating and time-dependent solution excluded by the linear term.

The last chapter, Chap. 8, is devoted to the cosmological implications of CCGG. We analyze the impact of the modified gravity ansatz within the FLRW geometry, aligned with Refs. [22–25]. Thereby torsion is shown to facilitate a dark energy effect in the late Universe, substituting the cosmological constant.

References

1. H. Weyl, Eine neue Erweiterung der Relativitätstheorie. Annalen der Physik IV Folge **59**, 101 (1919). https://doi.org/10.1002/andp.19193641002
2. Albert Einstein, *The Meaning of Relativity* (Princeton University Press, Princeton, 1955)
3. Yang, Mills, Conservation of isotopic spin and isotopic gauge invariance. Phys. Rev. **96**, 191 (1954)
4. D.W. Sciama, The analogy between charge and spin in general relativity. *Recent Developments in Festschrift for Infield*. (Pergamon Press, Oxford; PWN, Warsaw, 1962), pp. 415–439
5. T.W.B. Kibble, Lorentz invariance and the gravitational field. J. Math. Phys. **2**, 212–221 (1961). (Mar.)
6. R. Utiyama, Invariant theoretical interpretation of interaction. Phys. Rev. **101**(5), 1597 (1956)
7. F.W. Hehl, Four lectures on Poincaré gauge field theory, in *Cosmology and Gravitation: Spin*, ed. by P.G. Bergmann, V. De Sabbata. (Springer, Boston, MA, US, 1980), pp. 5–61. https://doi.org/10.1007/978-1-4613-3123-0_2. arXiv: 2303.05366 [gr-qc]
8. K. Hayashi, T. Shirafuji, Gravity from the Poincaré gauge theory of fundamental interactions. Prog. Theor. Phys. **64**(3), 866, 883, 1435, 2222 (1980). https://doi.org/10.1143/PTP.64.866
9. K. Hayashi, T. Shirafuji, Gravity from Poincaré gauge theory of the fundamental particles II. Prog. Theor. Phys. **64**(3), 883 (1980)
10. K. Hayashi, T. Shirafuji Gravity from Poincaré Gauge theory of the fundamental particles III. Prog. Theor. Phys. **64**(3), 866 (1980)
11. K. Hayashi, T. Shirafuji, Gravity from Poincaré Gauge theory of the fundamental particles IV. Prog. Theor. Phys. **64**(3), 2222 (1980)
12. K. Hayashi, T. Shirafuji, Gravity from Poincaré Gauge theory of the fundamental particles V. Prog. Theor. Phys. **65**(3), 525 (1981)

13. J. Struckmeier, A. Redelbach, Covariant hamiltonian field theory. Int. J. Mod. Phys. E **17**, 435–491 (2008). https://doi.org/10.1142/s0218301308009458. (arXiv: 0811.0508)
14. J. Struckmeier, H. Reichau, General U(N) gauge transformations in the realm of covariant Hamiltonian field theory, in *Exciting Interdisciplinary Physics*, ed. by W. Greiner. FIAS Interdisciplinary Science Series. (Springer International Publishing Switzerland, 2013, p. 367). ISBN: 978-3-319-00046-6. https://doi.org/10.1007/978-3-319-00047-3_31. arXiv: 1205.5754. Accessed from 29 May 2013
15. J. Struckmeier, D. Vasak, H. Stoecker, Covariant Hamiltonian representation of Noether's therorem and its application to SU(N) gauge theories, in *New Horizons in Fundamental*, ed. by S. Schramm, M. Schaefer. FIAS Interdisciplinary Science Series. (Springer International Publishing Switzerland, 2017). ISBN: 978-3-319-44165-8. https://doi.org/10.1007/978-3-319-44165-8. https://arxiv.org/abs/1608.01151
16. J. Struckmeier et al., Canonical transformation path to gauge theories of gravity. Phys. Rev. D **95**, 124048 (2017). https://doi.org/10.1103/PhysRevD.95.124048. arXiv: 1704.07246
17. Albert Einstein, *Private Letter to Hermann Weyl* (ETH Zürich Library, Archives and Estates, 1918)
18. D. Kehm et al., Violation of Birkhoff's theorem for pure quadratic gravity action. Astron. Nachr./AN **338**(9–10), 1015–1018 (2017). https://doi.org/10.1002/asna.201713421
19. F.W. Hehl, Gauge theory of gravity and spacetime (2014). https://doi.org/10.1007/978-1-4939-3210-8_5. arXiv:1204.36722 [gr-qc]
20. C-M. Chen, J.M. Nester, R-S. Tung, Gravitational energy for GR and Poincaré gauge theories: a covariant Hamiltonian approach. Int. J. Mod. Phys. D **24**(11), 1530026 (2015). https://doi.org/10.1142/S0218271815300268. arXiv: 1507.07300 [gr-qc]
21. J. Struckmeier, D. Vasak, J. Kirsch, Generic theory of geometrodynamics from Noether's theorem for the Diff(M) symmetry group, in *Discoveries at the Frontiers*, ed. by J. Kirsch et al. (Springer Nature Switzerland AG, 2020, pp. 143–181). https://doi.org/10.1007/978-3-030-34234-0_12. arXiv: 1807.03000 [gr-qc]
22. D. Vasak, J. Kirsch, J. Struckmeier, Dark energy and inflation invoked in covariant canonical gauge theory of gravity (CCGG) by locally contorted space-time. Eur. Phys. J. Plus **135**, 404 (2020). https://doi.org/10.1140/epjp/s13360-020-00415-7
23. D. Vasak et al., On the cosmological constant in the deformed Einstein-Cartan gauge gravity in De Donder-Weyl Hamiltonian formulation. Astron. Nachr. (2022). https://doi.org/10.1002/asna.20220069,2022. arXiv: 2209.00501 [gr-qc]
24. A. van de Venn et al. , Torsional dark energy in quadratic gauge gravity. Eur. Phys. J. (2023). https://doi.org/10.1140/epjc/s10052-023-11397-y. arXiv: 2211.11868 [gr-qc]. Accessed from 11 Apr 2023
25. J. Kirsch et al., Torsion driving cosmic expansion. Eur. Phys. J. C (2023). https://doi.org/10.1140/epjc/s10052-023-11571-2. https://link.springer.com/content/pdf/10.1140/epjc/s10052-023-11571-2.pdf

Chapter 2
Relativistic Space-Times

Our formulation of the field theory of matter in curved dynamical spacetime relies on the vierbein concept that is needed to accommodate matter with half-integer spin. The term tetrad or vierbein goes back to the German expression *begleitendes Vierbein* (accompanying four-leg) for a system of orthonormal basis vectors following a curve embedded in a curvilinear four-dimensional[1] Euclidian space. Applying this concept in pseudo-Riemannian and Riemann-Cartan geometries for an elegant description of spacetime requires insight into its foundations and rigorous definitions as laid out in the framework of differential geometry. Hence, in order to settle on a common understanding of the concept of curved spacetime, we first revisit the relevant definitions and relations. For more details we have to refer to textbooks on differential geometry (for example [1, 2]).

Our conventions in this paper are as follows: The metric's signature is defined as $(1, 3)$, hence the signs of its eigenvalues are $(+, -, -, -)$, and natural units with $\hbar = c = 1$ are used. For the definitions of connection, curvature, torsion etc. we follow the prescription of Misner, Thorne, Wheeler [3]. Denoting the dimension of physical quantities by the symbol [.], the dimensions of mass m, length L, time T, and energy E are then related as follows:

$$[m] = [L]^{-1} = [T]^{-1} = [E].$$

[1] In n dimensions the vierbein is generalized to "n-bein" or "vielbein" where "viel" stands for "many" in German.

D. Vasak et al., *Covariant Canonical Gauge Gravity*, FIAS Interdisciplinary Science Series, https://doi.org/10.1007/978-3-031-43717-5_2

2.1 Basic Concepts and Relations

In this chapter we summarize the basic concepts linking differential geometry, tensor calculus and physics. We review the definition of spacetime, points, coordinates, frames, vectors, velocities, and tensors, and show what straight lines are and how acceleration and force are defined in curvilinear spaces. This allows to generalize the notion of particle trajectories with respect to arbitrary observers and thus provides a generalized understanding of Newton's laws of kinematics and dynamics.

2.1.1 *Time-Orientable Lorentzian Manifold*

Our starting point will be the following definition of a general relativistic (Riemann-Cartan-Weyl) spacetime in which neither torsion nor non-metricity are *a priori* excluded. It is the so-called *Lorentzian manifold* $(M, \mathcal{O}, \mathcal{A}^{\uparrow}, \nabla, \mathfrak{g}, \mathbb{T})$, consisting of the set M of "points" equipped with several structural elements:

$\mathcal{O}$ is a (Hausdorff) topology enabling the definition of continuity of functions on M.

$\mathcal{A}^{\uparrow}$ denotes a smooth atlas that consists of charts (U_i, x_i) with open sets $U_i \subseteq M$ and maps $x_i : U_i \to \mathbb{R}^d$ that cover the entire manifold, i.e. $\bigcup_i U_i = M$. The integer d denotes the dimension of the manifold M, hence $d = 4$ for a four-dimensional spacetime. Coordinate transformations, hence transition maps between any two overlapping charts, $x_i \circ x_j^{-1} : \mathbb{R}^4 \to \mathbb{R}^4$, are differentiable and invertible. Here the indices i, j refer to the charts and maps. At each point $p \in M$ one defines the T_pM as the space of all tangents of curves ζ_p through p, see Fig. 2.1 illustrating a 2-dimensional example. The disjoint union of all tangent spaces, $TM := \dot{\bigcup}_p T_pM$ on M, equipped with a projection map $\pi : TM \to M$, is referred to as the . The arrow in $\mathcal{A}^{\uparrow}$ indicates that for this special manifold we require the atlas to be "oriented", i.e. restricted to charts (U_i, x_i) with transition maps satisfying a positive Jacobi determinant $\det\left(\partial x_i / \partial x_j\right) > 0$ to ensure a unique definition of an integral over the entire manifold.

∇ is a covariant derivative enabling the definition of (auto)parallel transport and of "straight" lines, and is unambiguously defined by the choice of an "" γ.

$\mathfrak{g}$ is the non-degenerate, symmetric , a smooth, non-singular $(0, 2)$-tensor field defined everywhere on the tangent bundle, $\mathfrak{g} : TM \times TM \to \mathbb{R}$, enabling the definition of length, angle, distance and geodesics. (A relation between the independent structural elements metric and connection is enforced only by assigning a length to a straight line.)

$\mathbb{T}$ is a smooth vector field that does not vanish anywhere on the bundle TM and defines the global "time" direction by the requirement $\mathfrak{g}(\mathbb{T}, \mathbb{T}) > 0$. This structural

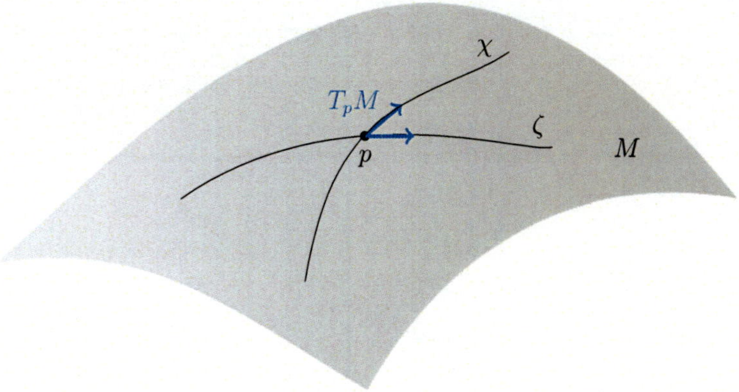

Fig. 2.1 The tangent vectors at different curves at a point p span the tangent space T_pM.

element provides *time-orientability* or *stable causality* [4, 5], i.e. ensures that a time direction can be defined and past and future stably distinguished.[2]

The elements of M are points. In the underlying philosophy, points are physical entities, their coordinates are not. They are just labels assigned to points via a specific choice of an atlas. And since that choice is arbitrary, a physical theory must be invariant with respect to that choice. (See Sect. 2.3 for the discussion of coordinate transformations.)

2.1.2 Velocities, Vectors and Co-vectors

A key concept in differential geometry is the along a parametrized curve $\zeta(\lambda)$ within a manifold. The *Velocity* alongside a curve $\zeta(\lambda)$ through the point $p \in M$ is the map

$$v_{\zeta,p} : C^\infty(M) \to \mathbb{R}$$

from the vector space of smooth real functions on M,

$$C^\infty(M) = \{f : M \to \mathbb{R}, \text{ smooth}\},$$

given by

$$f \mapsto v_{\zeta,p}(f) := (f \circ \zeta)'(0).$$

[2] This field (coined aether in the Einstein-Aether gravity, or khronon in a scalar field realization in the Hořava-Lifshitz gravity) can always be constructed in globally hyperbolic spacetimes where a global foliation is possible. For the discussion of the various implementations of causality see also [5–7].

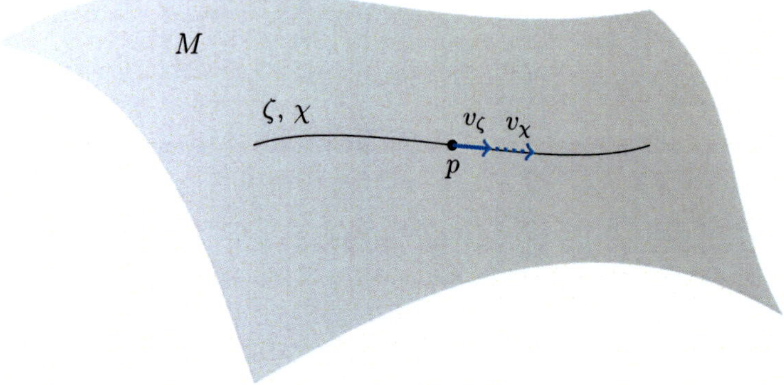

Fig. 2.2 An example demonstrating the relationship of the parameterization with the velocity: $v_\chi(f)|_p = (f \circ \chi)'(0) = 2(f \circ \zeta)'(0) = 2v_\zeta(f)|_p$, where we set $\chi(\lambda) = \zeta(2\lambda)$.

The prime denotes the derivative with respect to the curve parameter λ, whereby $\zeta(0) \equiv p$. Obviously, the definition of velocity is not unambiguous as it depends on the choice of the parametrization, see Fig. 2.2. Interpreting the functions f on M as some kind of measurable "signals", the physical "world-line" of a particle is identified through *changes* of any possible signals along the points on its trajectory. Let now (U, x) be a chart in $\mathcal{A}^\uparrow$ on M with $p \in U$ and $x : M \supseteq U \to \mathbb{R}^4$, mapping an open set of points in the manifold on an open set in $\mathbb{R}^4$. On a specific chart we can now identify the velocity vector with the directional derivative along the curve $\zeta(\lambda)$ at p:

$$\begin{aligned} v_{\zeta,p}(f) &= \frac{}{\lambda}(f \circ \zeta(\lambda)) = \frac{}{\lambda}\left((f \circ x^{-1}) \circ (x \circ \zeta(\lambda))\right) \\ &= \frac{}{\lambda}[f(x^\mu(\zeta(\lambda)))] =: [v^\mu(x(\lambda))\,\partial_\mu] f(x), \end{aligned} \tag{2.1}$$

where it is understood that $\lambda = 0$ is set after the differentiation. In the last step we abbreviate $x^\mu(\lambda) \equiv x^\mu(\zeta(\lambda))$ and apply the chain rule of standard differential calculus on $\mathbb{R}^4$. $\partial_\mu \equiv \partial/\partial x^\mu$ with $\mu = 0, 1, 2, 3$ constitute here the so called on T_pM.

An arbitrary vector $V \in T_pM$, can be expanded in that standard coordinate basis of T_pM at that point by

$$V = V^\mu \partial_\mu.$$

$V^\mu(x)$ are the vector's components at $x = x(p)$ on the chart (U, x). For the $\mathfrak{g}$, which is a (0, 2) tensor field, a similar expansion is defined on the chart (U, x) by

$$\mathfrak{g}_{(x)}\left(\partial_\mu, \partial_\nu\right) =: g_{\mu\nu}(x).$$

$g_{\mu\nu}(x) \in \mathbb{R}$ are the (dimensionless) components of the metric on the chart in question. The conditions for a world-line to be "time-like" and "future-directed" is expressed in chart coordinates ($\mu, \nu = 0, 1, 2, 3$, and the Einstein summation rule applies) as

$$\mathrm{g}_{(x)}\left(v_p, v_p\right) = \mathrm{g}_{(x)}\left(v_p^\mu\, \partial_\mu, v_p^\nu\, \partial_\nu\right) = v_p^\mu\, v_p^\nu\, \mathrm{g}_{(x)}\left(\partial_\mu, \partial_\nu\right) = v^\mu(x)\, v^\nu(x)\, g_{\mu\nu}(x) > 0 \tag{2.2a}$$

$$\mathrm{g}_{(x)}\left(\mathbb{T}, v_p\right) = T^\mu(x)\, v^\nu(x)\, g_{\mu\nu}(x) > 0. \tag{2.2b}$$

The basis vectors $\{e_0, e_1, e_2, e_3\}$ attached to each point of the manifold constitute an alternative, orthonormal frame in TM, the *Inertial space*. Obviously, as "orthonormal" refers to the given metric g,

$$\mathrm{g}\left(e_i, e_j\right) = \mathrm{g}_{(x)}\left(e_i^{\ \mu}\, \partial_\mu, e_j^{\ \nu}\, \partial_\nu\right) = e_i^{\ \mu}(x)\, e_j^{\ \nu}(x)\, g_{\mu\nu}(x) = \eta_{ij}, \tag{2.3}$$

with the Minkowski metric $\eta_{ij} = \mathrm{diag}(1, -1, -1, -1)$. The (dimensionless) coefficients $e_j^{\ \nu}$, referred to as *Vierbeins* or *tetrads*, are the components of the orthonormal frame vectors in the coordinate basis. By this choice of the basis, the postulated global metric g is, via (2.3), pushed forward to the inertial frame and the standard Minkowski metric. The vierbeins thus transform in the Greek index as a vector in the coordinate frame, while the Latin index refers to the flat inertial frame. They are also called "soldering forms", and contribute an additional structural element to the Lorentzian spacetime. By this definition the Lorentzian manifold is a so called *time-orientable orthonormal frame bundle.*[3]

The dual (co-tangent) space

$$T_p^* M = \left\{V : T_p M \to \mathbb{R}\right\}$$

is isomorphic to $T_p M$, and the co-vectors $\mathrm{d}x^\mu$ constitute a basis of the . They satisfy the condition

$$\mathrm{d}x^\mu(\partial_\nu) = \delta_\nu^\mu. \tag{2.4}$$

Any co-vector (or "1-form") $V \in T_p^* M$ can be expanded in that basis via $V = V_\mu\, \mathrm{d}x^\mu$. The dual basis of the frame bundle $\{\varepsilon^0, \varepsilon^1, \varepsilon^2, \varepsilon^3\}$ satisfies similarly the condition

$$\varepsilon^j(e_i) = \delta_i^j. \tag{2.5}$$

[3] The frame bundle is a principal G-bundle with the group G being the Lorentz group. The frame induces a global section $M \mapsto TM$, which makes the bundle to a trivial bundle, i.e. a direct product of the base manifold (the spacetime), and a local orthonormal frame $\{e_i^{\ \mu}\}$, the inertial observer's spacetime of Special Relativity with Minkowski metric at each point $p \in M$. For the description of matter fields on curvilinear spacetime, the bundle is extended to an associated frame bundle that has a fiber F attached to every point of the bundle. That fiber is a complex tensor space. Its basis can be arranged in such a way that the group elements will be in general composed of box-diagonal matrices where each $n \times n$ box corresponds to an irreducible representation of the Lorentz group. For $n = 1$ the corresponding representation of the Lorentz group is $\mathbb{1}$, i.e. the corresponding subspace of the fiber represents complex scalar fields, for $n = 2$ we deal with spinors, etc. Notice that in [8] the construction of local inertial frames and gauge invariance has been linked to translations in the *affine* bundle AM on a manifold M, and the diffeomorphisms on M.

The co-vector ε^a, a 1-form, maps the vector e_a on a scalar, i.e. $\varepsilon^a : TM \rightarrow \mathbb{R}$. The components of that basis in the co-tangent basis are the *dual vierbeins* $\varepsilon^i{}_\mu(x)$:

$$\varepsilon^i(x) = \varepsilon^i{}_\mu(x)\ \mathrm{d}x^\mu.$$

From Eqs. (2.5) and (2.4) follows

$$\delta_i^j = \varepsilon^j{}_\mu\, e_i{}^\mu.$$

Equation (2.3) can now be reversed by multiplying it twice with the dual vierbeins:

$$g_{\mu\nu}(x) = \varepsilon^i{}_\mu(x)\, \varepsilon^j{}_\nu(x)\, \eta_{ij}. \tag{2.6}$$

It is straightforward to see that the dual vierbeins can be constructed from the vierbeins via:

$$\varepsilon^i{}_\mu(x) = g_{\mu\nu}(x)\, \eta^{ij}\, e_j{}^\nu(x). \tag{2.7}$$

As we will see below, it is equivalent and more convenient to use the dual vierbein field for the orthonormal basis of the frame bundle. With the notation $e_\mu{}^i \equiv \varepsilon^i{}_\mu$, the spacetime manifold from Sect. 2.1.1 is then actually equivalent to $(M, \mathcal{O}, \mathcal{A}^\uparrow, \{e^i\}, \eta, \nabla, \mathbb{T})$.

In the above discussion we have referred to vectors, co-vectors and tensors without a proper definition. Formally a (n, m)-tensor is the map

$$\mathcal{T} : \underbrace{T^*M \times T^*M \times \ldots \times T^*M}_{\text{n times}} \times \underbrace{TM \times TM \times \cdots \times TM}_{\text{m times}} \rightarrow \mathbb{R}.$$

The (n, m)-tensor maps n co-vectors and m vectors on a real number. Its components are then defined as

$$T^{\nu_1,\nu_2,\ldots,\nu_n}{}_{\mu_1,\mu_2,\ldots,\mu_m} := \mathcal{T}(\mathrm{d}x^{\nu_1}, \mathrm{d}x^{\nu_2}, \ldots, \mathrm{d}x^{\nu_n}, \partial_{\mu_1}, \partial_{\mu_2}, \ldots, \partial_{\mu_m}) \in \mathbb{R}.$$

This gives then due to the orthonormality relation (2.4):

$$\mathcal{T} = T^{\nu_1,\ldots,\nu_n}{}_{\mu_1,\ldots,\mu_m}\ \partial_{\nu_1}, \ldots, \partial_{\nu_n}\ \mathrm{d}x^{\mu_1}, \ldots, \mathrm{d}x^{\mu_m}.$$

This is the reason why vectors with the components V^μ, also called contravariant vectors, are $(1, 0)$-tensors, and the components V_μ, also called covariant vectors or 1-forms, are $(0, 1)$-tensors.

For later use we list the following relations:

$$e_\mu{}^i e_i{}^\nu = \delta^\nu_\mu \tag{2.8a}$$

$$e_\alpha{}^j e_k{}^\alpha = \delta^j_k \tag{2.8b}$$

$$e_\alpha{}^k \frac{\partial}{\partial x^\mu} e_j{}^\alpha = \frac{\partial}{\partial x^\mu} \left(e_\alpha{}^k e_j{}^\alpha\right) - e_j{}^\alpha \frac{\partial}{\partial x^\mu} e_\alpha{}^k = -e_j{}^\alpha \frac{\partial}{\partial x^\mu} e_\alpha{}^k \tag{2.8c}$$

$$\frac{\partial}{\partial e_i{}^\alpha} e_\beta{}^k = -e_\beta{}^i e_\alpha{}^k \tag{2.8d}$$

$$\frac{\partial}{\partial e_\alpha{}^i} e_k{}^\beta = -e_k{}^\alpha e_i{}^\beta \tag{2.8e}$$

$$\varepsilon := \det(e_\mu{}^i). \tag{2.8f}$$

To calculate the derivatives of ε we utilize the definition of the determinant to get

$$\frac{\partial \varepsilon}{\partial e_\mu{}^k} e_\nu{}^k = \delta^\mu_\nu \varepsilon \quad \Rightarrow \quad \frac{\partial \varepsilon}{\partial e_\mu{}^k} = e_k{}^\mu \varepsilon \quad \frac{\partial}{\partial e_\mu{}^k} \frac{1}{\varepsilon} = -e_k{}^\mu \frac{1}{\varepsilon}. \tag{2.8g}$$

2.1.3 Lorentz Connection and the "Vierbein Equation"

In the following we adopt the convention that so far emerged naturally from the definitions and postulates in differential geometry: Latin indices $i, j, k, \ldots = 0, 1, 2, 3$ refer to the observer located in the inertial space with the Lorentz (also called non-coordinate or anholonomic) basis $\{e_i\}$, and the Greek indices $\mu, \nu, \lambda, \ldots = 0, 1, 2, 3$ refer to a chart on the base manifold M, and the coordinate basis $\{\partial_\mu\}$. The objects under consideration are tensor fields or sections defined over the entire bundle, involving the basis vectors according to Eqs. (2.3) and (2.6). The vierbeins at a point $p \in M$ are constructed such that swapping the indices of any tensor field components between the coordinate and inertial frame representations is made possible. For the contravariant vector component, for example, this means

$$a^j = a^\alpha e_\alpha{}^j, \qquad a^\mu = a^i e_i{}^\mu, \tag{2.9}$$

implying that also the length of vectors (and co-vectors) and the angles between them are retained.

The of contravariant and covariant vector components is defined by the given covariant derivative on the manifold. In the chart representation this is, respectively,

$$\nabla_\nu a^\mu := \frac{\partial a^\mu}{\partial x^\nu} + \gamma^\mu{}_{\alpha\nu} a^\alpha \tag{2.10a}$$

$$\nabla_\nu a_\mu := \frac{\partial a_\mu}{\partial x^\nu} - \gamma^\alpha{}_{\mu\nu} a_\alpha, \tag{2.10b}$$

with $\gamma^{\alpha}{}_{\mu\nu}$ being some given . (Notice that in our convention the differentiation index is the rightmost one of the connection.) By demanding that angles and lengths are globally retained after swapping coordinate and Lorentz indices, the parallel transport on the base manifold must determine the parallel transport across the inertial frames of the bundle where the metric is everywhere the Minkowski metric, i.e. it is "flat" everywhere. Inserting the first relation of Eq. (2.9) into Eq. (2.10b), we obtain for a contravariant vector component

$$\frac{\partial a^{\mu}}{\partial x^{\nu}} = \frac{\partial(a^{i}\, e_{i}{}^{\mu})}{\partial x^{\nu}} = a^{i}\, \frac{\partial e_{i}{}^{\mu}}{\partial x^{\nu}} + e_{i}{}^{\mu}\, \frac{\partial a^{i}}{\partial x^{\nu}}.$$

Adding to both sides of this equation $\gamma^{\mu}{}_{\alpha\nu} a^{\alpha}$, multiplying both sides by $e_{\mu}{}^{k}$ and contracting over μ, yields

$$e_{\mu}{}^{k}\left(\frac{\partial a^{\mu}}{\partial x^{\nu}} + \gamma^{\mu}{}_{\alpha\nu} a^{\alpha}\right) = \frac{\partial a^{k}}{\partial x^{\nu}} + a^{i}\left(e_{\mu}{}^{k}\, \nabla_{\nu}\, e_{i}{}^{\mu}\right) =: \frac{\partial a^{k}}{\partial x^{\nu}} + \omega^{k}{}_{i\nu}\, a^{i}, \tag{2.11}$$

defining new functions called *Lorentz connection coefficients* that play a similar role for the parallel transport in the inertial frame as the affine connection does in the coordinate frame:

$$\omega^{k}{}_{i\nu} := e_{\mu}{}^{k}\, \frac{\partial e_{i}{}^{\mu}}{\partial x^{\nu}} + e_{\mu}{}^{k}\, \gamma^{\mu}{}_{\alpha\nu}\, e_{i}{}^{\alpha} = e_{\mu}{}^{k}\, \nabla_{\nu}\, e_{i}{}^{\mu}. \tag{2.12}$$

With the basis vector and co-vector swapped, the Lorentz connection can also be expressed as:

$$\omega^{k}{}_{i\nu} \equiv -e_{i}{}^{\mu}\, \frac{\partial e_{\mu}{}^{k}}{\partial x^{\nu}} + e_{i}{}^{\mu}\, \gamma^{\alpha}{}_{\mu\nu}\, e_{\alpha}{}^{k} = -e_{i}{}^{\mu}\, \nabla_{\nu}\, e_{\mu}{}^{k}. \tag{2.13}$$

Note that the sign change in Eq. (2.13) is derived from (Eq. (2.8b)). The Lorentz connection is often called *Spin connection* because, as we will see in Sect. 4.1, it also appears as a gauge field for fermions. The definition Eq. (2.12) can obviously be rewritten in the form called the *Vierbein Equation*:

$$\frac{\partial e_{i}{}^{\mu}}{\partial x^{\nu}} - \omega^{k}{}_{i\nu}\, e_{k}{}^{\mu} + \gamma^{\mu}{}_{\alpha\nu}\, e_{i}{}^{\alpha} = 0, \tag{2.14}$$

or (with the dual basis)

$$\frac{\partial e_{\mu}{}^{i}}{\partial x^{\nu}} + \omega^{i}{}_{k\nu}\, e_{\mu}{}^{k} - \gamma^{\alpha}{}_{\mu\nu}\, e_{\alpha}{}^{i} = 0. \tag{2.15}$$

The Vierbein Equation can be reversed to express the affine connection coefficients in terms of the Lorentz connection coefficients:

$$\gamma^{\mu}{}_{\alpha\nu} = -e_{\alpha}{}^{i}\left(\frac{\partial e_{i}{}^{\mu}}{\partial x^{\nu}} - \omega^{k}{}_{i\nu}\, e_{k}{}^{\mu}\right) = e_{k}{}^{\mu}\left(\frac{\partial e_{\alpha}{}^{k}}{\partial x^{\nu}} + \omega^{k}{}_{i\nu}\, e_{\alpha}{}^{i}\right), \tag{2.16}$$

and thus defines a global invertible map of the parallel transport in the Lorentz frame into the basic manifold[4] with lengths and angles retained.

It is useful for a more compact notation to abbreviate the partial derivative by a comma, ",", and define a covariant derivative on the frame bundle denoted by ";", that acts on both the Lorentz and coordinate indices, as shown in the following examples:

$$a^{\mu}{}_{;\nu} := \frac{\partial a^{\mu}}{\partial x^{\nu}} + \gamma^{\mu}{}_{\alpha\nu}a^{\alpha} = a^{\mu}{}_{,\nu} + \gamma^{\mu}{}_{\alpha\nu}a^{\alpha} \equiv \nabla_{\nu}a^{\mu} \tag{2.17a}$$

$$a_{\mu;\nu} := \frac{\partial a_{\mu}}{\partial x^{\nu}} - \gamma^{\alpha}{}_{\mu\nu}a_{\alpha} = a_{\mu,\nu} - \gamma^{\alpha}{}_{\mu\nu}a_{\alpha} \equiv \nabla_{\nu}a_{\mu} \tag{2.17b}$$

$$a^{k}{}_{;\nu} := \frac{\partial a^{k}}{\partial x^{\nu}} + \omega^{k}{}_{i\nu}\, a^{i} = e_{\mu}{}^{k}\left(a^{k}{}_{;\nu} + \gamma^{\mu}{}_{\alpha\nu}a^{\alpha}\right) = e_{\mu}{}^{k}\,\nabla_{\nu}a^{\mu} \tag{2.17c}$$

$$a_{k;\nu} := \frac{\partial a_{k}}{\partial x^{\nu}} - \omega^{i}{}_{k\nu}\, a_{i} = e_{k}{}^{\mu}\left(a_{k,\nu} - \gamma^{\alpha}{}_{\mu\nu}a_{\alpha}\right) = e_{k}{}^{\mu}\,\nabla_{\nu}a_{\mu} \tag{2.17d}$$

$$t^{\mu\nu}{}_{i;\beta} = \frac{\partial t^{\mu\nu}{}_{i}}{\partial x^{\beta}} - \omega^{k}{}_{i\beta}\, t^{\mu\nu}{}_{k} + \gamma^{\mu}{}_{\alpha\beta}\, t^{\alpha\nu}{}_{i} + \gamma^{\nu}{}_{\alpha\beta}\, t^{\mu\alpha}{}_{i}\,. \tag{2.17e}$$

Notice that here and in the following the affine connection is actually just a placeholder for the right-hand side of the definition Eq. (2.16) where the sign in the connection terms is adjusted such that it aligns in both frames. With this definition of the frame-bundle covariant derivative, the notation of the Vierbein Equation simplifies to

$$e_{\mu}{}^{i}{}_{;\nu} = \frac{\partial e_{\mu}{}^{i}}{\partial x^{\nu}} + \omega^{i}{}_{k\nu}\, e_{\mu}{}^{k} - \gamma^{\alpha}{}_{\mu\nu}\, e_{\alpha}{}^{i} \equiv 0. \tag{2.18}$$

As the index ν in $\omega^{i}{}_{k\nu}$ is a coordinate index, we can swap the frames as in Eq. (2.9) and re-define the following quantities in the inertial frame that will turn out useful later on:

$$\omega_{ik\nu} := \eta_{il}\,\omega^{l}{}_{k\nu} \tag{2.19a}$$

$$\omega^{i}{}_{kn} := e_{n}{}^{\nu}\,\omega^{i}{}_{k\nu} \tag{2.19b}$$

$$X^{\cdots}{}_{\cdots;n_1\ldots n_k} := e_{n1}{}^{\nu_1}\ldots e_{n_k}{}^{\nu_k}\, X^{\cdots}{}_{\cdots;\nu_1\ldots\nu_k} \tag{2.19c}$$

for an arbitrary tensor $X^{\cdots}{}_{\cdots;\nu_1\ldots\nu_k}$.

[4] We will see below that the Lorentz connection emerges as the independent field, and that the affine connection becomes a derived quantity *defined* by Eq. (2.16) that is then called the Vierbein *Postulate*.

2.1.4 The Postulates of Kinematics

The above definition of the Lorentzian manifold allows the following physical postulates to be formulated that generalize Newton's laws of physics[5]:

(P1) The *World-line* $\zeta(\lambda) : \mathbb{R} \mapsto M$ of a massive (massless) particle satisfies for all $\lambda \in \mathbb{R}$

$$\begin{aligned} 1.\quad & \mathfrak{g}_{\zeta(\lambda)}\left(\upsilon_{\zeta,\zeta(\lambda)}, \upsilon_{\zeta,\zeta(\lambda)}\right) > 0 \text{ for massive particles (time-like)} \\ & \mathfrak{g}_{\zeta(\lambda)}\left(\upsilon_{\zeta,\zeta(\lambda)}, \upsilon_{\zeta,\zeta(\lambda)}\right) = 0 \text{ for massless particles (light-like)} \\ 2.\quad & \mathfrak{g}_{\zeta(\lambda)}\left(\mathbb{T}, \upsilon_{\zeta,\zeta(\lambda)}\right) > 0 \quad \text{(future-directed)} \end{aligned}$$

where $\mathfrak{g}_{\zeta(\lambda)}(.,.)$ denotes the value of the metric along the curve ζ, and $\upsilon_{\zeta,\zeta(\lambda)}$ is the tangent vector ("particle velocity") at $p = \zeta(\lambda)$.

(P2) An *Observer* is a world-line that satisfies the postulates P1 for a massive particle and has attached—for each λ and $T_{\zeta(\lambda)}M$ the world-line passes—a basis $\{e_i(\lambda)\}$ with the properties

$$e_0 = \upsilon_{\zeta,\zeta(\lambda)} \tag{2.20a}$$

$$\mathfrak{g}_{\zeta(\lambda)}\left(e_i(\lambda), e_j(\lambda)\right) = \eta_{ij}. \tag{2.20b}$$

This ensures a of the 0-component of the observer's vierbein system, and the orthonormality of the inertial frame basis w.r.t. the Lorentzian metric. The "global time" direction can be interpreted as determining the reference frame of a "distinguished observer" with the basis vector, say E_0, defined by the specific world-line $\zeta(\lambda)$ of that observer. Setting $T^{\mu} = E_0{}^{\mu}$ means, via the second relation of postulate P1, that other observers' world-lines must remain within the light cone of that fiducial observer. With this identification the vector field $\mathbb{T}$ is *implicitly* incorporated in the discussion below.

This does not introduce any symmetry breaking of the theory as the dynamics of the fiducial observer is not a priori restricted. However, selecting the observer frame, and enabling slicing of spacetime geometry, will a posteriori introduce a natural spontaneous breaking of the solution's covariance that is, though, necessary for observational work. The requirements (2.20a) and (2.20b) then completely define via Eq. (2.6) the metric tensor in the observer frame.

We can now use this abstract, chart independent framework to define coordinate independent measurements of elapsed time and other observables in each observer's "frame" $\{e_i\}$ as follows:

(P3) A clock that the observer carries will measure the "time" τ (*Eigentime*) elapsed between two "events" at $\zeta(\lambda_1)$ and $\zeta(\lambda_2)$ as

[5] The gist behind this set of postulates can be found in [9].

$$\tau := \int_{\lambda_1}^{\lambda_2} \mathrm{d}\lambda \sqrt{\mathfrak{g}_{\zeta(\lambda)}\left(\upsilon_{\zeta,\zeta(\lambda)}, \upsilon_{\zeta,\zeta(\lambda)}\right)} \tag{2.21}$$

(P4) At the intersection point p of the world-lines ζ of the observer and χ of a massive particle with $p \equiv \zeta(\tau_1) = \chi(\tau_2)$, the observer measures the spatial velocity of that particle as the "projection" with the spatial basis co-vector e^a, $a = 1, 2, 3$, acting on the velocity vector υ:

$$v^a(\tau_2) = e^a(\upsilon_{\chi,\chi(\tau_2)}) = e_\mu{}^a \, \mathrm{d}x^\mu \, (v^\nu \partial_\nu) = e_\mu{}^a \, \delta^\mu_\nu \, v^\nu = e_\mu{}^a \, v^\mu .$$

Similarly, the measurement of any observable will be the value of the corresponding tensor projected onto the observer's frame. For example the electrical field derives from the Faraday (0,2) tensor $\mathbb{F}$ (a 2-form, i.e. in coordinate notation with two lower indices) by

$$E_a := \mathbb{F}(e_0, e_a).$$

In chart representation this gives for the electrical field:

$$E_a := F_{0a}(x) = e_0{}^\mu(x) \, e_a{}^\nu(x) \, F_{\mu\nu}(x) \qquad a = 1, 2, 3.$$

2.1.5 Straight Lines, Acceleration and Force

Be now υ_ζ a future-directed particle world-line in the sense of postulate P1. A curve is said to be autoparallel when the covariant derivative of the velocity along the direction of its tangent vector vanishes, i.e. (in sloppy notation)

$$\nabla_{\upsilon_\zeta} \upsilon_\zeta = 0. \tag{2.22}$$

The left-hand side of the equation is physically the particle's acceleration along the curve ζ,

$$a_\zeta \equiv \nabla_{\upsilon_\zeta} \upsilon_\zeta, \tag{2.23}$$

and in the sense of Newton's first postulate this equation describes a free particle: The world-line of a particle under the influence of no force (i.e. a "") is a future directed autoparallel. In terms of the chart dependent affine connection $\gamma^\nu{}_{\alpha\mu}$—representing a specific choice of the structural element "covariant derivative" of the Lorentzian manifold—that world-line is given in coordinate notation by the autoparallel-transport equation

$$\left(\nabla_{\upsilon_\zeta} \upsilon_\zeta\right)^\nu = \left(v^\mu \nabla_\mu\right) v^\nu = \left(v^\mu \frac{\partial}{\partial x^\mu}\right) v^\nu + \gamma^\nu{}_{\alpha\mu} \, v^\alpha \, v^\mu = \dot{v}^\nu + \gamma^\nu{}_{\alpha\mu} \, v^\alpha \, v^\mu = 0, \tag{2.24}$$

where

$$\dot{v}^\nu \equiv \underset{\tau}{-} v^\nu := \left(v^\mu \frac{\partial}{\partial x^\mu} \right) v^\nu$$

defines the derivative with respect to the particle's eigentime τ. The dynamical effect of gravitation thus enters only via the *symmetric* portion of the affine connection.

In the presence of a force, equation (2.23) is amended to

$$a_\zeta = \nabla_{v_\zeta} v_\zeta = \frac{\mathbb{K}}{m_{\text{inert}}}, \tag{2.25}$$

where the force $\mathbb{K}$ is a spatial vector field with $g(\mathbb{T}, \mathbb{K}) = 0$, and m_{inert} is the particle's *inertial* mass. This, of course, is Newton's second postulate expressed in the language of differential geometry. Gravitation, which is not a force anywhere in the sense of classical mechanics, and also Newton's intuitive concept of a "straight line" receive a mathematical underpinning only with the structural elements of a Lorentzian manifold. Notice that minimizing the eigentime elapsed as given by Eq. (2.21) defines the *shortest line* between two events, that might differ from the autoparallel (straight) line.

2.2 Metric Compatibility

The stipulation that an observer's frame is everywhere inertial (flat) means that the frame is endowed with the constant global metric $\eta_{ij} = \text{diag}(1, -1, -1, -1)$. It thus remains invariant under parallel transport, hence its covariant derivative must vanish everywhere. This is indeed the case if the Lorentz connection is anti-symmetric:

$$\eta_{ij;\alpha} = \frac{\partial \eta_{ij}}{\partial x^\alpha} - \omega^k{}_{i\alpha}\, \eta_{kj} - \omega^k{}_{j\alpha}\, \eta_{ik} = -\left(\omega_{ji\alpha} + \omega_{ij\alpha}\right) = 0. \tag{2.26}$$

Even more, we have already seen that, provided the Vierbein Equation holds, this is equivalent to requesting that the metric $g_{\mu\nu}$ on the tangent bundle of the base manifold is "invariant under parallel transport", i.e. that globally the length of vectors and the angles between them remain constant:

$$g_{\mu\nu;\alpha}(x) = e_\mu{}^i\, e_\nu{}^j\, \eta_{ij;\alpha} = 0. \tag{2.27}$$

This is called *Metric compatibility*. If the Vierbein Equation does not hold, or/and $\omega_{(ji)\alpha} \neq 0$, then the so called *Non-metricity tensor*

$$N_{\alpha\mu\nu}(x) := g_{\mu\nu;\alpha}(x) \equiv \frac{\partial g_{\mu\nu}}{\partial x^\alpha} - \gamma^\lambda{}_{\mu\alpha}\, g_{\nu\lambda} - \gamma^\lambda{}_{\nu\alpha}\, g_{\lambda\mu} \neq 0 \tag{2.28}$$

carries additional degrees of freedom. It is symmetric in its last two indices.

2.3 Frame Transitions

The Lorentz connection is a group-valued 1-frame that "rests on three legs": two Latin indices in the Lorentz space (group indices), and one Greek index in the coordinate space (coordinate index). How does it change under arbitrary frame and variable transformations?

There are two kinds of transformations:

Chart transitions are local (passive) diffeomorphisms. In the coordinate space (hence denoted by Greek indices), where the chart transition $x \mapsto X(x)$ is a bijective differentiable map $\mathbb{R}^4 \mapsto \mathbb{R}^4$, the components of an arbitrary tensor $\mathcal{T}$ transform as

$$\mathcal{T}^{\alpha\ldots}{}_{\beta\ldots}(X) = \frac{\partial X^\alpha}{\partial x^\mu} \cdots \frac{\partial x^\nu}{\partial X^\beta} \cdots T^{\mu\ldots}{}_{\nu\ldots}(x). \tag{2.29}$$

Demanding covariance with respect to all analytical coordinate (not point) transformations is here identified with the . In the mathematics of manifolds and bundles that Principle is thus merely a "hygiene" issue as the independence of calculations from the selected chart is a necessary requirement for the consistency of any theory.[6]

It is important to stress here, that by passive diffeomorphisms the points (events) in the manifold (spacetime) retain their individuality given by the metric. This is not the case when identifying coordinate relativity with active diffeomorphisms[7] as it is often understood when generalizing the (Poincare) symmetry transformations of Special Relativity from inertial frames to curved spacetimes (Fig. 2.3).

Local Lorentz transformations relate two observer frames as defined in postulate $P3$. According to that postulate observers live in inertial frames that are constructed such that locally the *Principle of Special Relativity* holds. Let (ζ, e) and $(\tilde{\zeta}, E)$ be two observers meeting at the point $p = \zeta(\tau = 0) = \tilde{\zeta}(\tilde{\tau} = 0)$. Between the two bases, $\{e_i\} \in T_pM$ of observer 1 and $\{E_I\} \in T_pM$ of observer 2, must exist an invertible linear transformation (matrix) $\Lambda \in \mathrm{GL}(4)$ such that

$$E_A(x) = \Lambda_A{}^b(x)\, e_b(x). \tag{2.30}$$

For vector components in the frame of observer 2 we use in the following capital letters and capital Latin indices, for example $A^K(x)$ with the index K referring to the basis $\{E_K\}(x)$. Now according to the definition of the observer's frame we find (in simplified notation)

[6] In the various versions of a coordinate-free notation of differential geometry this demand is automatically incorporated, but we prefer to stick here with the "workhorse" of theoretical physics, the tensor calculus.

[7] For a discussion of the significance of passive versus active diffeomorphisms in theories of gravitation see, for example, [10].

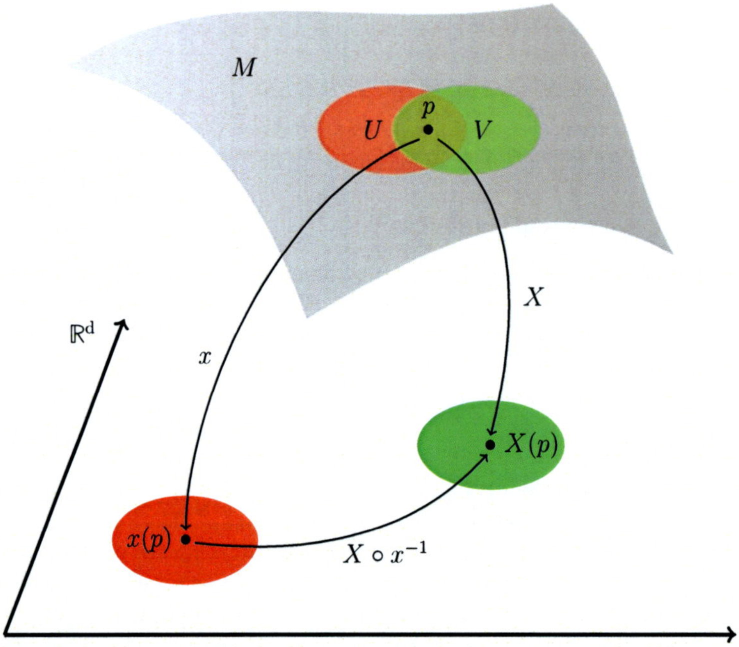

Fig. 2.3 If two charts (U, x) and (V, X) overlap, two sets of coordinates x_μ resp. X_μ can be assigned each point p with $p \in U \cap V$. The map $X \circ x^{-1}$ then provides the transition $x_\mu \mapsto X_\mu$

$$\begin{aligned}\eta_{AB} &= \mathfrak{g}(E_A, E_B) = \mathfrak{g}\left(\Lambda_A{}^m\, e_m, \Lambda_B{}^n\, e_n\right)\\ &= \Lambda_A{}^m\, \Lambda_B{}^n\, \mathfrak{g}\,(e_m, e_n) = \Lambda_A{}^m\, \Lambda_B{}^n\, \eta_{mn} = \Lambda_A{}^m\, (\Lambda^T)_{mB}\end{aligned}$$

i.e. the transformation (matrix) must satisfy

$$\Lambda \cdot \Lambda^T = \eta, \tag{2.31}$$

which defines the group of Lorentz transformations, i.e. $\Lambda \in \mathrm{SO}(1, 3)$. Actually, because of the preservation of the time direction dictated by the postulate P1, we find $\Lambda \in \mathrm{SO}^+(1, 3)$, i.e. in the orthochronous subgroup of the .

The conclusion is that the Lorentz transformation relates the so called Lorentz (inertial) frames of any two observers at the same point. Einstein's Principle of General Relativity, stipulating that the dynamics of a physical system is independent of the observer's frame of reference, thus requires not only the passive diffeomorphism covariance on the base manifold, but in addition the covariance under local orthochronous Lorentz transformations. In passing we also notice that

$$E_\mu{}^A\, e_b{}^\mu = \Lambda^A{}_b.$$

2.3.1 Interrelationships of the Affine and Lorentz Connections

In our notification chart transitions act only on Greek indices, and Lorentz transformations on Latin indices. However, each acts, at some point p, on a coordinate representation of a tensor field in a particular frame. The pertaining connections at that point are related via the definition (2.12) and its inverse, Eq. (2.16). In combination, these two transformations make the Lorentz space affine as they include an arbitrary shift of the origin of the inertial frame.

According to the definition (2.12), the $\omega^l{}_{i\nu}(x)$ is a co-vector field with respect to the coordinate index ν, and transforms via

$$\omega^l{}_{i\nu}(X) = \frac{\partial x^\mu}{\partial X^\nu}\,\omega^l{}_{i\mu}(x),$$

where $\partial x^\mu/\partial X^\nu$ is the transition map between the charts (U, x) and (V, X) on $U \cap V$ at the point $p \in U \cap V$. Also the vierbeins transform as vectors with respect to the coordinate index. In contrast, the transformation of an arbitrary vector components $a^j(x)$ at point x from the local frame $\{e_j\}(x)$ of observer 1 to the local frame $\{E_I\}(x)$ of observer 2 is mediated by the local Lorentz group element $\Lambda^I{}_j(x)$. A vector component in the coordinate frame can be expressed in either of the observers' frames

$$a^\mu = a^i\, e_i{}^\mu = A^I\, E_I{}^\mu.$$

Obviously using Eq. (2.17c) and Eq. (2.18) we get:

$$A^K{}_{;\nu} = \left(E_\mu{}^K\, a^\mu\right)_{;\nu} = E_\mu{}^K\, a^\mu{}_{;\nu} = E_\mu{}^K\left(e_l{}^\mu\, a^l\right)_{;\nu} = E_\mu{}^K\, e_l{}^\mu\left(a^l{}_{,\nu} + \omega^l{}_{i\nu}\, a^i\right) = \Lambda^K{}_l\, a^l{}_{;\nu}.$$

Now the partial derivative of Eq. (2.30), with Eq. (2.18), yields

$$\left(\frac{\partial \Lambda^l{}_K}{\partial x^\nu} - \Omega^I{}_{K\nu}\,\Lambda^l{}_I\right) e_l = \Lambda^m{}_K\,\omega^l{}_{m\nu}\, e_l.$$

Notice that this equation holds at any point p with the chart coordinate x, and that the Lorentz transformation $\Lambda^l{}_K$ is of course local to that point. Hence its derivative is in general non-vanishing. As the above equation is valid for any basis vector component e_l, we find for the transformation of the Lorentz connection at x:

$$\Omega^I{}_{K\nu} = \Lambda^I{}_l\left(\Lambda^i{}_K\,\omega^l{}_{i\nu} + \frac{\partial \Lambda^l{}_K}{\partial x^\nu}\right) = \Lambda^i{}_K\left(\Lambda^I{}_l\,\omega^l{}_{i\nu} - \frac{\partial \Lambda^I{}_i}{\partial x^\nu}\right). \tag{2.32}$$

This transformation relation is inhomogeneous. From Eq. (2.31) it follows that

$$\Lambda^i{}_K\,\frac{\partial \Lambda^I{}_i}{\partial x^\nu} = -\Lambda^I{}_i\,\frac{\partial \Lambda^i{}_K}{\partial x^\nu},$$

and we conclude that the Lorentz connection must not be symmetric in the Lorentz indices, $\omega_{ki\nu} + \omega_{ik\nu} \neq 0$. The index ν is a chart component index, and is not affected by transformations in the inertial frame.

Note that using Eqs. (2.19) and (2.30), we can swap that index into the inertial frame,

$$\omega^{k}{}_{jn} := e_{n}{}^{\nu}\,\omega^{k}{}_{j\nu}$$
$$\Omega^{K}{}_{JN} := E_{N}{}^{\nu}\,\Omega^{K}{}_{J\nu} = \Lambda^{m}{}_{N}\,e_{m}{}^{\nu}\,\Omega^{K}{}_{J\nu},$$

to obtain

$$\Omega^{I}{}_{KN} = \Lambda^{m}{}_{N}\,\Lambda^{i}{}_{K}\left(\Lambda^{I}{}_{l}\,\omega^{l}{}_{im} - \frac{\partial\Lambda^{I}{}_{i}}{\partial x^{m}}\right). \tag{2.33}$$

Substituting Eq. (2.13) for $\omega^{l}{}_{i\nu}$ into the above equation gives, after a lengthy calculation, the transformation law $\gamma \to \Gamma$ of the affine connection

$$\Gamma^{\alpha}{}_{\nu\beta} = \gamma^{\sigma}{}_{\eta\mu}\,\frac{\partial x^{\eta}}{\partial X^{\nu}}\,\frac{\partial x^{\mu}}{\partial X^{\beta}}\,\frac{\partial X^{\alpha}}{\partial x^{\sigma}} - \frac{\partial x^{\eta}}{\partial X^{\nu}}\,\frac{\partial x^{\mu}}{\partial X^{\beta}}\,\frac{\partial^{2} X^{\alpha}}{\partial x^{\mu}\partial x^{\eta}}. \tag{2.34}$$

2.3.2 *Relative Tensors and Their Transformation Rules*

If we wish to apply the Hamiltonian formalism of field theory to action integrals involving chart transitions, the transformation rules (2.29) of *absolute tensors* encountered so far has to be generalized to those of *relative tensors*. If a the matrix field $t^{\alpha_1\ldots\alpha_n}{}_{\beta_1\ldots\beta_m}$ at x transforms under a chart transition $x \mapsto X$ according to

$$T^{\xi_1\ldots\xi_n}{}_{\eta_1\ldots\eta_m}(X) = t^{\alpha_1\ldots\alpha_n}{}_{\beta_1\ldots\beta_m}(x)\frac{\partial X^{\xi_1}}{\partial x^{\alpha_1}}\cdots\frac{\partial X^{\xi_n}}{\partial x^{\alpha_n}}\frac{\partial x^{\beta_1}}{\partial X^{\eta_1}}\cdots\frac{\partial x^{\beta_m}}{\partial X^{\eta_m}}\left|\frac{\partial x}{\partial X}\right|^{w} \tag{2.35}$$

with $\left|\frac{\partial x}{\partial X}\right|$ the determinant of the Jacobi matrix (Jacobian) of the transformation $x \mapsto X$,

$$\left|\frac{\partial x}{\partial X}\right| = \frac{\partial\left(x^{0},\ldots,x^{3}\right)}{\partial\left(X^{0},\ldots,X^{3}\right)}, \tag{2.36}$$

then it is referred to as a relative tensor of weight w. For $w = 0$, the rule (2.35) includes the transformation rule (2.29) for absolute tensors. The particular case of a relative tensor of weight $w = 1$ is also referred to as a *Tensor density*. With *scalars* denoting the particular class of tensors of rank zero, Eq. (2.35) also defines the transformation rule for scalars of weight w. Accordingly, a scalar of weight $w = 1$ is called a *Scalar density*. For instance, the square root of the determinant of a (0, 2) tensor transforms as a scalar density as the tensor itself transforms as

$$G_{\mu\nu}(X) = g_{\alpha\beta}(x) \frac{\partial x^\alpha}{\partial X^\mu} \frac{\partial x^\beta}{\partial X^\nu},$$

hence

$$(\det G_{\mu\nu})(X) = (\det g_{\alpha\beta})(x) \left| \frac{\partial x}{\partial X} \right|^2 \quad \Leftrightarrow \quad G(X) = g(x) \left| \frac{\partial x}{\partial X} \right|^2 .$$

If in particular $g \equiv \det g_{\alpha\beta} < 0$ and $G \equiv \det G_{\mu\nu} < 0$ denote the determinants of the covariant form of the metric tensors $g_{\mu\nu}$ and $G_{\mu\nu}$, the transformation rule $g(x) \mapsto G(X)$ for the scalar density follows as

$$\sqrt{-G} = \sqrt{-g} \left| \frac{\partial x}{\partial X} \right| . \tag{2.37}$$

$\sqrt{-g}$ thus represents a relative scalar of weight $w = 1$, i.e. a scalar density.

Correspondingly, the determinant of the contravariant representation of the metric tensor is a relative scalar of weight $w = -2$

$$G^{\mu\nu}(X) = g^{\alpha\beta}(x) \frac{\partial X^\mu}{\partial x^\alpha} \frac{\partial X^\nu}{\partial x^\beta} \quad \Rightarrow \quad (\det G^{\mu\nu})(X) = (\det g^{\alpha\beta})(x) \left| \frac{\partial x}{\partial X} \right|^{-2} .$$

From Eq. (2.35) one concludes directly that the product of a relative (i, n) tensor t of weight w_1 with a relative (j, m) tensor s of weight w_2 yields a relative $(i + j, n + m)$ tensor $t \otimes s$ of weight $w_1 + w_2$.

The volume form $\mathrm{d}^4 x$ transforms as a relative scalar of weight $w = -1$

$$\mathrm{d}^4 X = \frac{\partial(X^0, \ldots, X^3)}{\partial(x^0, \ldots, x^3)} \, \mathrm{d}^4 x = \left| \frac{\partial X}{\partial x} \right| \, \mathrm{d}^4 x = \left| \frac{\partial x}{\partial X} \right|^{-1} \, \mathrm{d}^4 x . \tag{2.38}$$

In conjunction with Eq. (2.37), one concludes that the product $\sqrt{-g}\, \mathrm{d}^4 x$ transforms as a scalar of weight $w = 0$, hence as an absolute scalar

$$\sqrt{-G}\, \mathrm{d}^4 X = \sqrt{-g}\, \mathrm{d}^4 x . \tag{2.39}$$

$\sqrt{-g}\, \mathrm{d}^4 x$ is thus referred to as the *Invariant volume form.*

2.3.3 *Invariant Integrals*

As both the definition and the selection of charts are arbitrary, we always have to prove that physical observables are results of calculations that are chart independent. This means that any observable quantity Q calculated at point p in chart coordinates (U, x) must remain unchanged if instead the chart (V, X) with $p \in U \cap V$ is used.

That requirement, of course, must also apply for the Lagrangian or Hamiltonian from which the observables are derived.

The transformation of tensors under chart transitions are determined by the invertible and differentiable chart transition map $X(x)$—a diffeomorphism on M—via the coefficient matrix $\partial X^\mu/\partial x^\nu$ and its inverse (c.f. Eq. (2.29)). That matrix also makes the value of an integral chart dependent. The naïve integral definition on a chart $(U, x) \cap (V, X)$

$$I_x = \int\limits_{x(U\cap V)} \mathrm{d}^4\alpha\, f_x(\alpha).$$

is under the substitution of variables by the transition map $X(x)$ between (U, x) and (V, X) not equal to I_X,

$$I_{x\mapsto X} = \int\limits_{X(U\cap V)} \mathrm{d}^4\beta\, f_X(\beta) \left|\det\left(\frac{\partial\alpha^\mu}{\partial\beta^\nu}\right)\right| \neq I_X = \int\limits_{X(U\cap V)} \mathrm{d}^4\alpha\, f_X(\alpha),$$

as the determinant of the transition coefficient (Jacobi) matrix is in general not equal to 1. For the calculus to be consistent, a chart independent integral, $\int\limits_{U\cap V} f$, of a function $f : M \mapsto \mathbb{R}^4$ over an open set $(U \cap V) \in M$ must be constructed. The volume element d^4x must hence be modified such that the Jacobi determinant is canceled under chart transitions and $I_{x\mapsto y} = I_y$, which is accomplished by introducing the from Eq. (2.39). This is remarkable as this construct, in which $g_{\mu\nu}(x)$ is treated just as a real matrix, is chart specific. For variational analysis of action integrals, the notation

$$I = \int \mathrm{d}^4x\, \sqrt{-g(x)}\, f(x) \equiv \int \mathrm{d}^4x\, \tilde{f}(x) \tag{2.40}$$

will be applied throughout this book, where $\tilde{f}(x) := \sqrt{-g(x)}\, f(x)$ is a scalar density.

For the application of the vierbein formalism we note that a similar effect will be achieved if in the volume element we replace $\sqrt{-g}$ by $\varepsilon \equiv \det\left(e_\mu{}^i\right)$.[8] This is reason why the (dual!) vierbein is sometimes considered the "square root of the metric".

This construction can now be extended to define the integral over the entire manifold,

$$\int\limits_M f = \sum_{i=1}^{N} \int\limits_{U_i} (\rho_i \cdot f),$$

[8] In principle, invariant volume elements can be created using any (0, 2) tensor as long as its determinant does not vanish anywhere. Guendelman et al. [11] for example introduce two volume elements with independent degree of freedom for dark matter and dark energy.

where $\rho_i : U^i \mapsto \mathbb{R}$ is the so called partition of unity consisting of continuous functions such that for any p sitting in the overlap of a finite number of charts $U_i, i = 1, \ldots, N$,

$$\sum_{i=1}^{N} \rho_i(p) = 1$$

holds. We will in the following adhere to the general practice and assume that the partition, implicit in any integral definition over a manifold, is well defined such that its explicit form remains irrelevant for concrete calculations.

2.4 Vierbein Fields and Coefficients of Anholonomy

In order to derive key tensor expressions in curved geometries we explore the fact that vectors (and of course tensors in general) on a manifold are operators defined on the manifold's tangent space that act upon functions $f : M \mapsto \mathbb{R}$, see Sect. 2.1.2. For the vierbein fields this fact, expressed in the chart representation, means

$$e_i\, f = e_i^{\ \mu} \frac{\partial f}{\partial x^\mu}.$$

These vierbein operators constitute a Lie algebra as follows:

$$\begin{aligned}
[e_i\,,\, e_j]\, f &= \left[e_i^{\ \mu}\frac{\partial}{\partial x^\mu}\,,\, e_j^{\ \nu}\frac{\partial}{\partial x^\nu}\right] f \\
&= e_i^{\ \mu}\, e_j^{\ \nu}\left(\frac{\partial^2}{\partial x^\nu \partial x^\mu} - \frac{\partial^2}{\partial x^\mu \partial x^\nu}\right) f + e_i^{\ \mu}\frac{\partial e_j^{\ \nu}}{\partial x^\mu}\frac{\partial f}{\partial x^\nu} - e_j^{\ \nu}\frac{\partial e_i^{\ \mu}}{\partial x^\nu}\frac{\partial f}{\partial x^\mu} \\
&= 0 + e_i^{\ \mu}\, e_k^{\ \nu}\, e_\lambda^{\ k}\frac{\partial e_j^{\ \lambda}}{\partial x^\mu}\frac{\partial f}{\partial x^\nu} - e_j^{\ \nu}\, e_k^{\ \mu}\, e_\lambda^{\ k}\frac{\partial e_i^{\ \lambda}}{\partial x^\nu}\frac{\partial f}{\partial x^\mu} \\
&= -e_i^{\ \mu}\, e_k^{\ \nu}\, e_j^{\ \lambda}\frac{\partial e_\lambda^{\ k}}{\partial x^\mu}\frac{\partial f}{\partial x^\nu} + e_j^{\ \nu}\, e_k^{\ \mu}\, e_i^{\ \lambda}\frac{\partial e_\lambda^{\ k}}{\partial x^\nu}\frac{\partial f}{\partial x^\mu} \\
&= e_i^{\ \mu}\, e_j^{\ \nu}\left(\frac{\partial e_\mu^{\ k}}{\partial x^\nu} - \frac{\partial e_\nu^{\ k}}{\partial x^\mu}\right) e_k^{\ \lambda}\frac{\partial f}{\partial x^\lambda} = \left(C^k_{\ ij}\, e_k\right) f.
\end{aligned}$$

The structure constants, or better structure functions, of the vierbein fields,

$$C^k_{\ ij} := e_i^{\ \mu}\, e_j^{\ \nu}\left(\frac{\partial e_\mu^{\ k}}{\partial x^\nu} - \frac{\partial e_\nu^{\ k}}{\partial x^\mu}\right), \tag{2.41}$$

are called *Coefficients of anholonomy*, and the Lorentz basis *anholonomic*. In contrast, the natural basis $\{\partial_\mu\}$ on the tangent space is *holonomic* as these basis vectors commute, $[\partial_\mu, \partial_\nu] = 0$, and the pertaining structure coefficients vanish.[9]

Notice that by definition the coefficients of anholonomy are anti-symmetric in their lower indices,

$$C^k{}_{ij} = -C^k{}_{ji}.$$

Their interpretation is as follows. For two arbitrary vector fields A and B, defined on a smooth manifold, the commutator is identical to the [12, 13]

$$\mathcal{L}_A B := [A, B] \tag{2.42}$$

of the vector field B along the integral curve of vector field A where holds $\zeta'(\lambda) = A(\zeta(\lambda)$ holds. Hence by $\left[e_i, e_j\right] = C^k{}_{ij}\, e_k$ the coefficient $C^k{}_{ij}$ can be seen as the component in direction of e_k of the Lie derivative of the basis[10] vector e_j along the curve of e_i:

$$\mathcal{L}_{e_i} e_j = [e_i, e_j].$$

2.5 Curvature and Torsion

The Riemann-Cartan curvature tensor is a $(1, 3)$ tensor intimately connected to the affine connection and can, via Eq. (2.16), be equivalently expressed in terms of the Lorentz connection:

$$\begin{aligned} R^\lambda{}_{\sigma\mu\nu} &:= \frac{\partial \gamma^\lambda{}_{\sigma\nu}}{\partial x^\mu} - \frac{\partial \gamma^\lambda{}_{\sigma\mu}}{\partial x^\nu} + \gamma^\lambda{}_{\delta\mu}\gamma^\delta{}_{\sigma\nu} - \gamma^\lambda{}_{\delta\nu}\gamma^\delta{}_{\sigma\mu} \\ &= e_a{}^\lambda e_\sigma{}^b \left(\frac{\partial \omega^a{}_{b\nu}}{\partial x^\mu} - \frac{\partial \omega^a{}_{b\mu}}{\partial x^\nu} + \omega^a{}_{c\mu}\,\omega^c{}_{b\nu} - \omega^a{}_{c\nu}\,\omega^c{}_{b\mu} \right) \\ &=: e_a{}^\lambda e_\sigma{}^b R^a{}_{b\mu\nu}. \end{aligned} \tag{2.43}$$

The Riemann-Cartan curvature tensor is by construction anti-symmetric in its last two indices, and with the anti-symmetry of the Lorentz connection (hence in metric-compatible geometries), also in the first index pair:

[9] The attribute "anholonomic" means that the relation between the two bases and coordinate sets is not integrable. In particular it is not possible to set $e_\mu{}^k = \frac{\partial x^k}{\partial y^\mu}$.

[10] The Lie derivative is a construct based on an infinitesimal change of the chart, c.f. Sect. 6.2.1 for its definition, and is thus independent of the definition of the covariant derivative and the connection. Hence care must be taken when looking at the interpretations of the physical meanings of the structure constants of the Lorentz basis vectors as compared to the definition of the parallel transport and acceleration via the covariant derivative in the observer's inertial frame using the Lorentz connection.

$$R_{\lambda\sigma\mu\nu} \equiv -R_{\lambda\sigma\nu\mu} = -R_{\sigma\lambda\mu\nu}. \tag{2.44}$$

The is derived from $R^{\lambda}{}_{\sigma\mu\nu}$ by the contraction

$$R_{\sigma\nu} := R^{\lambda}{}_{\sigma\lambda\nu}, \tag{2.45}$$

and the is given by

$$R := R_{\alpha\nu}\, g^{\alpha\nu}. \tag{2.46}$$

Notice that neither the Riemann-Cartan nor the Ricci tensor are affected by the signature of the metric, in contrast to the Ricci scalar.

The (Cartan) ,

$$S^{\lambda}{}_{\mu\nu} := \tfrac{1}{2}\left(\gamma^{\lambda}{}_{\mu\nu} - \gamma^{\lambda}{}_{\nu\mu}\right), \tag{2.47}$$

can according to Eq. (2.16) be expressed as

$$S^{\lambda}{}_{\mu\nu} \equiv \tfrac{1}{2} e_i{}^{\lambda} \left(\frac{\partial e_\mu{}^i}{\partial x^\nu} - \frac{\partial e_\nu{}^i}{\partial x^\mu} + \omega^i{}_{k\nu}\, e_\mu{}^k - \omega^i{}_{k\mu}\, e_\nu{}^k \right). \tag{2.48}$$

With Eq. (2.41) this identity can be rewritten in terms of the Lorentz connections and the structure coefficients:

$$S^{\lambda}{}_{\mu\nu} = \tfrac{1}{2} e_l{}^{\lambda}\, e_\mu{}^m\, e_\nu{}^n \left(C^l{}_{mn} + \omega^l{}_{mn} - \omega^l{}_{nm}\right) =: e_l{}^{\lambda}\, e_\mu{}^m\, e_\nu{}^n\, S^l{}_{mn}. \tag{2.49}$$

Both, the Riemann-Cartan curvature and the Cartan torsion contribute to the parallel transport of a vector along a closed loop:

$$\begin{aligned}
&(\nabla_\beta \nabla_\alpha - \nabla_\alpha \nabla_\beta)\, v^\mu = v^\mu{}_{;\alpha;\beta} - v^\mu{}_{;\beta;\alpha} \\
&= \left(v^\mu{}_{,\beta} + \gamma^\mu{}_{\sigma\beta} v^\sigma\right)_{;\alpha} - \left(v^\mu{}_{,\alpha} + \gamma^\mu{}_{\sigma\alpha} v^\sigma\right)_{;\beta} \\
&= v^\mu{}_{,\alpha,\beta} + v^\sigma \gamma^\mu{}_{\sigma\beta,\alpha} + \gamma^\mu{}_{\sigma\beta} v^\sigma{}_{,\alpha} + \gamma^\mu{}_{\sigma\alpha}\left(v^\sigma{}_{,\beta} + \gamma^\sigma{}_{\rho\beta} v^\rho\right) - \gamma^\sigma{}_{\beta\alpha}\left(v^\mu{}_{,\sigma} + \gamma^\mu{}_{\rho\sigma} v^\rho\right) \\
&\quad - (\alpha \leftrightarrow \beta) \\
&= v^\sigma\left(\gamma^\mu{}_{\sigma\beta,\alpha} - \gamma^\mu{}_{\sigma\alpha,\beta} + \gamma^\rho{}_{\sigma\beta}\gamma^\mu{}_{\rho\alpha} - \gamma^\rho{}_{\sigma\alpha}\gamma^\mu{}_{\rho\beta}\right) + \left(\gamma^\sigma{}_{\alpha\beta} - \gamma^\sigma{}_{\beta\alpha}\right)\left(v^\mu{}_{,\sigma} + \gamma^\mu{}_{\rho\sigma} v^\rho\right) \\
&= R^\mu{}_{\sigma\alpha\beta} v^\sigma + 2S^\sigma{}_{\alpha\beta} v^\mu{}_{;\sigma}.
\end{aligned} \tag{2.50}$$

The impact of torsion is illustrated in Fig. 2.4. The metric-compatible spacetime with torsion is called Riemann-Cartan space [14], and in four dimensions denoted by U_4.

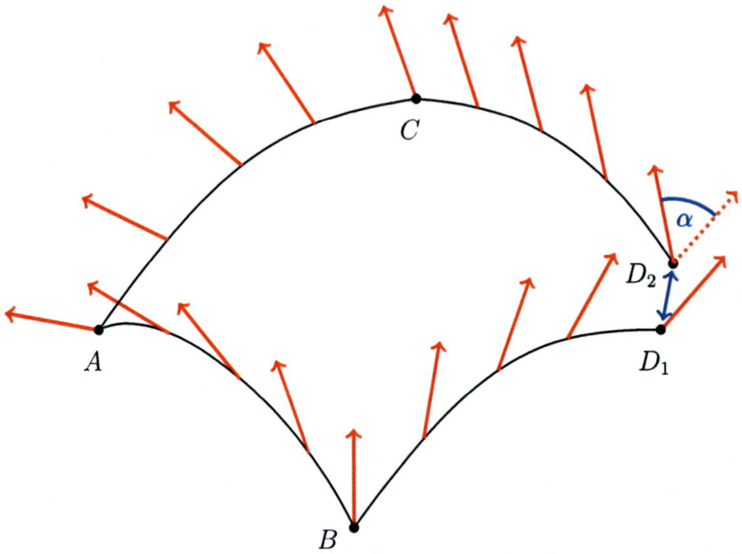

Fig. 2.4 In a space with torsion a parallelogram does not close, indicated by the blue arrow between the points D_1 and D_2. A vector parallel transported from A via B to D_1 or alternatively from A via C to D_2 points in the same direction at both endpoints if there is no curvature. Otherwise, the vector in D_1 is rotated against the vector in D_2 by an angle α

2.6 Connection and Contortion

The following linear combination of the non-metricity tensor (2.28),

$$\begin{aligned} L_{\alpha\mu\nu} :&= N_{\alpha\mu\nu} - N_{\mu\nu\alpha} - N_{\nu\alpha\mu} = g_{\mu\nu;\alpha} - g_{\nu\alpha;\mu} - g_{\alpha\mu;\nu} \\ &= \frac{\partial g_{\mu\nu}}{\partial x^\alpha} - \frac{\partial g_{\nu\alpha}}{\partial x^\mu} - \frac{\partial g_{\alpha\mu}}{\partial x^\nu} \\ &\quad - \gamma^\lambda{}_{\mu\alpha}\, g_{\nu\lambda} - \gamma^\lambda{}_{\nu\alpha}\, g_{\lambda\mu} + \gamma^\lambda{}_{\nu\mu}\, g_{\nu\lambda} + \gamma^\lambda{}_{\alpha\mu}\, g_{\lambda\alpha} + \gamma^\lambda{}_{\alpha\nu}\, g_{\lambda\mu} + \gamma^\lambda{}_{\mu\nu}\, g_{\alpha\lambda}. \end{aligned} \tag{2.51}$$

can be resolved for $\gamma^\lambda{}_{\mu\nu}$ yielding the relation

$$\gamma^\lambda{}_{\mu\nu} = \begin{Bmatrix} \lambda \\ \mu\nu \end{Bmatrix} + K^\lambda{}_{\mu\nu} - L^\lambda{}_{\mu\nu}. \tag{2.52}$$

The first term on the right-hand side is the symmetric *Christoffel symbol* that in the Einstein-Hilbert theory is identified with the affine connection and also called the Levi-Civita connection. It is determined entirely by the metric via the *Levi-Civita relation* (explicitely derived in the box below)

$$\left\{ \begin{matrix} \lambda \\ \mu\nu \end{matrix} \right\} := \tfrac{1}{2} g^{\lambda\alpha} \left(-\frac{\partial g_{\mu\nu}}{\partial x^\alpha} + \frac{\partial g_{\nu\alpha}}{\partial x^\mu} + \frac{\partial g_{\alpha\mu}}{\partial x^\nu} \right). \tag{2.53}$$

✓ Explicit Calculation

Provided that the of the metric tensor vanishes

$$g_{\mu\nu;\alpha} \equiv \frac{\partial g_{\mu\nu}}{\partial x^\alpha} - g_{\xi\nu}\gamma^\xi{}_{\mu\alpha} - g_{\mu\xi}\gamma^\xi{}_{\nu\alpha} = 0, \tag{2.54}$$

it is possible to directly correlate *symmetric* connection coefficients to the metric. A vanishing covariant derivative of the metric can be expressed in three equivalent versions of its indices as follows:

$$\begin{aligned} \frac{\partial g_{\mu\nu}}{\partial x^\alpha} &= g_{\nu\xi}\gamma^\xi{}_{\mu\alpha} + g_{\mu\xi}\gamma^\xi{}_{\nu\alpha} \\ \frac{\partial g_{\mu\alpha}}{\partial x^\nu} &= g_{\alpha\xi}\gamma^\xi{}_{\mu\nu} + g_{\mu\xi}\gamma^\xi{}_{\alpha\nu} \\ -\frac{\partial g_{\nu\alpha}}{\partial x^\mu} &= -g_{\alpha\xi}\gamma^\xi{}_{\nu\mu} - g_{\nu\xi}\gamma^\xi{}_{\alpha\mu}. \end{aligned}$$

With the symmetry property of the connection coefficients (2.52) that applies for torsion-free spaces, the three equations can be summed up to yield

$$2g_{\mu\xi}\gamma^\xi{}_{\nu\alpha} = \frac{\partial g_{\mu\nu}}{\partial x^\alpha} + \frac{\partial g_{\mu\alpha}}{\partial x^\nu} - \frac{\partial g_{\nu\alpha}}{\partial x^\mu}.$$

Thus, for torsion-free spaces, the connection coefficient $\gamma^\alpha{}_{\mu\nu}$ is uniquely correlated to the contravariant metric $g^{\mu\nu}$ and the derivatives of the covariant metric $g_{\mu\nu}$ by

$$\gamma^\alpha{}_{\mu\nu} = \tfrac{1}{2} g^{\alpha\xi} \left(\frac{\partial g_{\mu\xi}}{\partial x^\nu} + \frac{\partial g_{\nu\xi}}{\partial x^\mu} - \frac{\partial g_{\mu\nu}}{\partial x^\xi} \right) := \left\{ \begin{matrix} \alpha \\ \mu\nu \end{matrix} \right\}. \tag{2.55}$$

This particular case of a connection is called the *Levi-Civita connection*. The contraction $\alpha = \mu$ follows as

$$\gamma^\alpha{}_{\alpha\nu} = \tfrac{1}{2} g^{\alpha\xi} \left(\frac{\partial g_{\alpha\xi}}{\partial x^\nu} + \frac{\partial g_{\xi\nu}}{\partial x^\alpha} - \frac{\partial g_{\alpha\nu}}{\partial x^\xi} \right) = \tfrac{1}{2} g^{\alpha\xi} \frac{\partial g_{\alpha\xi}}{\partial x^\nu}. \tag{2.56}$$

The tensor $K^\lambda{}_{\mu\nu} - L^\lambda{}_{\mu\nu}$ is called the "distortion" tensor. and denotes the difference between the Levi-Civita connection and the general affine connection with torsion and non-metricity. The tensor

$$K^{\lambda}_{\ \ \mu\nu} = g^{\lambda\alpha} K_{\alpha\mu\nu} = g^{\lambda\alpha}\left(S_{\alpha\mu\nu} - S_{\mu\alpha\nu} - S_{\nu\alpha\mu}\right) = S^{\lambda}_{\ \ \mu\nu} - V^{\lambda}_{\ \ \mu\nu} \tag{2.57}$$

is called *Contortion tensor*. It is easy to see that $K_{\alpha\mu\nu}$ is anti-symmetric in its first two indices as the torsion $S_{\alpha\mu\nu}$ is anti-symmetric in μ and ν. Then it also follows that the difference between torsion and contortion,

$$V^{\lambda}_{\ \ \mu\nu} := g^{\lambda\alpha}\left(S^{\beta}_{\ \ \alpha\nu}\, g_{\beta\mu} - S^{\beta}_{\ \ \mu\alpha}\, g_{\beta\nu}\right) = g^{\lambda\alpha}\left(S^{\beta}_{\ \ \alpha\nu}\, g_{\beta\mu} + S^{\beta}_{\ \ \alpha\mu}\, g_{\beta\nu}\right) \tag{2.58}$$

is symmetric in μ and ν. Obviously with a non-vanishing torsion but with vanishing $V^{\lambda}_{\ \ \mu\nu}$ we get from Eqs. (2.47) and (2.15):

$$\omega^{k}_{\ \ i\nu} - e_{\lambda}^{\ \ k}\, e_{i\ \ ,\nu}^{\ \ \lambda} = e_{\lambda}^{\ \ k}\, \gamma^{\lambda}_{\ \ \mu\nu}\, e_{i}^{\ \ \mu} = e_{\lambda}^{\ \ k}\, e_{i}^{\ \ \mu}\left(\begin{Bmatrix}\lambda\\ \mu\nu\end{Bmatrix} + S^{\lambda}_{\ \ \mu\nu}\right) \tag{2.59}$$

This is the case if in a metric-compatible Lorentzian manifold the torsion tensor $S_{\alpha\mu\nu}$ is totally anti-symmetric. Then the symmetric portion of the affine connection still satisfies the Levi-Civita relation which means that a straight line between two points on the manifold is also the geodesic (of shortest elapsed time or of extremal length) one [9] satisfying

$$\left(v^{\mu}\frac{\partial}{\partial x^{\mu}}\right) v^{\nu} + \begin{Bmatrix}\nu\\ \alpha\mu\end{Bmatrix} v^{\alpha}\, v^{\mu} = 0. \tag{2.60}$$

Then also a geodetic (also called "normal") reference system, in which the symmetric portion of the affine connection vanishes,[11] can be constructed (c.f. Sect. 2.10.2).

Because a totally anti-symmetric 3-tensor in four dimensions has only four independent elements, we can re-write it as

$$S_{\alpha\mu\nu} = \frac{1}{\sqrt{3!}}\, \epsilon_{\alpha\mu\nu\sigma}\, A^{\sigma} \tag{2.61}$$

where we use the totally anti-symmetric covariant [3]

$$\epsilon_{\alpha\beta\gamma\delta} = \sqrt{-g}\, \{\alpha\beta\gamma\delta\}\,.$$

$\{\alpha\beta\gamma\delta\}$ is the so called that is totally anti-symmetric with $\{\alpha\beta\gamma\delta\} = -1$. The contravariant Levi-Civita tensor is then

$$\epsilon^{\alpha\beta\gamma\delta} = g^{\alpha\mu}\, g^{\beta\nu}\, g^{\gamma\rho}\, g^{\delta\sigma}\, \sqrt{-g}\, \{\mu\nu\rho\sigma\} = -g^{-1}\sqrt{-g}\, \{\alpha\beta\gamma\delta\} = -\frac{1}{\sqrt{-g}}\, \{\alpha\beta\gamma\delta\}.$$

Because of the relation

$$\epsilon^{\mu\beta\gamma\delta}\, \epsilon_{\nu\beta\gamma\delta} = 3!\, \delta^{\mu}_{\nu}$$

[11] The totally anti-symmetric torsion and the existence of a geodetic frame and of the Newton (weak field) limit was addressed by M. Socolovsky [15].

Eq. (2.61) can be reversed:

$$A^\sigma := \frac{1}{\sqrt{3!}} \epsilon^{\sigma\alpha\mu\nu} S_{\alpha\mu\nu} . \tag{2.62}$$

For a general tensor $S_{\alpha\mu\nu}$ the axial vector A_σ represents just its totally anti-symmetric portion aka axial torsion.[12]

2.7 Dimensions

The basic construct on a manifold is the velocity as defined in Eq. (2.1) on page 8:

$$v_{\zeta,p}(f) := \frac{d}{d\lambda}(f \circ \zeta)\,|_{\lambda=0} .$$

Elapsed time—of length dimension L—is defined in the postulate P_2. With Eq. (2.2a) and the dimension of velocity $[v^\mu] = 1$ we conclude that the parameter λ corresponds to eigentime and thus also has the dimension L. With this input we infer:

$$\begin{aligned}
[\partial_\mu] &= \frac{1}{L}\\
[g_{\mu\nu}] &= [\eta_{ij}] = 1\\
[\mathfrak{g}] &= L^2\\
[e_\mu{}^i] &= [e_i{}^\mu] = 1\\
[e_i] &= [v] = \frac{1}{L}\\
[\varepsilon^i] &= L\\
[\omega^i{}_{k\mu}] &= [\gamma_\alpha{}^{\mu\nu}] = [C^i{}_{jk}] = \frac{1}{L}\\
[a^\mu] &= \frac{1}{L}\\
[a] &= \frac{1}{L^2}\\
[S^\alpha{}_{\mu\nu}] &= [K^\alpha{}_{\mu\nu}] = [N^\alpha{}_{\mu\nu}] = \frac{1}{L}\\
[R^\alpha{}_{\beta\mu\nu}] &= \frac{1}{L^2} .
\end{aligned}$$

A careful look at the dimensions of tensor expressions is important when checking the consistency of expressions and especially in constructing Lagrangian or Hamiltonian densities.

---

12 For a thorough discussion of irreducible components of the torsion and contortion tensors see [16].

2.8 Ricci Rotation Coefficients and Riemann-Cartan Geometry

We return to Eq. (2.59) and then split off the Lorentz connection in Eq. (2.52). For simplicity a metric-compatible spacetime is assumed in which the non-metricity tensor vanishes and $L^{\lambda}{}_{\mu\nu} \equiv 0$. Then

$$\omega^{k}{}_{i\nu} = \bar{\omega}^{k}{}_{i\nu} + e_{\lambda}{}^{k}\, e_{i}{}^{\mu}\, K^{\lambda}{}_{\mu\nu}, \tag{2.63}$$

where the bar over ω (and over other symbols in the following) means that in its definition the affine connection has been substituted by the Christoffel symbol. This gives

$$\bar{\omega}^{k}{}_{i\nu} := e_{\lambda}{}^{k} \frac{\partial e_{i}{}^{\lambda}}{\partial x^{\nu}} + e_{\lambda}{}^{k}\, e_{i}{}^{\mu} \begin{Bmatrix} \lambda \\ \mu\nu \end{Bmatrix} = e_{i}{}^{\mu} \left(-\frac{\partial e_{\mu}{}^{k}}{\partial x^{\nu}} + e_{\lambda}{}^{k} \begin{Bmatrix} \lambda \\ \mu\nu \end{Bmatrix} \right), \tag{2.64}$$

called the [17].

Now, expanding the metric in the expression (2.53) with (2.6) and expressing the Christoffel symbol in terms of the vierbein components and their derivatives (see also [18]) gives:

$$\begin{aligned} \begin{Bmatrix} \lambda \\ \mu\nu \end{Bmatrix} &= \tfrac{1}{2} e_{k}{}^{\lambda} e_{i}{}^{\rho} \eta^{ki}\, \eta_{mn} \left[\frac{\partial}{\partial x^{\mu}} (e_{\nu}{}^{n} e_{\rho}{}^{m}) + \frac{\partial}{\partial x^{\nu}} (e_{\mu}{}^{n} e_{\rho}{}^{m}) - \frac{\partial}{\partial x^{\rho}} (e_{\nu}{}^{n} e_{\mu}{}^{m}) \right] \\ &= \tfrac{1}{2} e_{j}{}^{\lambda} \left(\frac{\partial e_{\mu}{}^{j}}{\partial x^{\nu}} + \frac{\partial e_{\nu}{}^{j}}{\partial x^{\mu}} \right) \\ &\quad + \tfrac{1}{2} \eta^{ik}\, e_{k}{}^{\lambda} e_{i}{}^{\rho}\, \eta_{mn} \left[e_{\nu}{}^{n} \left(\frac{\partial e_{\rho}{}^{m}}{\partial x^{\mu}} - \frac{\partial e_{\mu}{}^{m}}{\partial x^{\rho}} \right) + e_{\mu}{}^{n} \left(\frac{\partial e_{\rho}{}^{m}}{\partial x^{\nu}} - \frac{\partial e_{\nu}{}^{m}}{\partial x^{\rho}} \right) \right]. \end{aligned}$$

Inserting this into

$$\bar{\omega}^{k}{}_{in} = \bar{\omega}^{k}{}_{i\nu}\, e_{n}{}^{\nu}$$

gives

$$\begin{aligned} \bar{\omega}^{k}{}_{in} &= \tfrac{1}{2} [\, e_{i}{}^{\mu} e_{n}{}^{\nu} \left(\frac{\partial e_{\nu}{}^{k}}{\partial x^{\mu}} - \frac{\partial e_{\mu}{}^{k}}{\partial x^{\nu}} \right) \\ &\quad + \eta^{ab}\, \eta_{cd}\, \delta^{k}_{a}\, \delta^{c}_{n}\, e_{i}{}^{\mu} e_{b}{}^{\rho} \left(\frac{\partial e_{\rho}{}^{d}}{\partial x^{\mu}} - \frac{\partial e_{\mu}{}^{d}}{\partial x^{\rho}} \right) + \eta^{ab}\, \eta_{cd}\, \delta^{k}_{a}\, \delta^{c}_{i}\, e_{b}{}^{\rho} e_{n}{}^{\nu} \left(\frac{\partial e_{\rho}{}^{d}}{\partial x^{\nu}} - \frac{\partial e_{\nu}{}^{d}}{\partial x^{\rho}} \right)] \\ &= \tfrac{1}{2} \left(-C^{k}{}_{in} + \eta^{ak}\, \eta_{cn}\, C^{c}{}_{ai} + \eta^{ak}\, \eta_{ci}\, C^{c}{}_{an} \right) = \tfrac{1}{2} \eta^{ak} \left(-C_{ain} + C_{nai} - C_{ina} \right), \end{aligned} \tag{2.65}$$

where the $C^{i}{}_{jk}$ are the defined in Eq. (2.41). Notice that this implies

$$\bar{\omega}^k{}_{ni} - \bar{\omega}^k{}_{in} = C^k{}_{in}$$
$$\bar{\omega}_{kin} + \bar{\omega}_{ikn} = 0$$
$$\omega_{kin} = \tfrac{1}{2}\left(-C_{kin} + C_{nki} - C_{ink}\right) + K_{kin}.$$

The first of these three equations is, according to Eq. (2.49), equivalent to $S^\alpha{}_{\mu\nu}(\bar{\omega}) \equiv \bar{S}^\alpha{}_{\mu\nu} = 0$, recovering the fact that for the Levi-Civita case the associated Cartan torsion vanishes.

Notice that since contortion is anti-symmetric in its first two indices, the corresponding anti-symmetry of the Lorentz connection is not violated by torsion.

Moreover, inserting the relation (2.63) into the definition (2.43) of the Riemann-Cartan curvature tensor we find after a straightforward calculation

$$\begin{aligned} R^\lambda{}_{\sigma\mu\nu}(\omega^k{}_{i\nu}) &= R^\lambda{}_{\sigma\mu\nu}(\bar{\omega}^k{}_{i\nu} + e_\lambda{}^k e_i{}^\mu K^\lambda{}_{\mu\nu}) \\ &= R^\lambda{}_{\sigma\mu\nu}(\bar{\omega}^k{}_{i\nu}) + \bar{\nabla}_\mu K^\lambda{}_{\sigma\nu} - \bar{\nabla}_\nu K^\lambda{}_{\sigma\mu} + K^\lambda{}_{\beta\mu} K^\beta{}_{\sigma\nu} - K^\lambda{}_{\beta\nu} K^\beta{}_{\sigma\mu} \\ &=: \bar{R}^\lambda{}_{\sigma\mu\nu} + P^\lambda{}_{\sigma\mu\nu} \end{aligned} \tag{2.66}$$

where we have set

$$R^\lambda{}_{\sigma\mu\nu} \equiv R^\lambda{}_{\sigma\mu\nu}(\omega), \qquad \bar{R}^\lambda{}_{\sigma\mu\nu} \equiv R^\lambda{}_{\sigma\mu\nu}(\bar{\omega})$$

for the Riemann-Cartan and Riemann tensors, respectively, and call

$$P^\lambda{}_{\sigma\mu\nu} := \bar{\nabla}_\mu K^\lambda{}_{\sigma\nu} - \bar{\nabla}_\nu K^\lambda{}_{\sigma\mu} + K^\lambda{}_{\beta\mu} K^\beta{}_{\sigma\nu} - K^\lambda{}_{\beta\nu} K^\beta{}_{\sigma\mu} \tag{2.67}$$

the . The Riemann-Cartan curvature tensor is anti-symmetric in its second pair of indices (by definition) and, for an anti-symmetric spin connection, i.e. in the absence of the non-metricity tensor, also in its first two indices. That follows directly from the anti-symmetry of the contortion tensor (2.57) in its two left-most indices. For the Ricci tensor and the curvature scalar we then obtain

$$R_{\mu\nu} = \bar{R}_{\mu\nu} + K_\sigma K^\sigma{}_{\mu\nu} - K^\sigma{}_{\beta\nu} K^\beta{}_{\mu\sigma} + \bar{\nabla}_\sigma K^\sigma{}_{\mu\nu} - \bar{\nabla}_\nu K_\mu \tag{2.68}$$
$$R = \bar{R} + g^{\mu\nu}\left(K_\sigma K^\sigma{}_{\mu\nu} - K^\sigma{}_{\beta\nu} K^\beta{}_{\mu\sigma} + \bar{\nabla}_\sigma K^\sigma{}_{\mu\nu} - \bar{\nabla}_\nu K_\mu\right) \tag{2.69}$$

with the contracted contortion being equivalent to the contracted torsion tensor:

$$K_\sigma := K^\lambda{}_{\sigma\lambda} = 2\, S^\lambda{}_{\sigma\lambda} \equiv 2\, S_\sigma.$$

Notice for later use that while $\bar{R}_{[\mu\nu]} = 0$, the identity

$$\begin{aligned} R_{[\mu\nu]} \equiv P_{[\mu\nu]} &= (\nabla_\lambda - 2\, S_\lambda)\, S^\lambda{}_{\mu\nu} - (\nabla_\mu S_\nu - \nabla_\nu S_\mu) \\ &= (\bar{\nabla}_\lambda - 2\, S_\lambda)\, S^\lambda{}_{\mu\nu} - (\bar{\nabla}_\mu S_\nu - \bar{\nabla}_\nu S_\mu) \end{aligned} \tag{2.70}$$

holds [19]. This can also be re-written as

$$R_{[\mu\nu]} = \left(\bar{\nabla}_\lambda - 2\,S_\lambda\right)\left(S^\lambda{}_{\mu\nu} - \delta^\lambda_\mu\,S_\nu + \delta^\lambda_\nu\,S_\mu\right). \tag{2.71}$$

Interestingly, even in an overall flat Riemann-Cartan spacetime ($R^\lambda{}_{\sigma\mu\nu} = 0$) the Riemann curvature $\bar{R}^\lambda{}_{\sigma\mu\nu}$ that is used in general relativity may still be nonzero due to the presence of torsion.

2.9 Teleparallel Gravity

A combination of zero curvature with non-zero torsion for a non-trivial spacetime has been scrutinized under the tag of "Fernparallelismus" or "Teleparallel Gravity" (TEGR). This goes back to the 1920s as Einstein set out to establish a unified field theory based on geometric principles that would describe both the gravitation and the electromagnetic interaction.[13] In his ansatz, Einstein postulated that the vierbein basis vectors remain parallel upon translation, i.e. that

$$\hat{\nabla}_\nu\, e_\mu{}^k = 0. \tag{2.72}$$

We put a hat on this specific, so called Weitzenböck connection, and quantities pertinent to this special choice of spacetime. Then also

$$\frac{\partial e_\mu{}^k}{\partial x^\nu} = \hat{\Gamma}^\alpha{}_{\mu\nu}\, e_\alpha{}^k \tag{2.73}$$

holds globally. According to the definition (2.12) the Lorentz connection then vanishes,

$$\hat{\omega}^k{}_{i\nu} := -e_i{}^\mu\,\hat{\nabla}_\nu\, e_\mu{}^k = 0.$$

For a to be non-trivial, the affine connection must not be symmetric. Otherwise, $\hat{\Gamma}^\mu{}_{\alpha\nu} = \hat{\Gamma}^\mu{}_{\nu\alpha}$ ensures that we can than find a chart map, $x \mapsto y$, such that the connection vanishes at point p. By the integrability condition (2.73) a linear transformation on the bundle then exists that extends $\hat{\Gamma}^\mu{}_{\nu\alpha}(x) = 0$ to $\hat{\Gamma}^\mu{}_{\nu\alpha}(y)$ such that it vanishes globally, and the spacetime under consideration is trivial, namely flat and free of torsion [9]. Indeed, with the globally vanishing Lorentz connection we find from Eqs. (2.43) and (2.48) directly

[13] See the overview article by Einstein [20]. The mathematical formalism was developed independently by Cartan [21] and Weitzenböck [22]. That approach was based on the requirement of distant or teleparallelism of vectors but was abandoned by Einstein in the early 1930s. Although fundamental objections exist, see also for example Grossmann [23], it is still discussed in the literature, e.g. in [24].

$$\hat{R}^{a}{}_{b\mu\nu} \equiv R^{a}{}_{b\mu\nu}(\hat{\omega}) = R^{a}{}_{b\mu\nu}(0) = 0.$$

Notice that the vanishing of the Weitzenböck connection ensures metric compatibility in the sense of Eq. (2.26). However, as by the requirement (2.72) of total parallelism the affine connection must not be symmetric, we can see from the integrability condition (2.73):

$$\hat{\Gamma}^{\lambda}{}_{\mu\nu} = \frac{\partial e_{\mu}{}^{k}}{\partial x^{\nu}} e_{k}{}^{\lambda}, \tag{2.74}$$

and calculate the from the coefficients of anholonomy (2.41):

$$\hat{S}^{\lambda}{}_{\mu\nu} \equiv S^{\lambda}{}_{\mu\nu}(\hat{\omega}) = S^{\lambda}{}_{\mu\nu}(0) = \tfrac{1}{2} e_{i}{}^{\lambda} \left(\frac{\partial e_{\mu}{}^{i}}{\partial x^{\nu}} - \frac{\partial e_{\nu}{}^{i}}{\partial x^{\mu}} \right) = \tfrac{1}{2} e_{i}{}^{\lambda} e_{\mu}{}^{k} e_{\nu}{}^{j} C^{i}{}_{kj}. \tag{2.75}$$

Obviously, a non-trivial spacetime—with vanishing curvature and non-vanishing torsion—is intimately linked to the anholonomy of the inertial frame.

Furthermore, Eq. (2.63) yields a relation between the Ricci rotation coefficients and the Weitzenböck contortion:

$$\bar{\omega}^{k}{}_{i\nu} + e_{\lambda}{}^{k} e_{i}{}^{\mu} \hat{K}^{\lambda}{}_{\mu\nu} = 0. \tag{2.76}$$

or

$$\hat{K}^{\lambda}{}_{\mu\nu} = -e_{k}{}^{\lambda} e_{\mu}{}^{i} \bar{\omega}^{k}{}_{i\nu}. \tag{2.77}$$

The correlation (2.69) of the Riemann curvature scalar $\bar{R}$ with the Weitzenböck contortion has been explored in TEGR by replacing the Einstein-Hilbert Lagrangian by a term quadratic in the contortion tensor, c.f., e.g. [18, 24]. Moreover, the so called $f(T)$ theories, where T stands for torsion, have been invoked to explain effects attributed to dark energy [25, 26].

2.10 The Equivalence Principle

The acceleration along the trajectory ζ of a particle moving with the velocity $u = v_{\zeta}$ is defined in Eq. (2.23) or, in the chart-dependent notation, in Eq. (2.24). According to Eq. (2.22), $a_u = 0$ defines the world-line of a free particle (autoparallely transported curve i.e. a straight line) in curved spacetime geometry. It is important to recall here that, according to our postulate, a straight trajectory between two points is defined by the affine connection. It is in general not equal to the trajectory of extremal length, the so called geodesic, which is determined by the metric via the geodetic equation (2.60). They coincide only in metric manifolds with vanishing or totally anti-symmetric torsion. However, for an observer whose world-line at a point p

is represented by the tangent vector e_0, defining the temporal direction, and the attached orthogonal inertial frame $\{e_i\}$, the of the time direction e_0 gives with the definition (2.12) the n-th component of the acceleration in temporal direction:

$$a_0^n := e_\mu{}^n \left(\nabla_{e_0} e_0\right)^\mu = e_\mu{}^n \, e_0{}^\alpha \left(\nabla_\alpha e_0{}^\mu\right) = e_0{}^\alpha \, \omega^n{}_{0\alpha} = \omega^n{}_{00}. \tag{2.78}$$

(Notice that in this notation also the rightmost index in $\omega^n{}_{00}$ is a Lorentz index!) The right-hand side of Eq. (2.78) vanishes for $n = 0$ due to the anti-symmetry of ω_{ij0} in i, j, retaining the observer's time direction in the inertial frame.

2.10.1 The Weak Equivalence Principle

For $n = i \neq 0$ the co-moving observer notices an acceleration in the spatial direction i as if "its" coordinate system were subject to an external force. That force, though, is of purely geometrical origin and thus a pseudo force. Hence the pseudo force, when formulated as a force in Eq. (2.25), gives $\omega^i{}_{00} = F^i/m_{\text{inert}}$, and that gravitational force appears proportional to the inertial mass, $F^i \equiv m_{\text{inert}}\,\omega^i{}_{00}$. The gravitational mass of a particle is thus identical to its inertial mass. This identity is the so called Weak Equivalence Principle (WEP).

Moreover, the transport of the spatial base vectors e_i with $i = 1, 2, 3$ of the tangent space perpendicular to the trajectory that determines e_0 is calculated similarly to Eq. (2.78) to yield

$$a_i^n := e_\mu{}^n \left(\nabla_{e_0} e_i\right)^\mu = e_\mu{}^n \left(e_0{}^\alpha \nabla_\alpha\right) e_i{}^\mu = \omega^n{}_{i0}. \tag{2.79}$$

The Lorentz connection $\omega^n{}_{i0}$ is the n-th component of the acceleration of the spatial base vector e_i. For $n = 0$ the components $\omega^0{}_{i0}$, $i = 1, 2, 3$ describe tidal forces.

Moreover, a non-vanishing $\omega^i{}_{j0}$ with spatial indices $i, j = 1, 2, 3$ has, due to its anti-symmetry in i, j, three independent entries, and can thus be rewritten as

$$\omega^k = \tfrac{1}{2}\epsilon^{kij}\,\omega_{ij0}.$$

ω^k is the angular velocity of the frame's rotation which is a purely inertial effect (Coriolis force [18]).

2.10.2 The Geodetic System and the Strong Equivalence Principle

For any point p in a Lorentzian manifold M, as defined in Sect. 2.1.1, a distinguished, so called can be chosen such that the symmetric portion of the affine connection,

$\gamma^{\alpha}{}_{(\beta\gamma)}(p)$ vanishes, and with it the gravity term in the autoparallel transport Eq. (2.24). In torsion-free geometries the connection is symmetric and hence vanishes then altogether.

The geodetic system is constructed as follows. Let (V, y) be a chart with $p \in V \subset M$ where in general $\gamma_{(y)}{}^{\alpha}{}_{\beta\gamma}(p) \neq 0$. Then consider a new chart, $(U, x) \ni p$, and a shift vector $\xi = (\xi^0, \ldots, \xi^3) \in V(p)$ such that $p \equiv y^{-1}(\xi = 0)$. Then we define the chart map $V \to U, \xi \mapsto x$, as

$$(x \circ y^{-1})^{\mu}(\xi) = \xi^{\mu} - \tfrac{1}{2}\gamma_{(y)}{}^{\mu}{}_{\beta\gamma}(p)\, \xi^{\beta}\, \xi^{\gamma}.$$

This gives the transformation coefficient matrices

$$\frac{\partial x^{\mu}}{\partial y^{\alpha}} = \frac{\partial}{\partial \xi^{\alpha}} (x \circ y^{-1})^{\mu}(\xi) = \delta^{\mu}_{\alpha} - \tfrac{1}{2}\left(\gamma_{(y)}{}^{\mu}{}_{\alpha\beta}(p) + \gamma_{(y)}{}^{\mu}{}_{\beta\alpha}(p)\right) \xi^{\beta}$$

and

$$\frac{\partial^2 x^{\mu}}{\partial y^{\alpha}\partial y^{\beta}} = -\tfrac{1}{2}\left(\gamma_{(y)}{}^{\mu}{}_{\alpha\beta}(p) + \gamma_{(y)}{}^{\mu}{}_{\beta\alpha}(p)\right) = -\gamma_{(y)}{}^{\mu}{}_{(\alpha\beta)}(p).$$

Using the transformation relation (2.34) of the connection coefficients we get

$$\begin{aligned}\gamma_{(x)}{}^{\mu}{}_{\alpha\beta}(\xi = 0) &= \delta^{\mu}_{\tau}\left(\delta^{\sigma}_{\alpha}\,\delta^{\rho}_{\beta}\,\gamma_{(y)}{}^{\tau}{}_{\sigma\rho}(p) - \gamma_{(y)}{}^{\tau}{}_{(\alpha\beta)}(p)\right)\\ &= \gamma_{(y)}{}^{\mu}{}_{\alpha\beta}(p) - \gamma_{(y)}{}^{\mu}{}_{(\alpha\beta)}(p) = \gamma_{(y)}{}^{\mu}{}_{[\alpha\beta]}(p) = S^{\mu}{}_{\alpha\beta}(p),\end{aligned}$$

i.e. we have transformed the affine connection to a residual torsion. That residual term does not contribute to the autoparallel-transport equation (2.24), as only the symmetric portion of the connection contributes. Hence in the geodetic system, in absence of other interactions, the observer moves on a force-free autoparallel trajectory.[14] This is called the strong equivalence principle (SEP).

However, with Eqs. (2.52) and (2.58) the symmetric portion of the connection in metric-compatible geometries with a general torsion,

$$\gamma_{(y)}{}^{\lambda}{}_{(\mu\nu)}(p) = \left\{ {\lambda \atop \mu\nu} \right\} - V^{\lambda}{}_{\mu\nu} = \left\{ {\lambda \atop \mu\nu} \right\} - g^{\lambda\alpha}\left(S^{\beta}{}_{\alpha\nu}\, g_{\beta\mu} - S^{\beta}{}_{\mu\alpha}\, g_{\beta\nu}\right) = 0, \tag{2.80}$$

will retain a torsion-dependent contribution. Then the autoparallel equation that determines force-free, straight world-lines, will not coincide with the geodetic equation (2.60) that describes a world-line with extremal length. It looks similarly to Eq. (2.24), but with the affine connection substituted by the Christoffel symbol. Only in geometries with a totally anti-symmetric or no torsion does the geodesic coincide with a force-free motion, and the extremal and straight world-lines are identical.

---

14 See also [27] for a discussion of the Equivalence Principle in U_4.

In any case, though, torsion and also the derivative of the connection contribute to the curvature tensor and give rise to tidal forces distorting adjacent trajectories (and extended bodies) in direction perpendicular to that free (autoparallel or geodetic) trajectory.[15] Except for idealized point-like test particles, gravity will thus in general not remain unnoticed by the observers.

References

1. C.J. Isham, Modern differential geometry for physicists (1999)
2. M. Nakahara, *Geometry, Topology and Physics*, ed. by D.F. Brewer, 2nd edn. (Taylor & Francis Ltd, 2003), p. 596. ISBN: 0750306068. http://www.ebook.de/de/product/3646467/mikio_nakahara_geometry_topology_and_physics.html
3. C.W. Misner, K.S. Thorne, J.A. Wheeler, *Gravitation* (W. H. Freeman and Company, New York, 1973)
4. S. Hawking, *The Theory of Everything* (New Millenium Press, 2003). ISBN: 978-1-59777-611-0
5. R. Carballo-Rubio et al., Causal hierarchy in modified gravity. JHEP **12**, 055 (2020). https://doi.org/10.1007/JHEP12(2020)055. (arXiv: 2005.08533 [gr-qc])
6. D.D. Sega, D. Galviz, Are the notions of past, present and future compatible with the General 2021. arXiv: 2005.10065 [gr-qc]
7. N. Swanson, On the Ostrogradski instability; or, why physics really uses second derivatives. British J. Philos. Sci. **73**(1), 23–46 (2022). https://doi.org/10.1093/bjps/axz042
8. F. Gronwald, Metric-affine gauge theory of gravity, I. Fundamental structure and field equations (1997), arXiv:gr-qc/9702034v1
9. E. Schrödinger, *Space-Time Structure* (Cambridge University Press, 2002), p. 128. ISBN: 0521315204. http://www.ebook.de/de/product/4284759/erwin_schrodinger_erwin_schr_dinger_space_time_structure.html
10. L. Lusanna, M. Pauri, General covariance and the objectivity of spacetime point events: the Physical role of gravitational and gauge degrees of freedom in general relativity (2003), arXiv:gr-qc/0301040
11. E. Guendelman, E. Nissimov, S. Pacheva, Int. J. Mod. Phys. A **30**, 1550133 (2015). https://doi.org/10.1142/S0217751X1550133X
12. P. Matteucci, Einstein-Dirac theory on gauge-natural bundles. Rep. Math. Phys. **52**, 115–139 (2003). https://doi.org/10.1016/S0034-4877(03)90007-3. arxiv.org/abs/gr-qc/0201079
13. M. Godina, P. Matteucci, The Lie derivative of spinor fields: theory and applications. Int. J. Geom. Methods Mod. Phys. **2**, 159–188 (2005). https://doi.org/10.1142/S0219887805000624. arxiv.org/abs/math/0504366
14. F.W. Hehl, Gauge theory of gravity and spacetime (2014). https://doi.org/10.1007/978-1-4939-3210-8_5. arXiv:1204.3672%20 [gr-qc]
15. M. Socolovsky, Annales Fond. Louis Broglie **37**, 73 (2012)
16. S. Capozziello, C. Stornaiolo, Torsion tensor and its geometric interpretation. Annales Fond. Broglie **32**(2–3), 196–214 (2007)
17. K. Hayashi, T. Shirafuji, Gravity from the Poincaré gauge theory of fundamental interactions. Prog. Theor. Phys. **64**(3) 866, 883, 1435, 2222 (1980). https://doi.org/10.1143/PTP.64.866
18. R. Aldrovandi, P.B. Barros, J.G. Pereira, Spin and anhomology in general relativity (2004). arXiv:gr-qc/0402022
19. R.T. Hammond, Strings in gravity with torsion. Gener. Relat. Gravit. **32**(10), 2007–2019 (2000). https://doi.org/10.1023/a:1001942301598

[15] The dynamical equation behind this effect is the so-called geodesic-deviation equation [3].

20. A. Einstein, Auf die Riemann-Metrik und den Fern-Parallelismus gegründete einheitliche Feldtheorie. Math. Ann. **102**, 685 (1930). https://doi.org/10.1007/BF01782370
21. E. Cartan, Notice historique sur la notion de parallélisme absolu. Math. Ann. **102**, 689 (1930)
22. R. Weitzenboeck, Differentialinvarianten in der Einsteinschen Theorie des Fernparallelismus, in *Sitzungsberichte der Preussischen Akademie der Wissenschaften* (1928), p. 466
23. M. Grossmann, Fernparallelismus?, in *Vierteljahresschrift der Naturforschenden Gesellschaft in Zürich*, vol. 76 (1931), p. 42
24. J.W. Maluf Teleparallel equivalent of general relativity. Phys. Rev. D **67**, 108501 (2003)
25. E.V. Linder, Einstein's other gravity and the acceleration of the universe **81** (2010). arXiv:1005.3039%20 [astro-ph.CO]
26. G.R. Bengochea, R. Ferraro, Dark torsion as the cosmic speed-up. Phys. Rev. D **79**, 24019 (2009). https://doi.org/10.1103/PhysRevD.79.124019. arXiv:0812.1205%20 [astro-ph]
27. P. von der Heyde, The equivalence principle in the U4 theory of gravitation. Nuovo Cimento **14**(7), 250–252 (1975)

Chapter 3
Theorem for Scalar-Valued Functions of Absolute and Relative Tensors

In this chapter we present the proof of a useful theorem for that constitutes the tensor analogue of Euler's theorem on homogeneous functions of calculus, following Ref. [1]. We first prove the theorem for absolute scalars constituted by contractions of (n, n)-tensors, which on their part may be tensor products of lower rank tensors. On that basis, the theorem is generalized for tensors of multiple index classes, such as the Lorentz indices of the vierbeins, or for spinor-tensors (e.g. Dirac matrices), and furthermore for relative tensors that contract to a relative scalar of weight w. Since the Lagrangians of classical field theories that describe the dynamics of classical scalar, vector, and spinor matter in dynamical geometries represent relative scalars of weight $w = 1$, which are commonly referred to as scalar densities, this theorem is well suited to derive identities between those tensor expressions. It will be applied in the following sections, for example in deriving correlations of the metric energy-momentum tensors to their canonical versions.

3.1 Absolute Tensors

Proposition 3.1 *Let $S = S(T) \in \mathbb{R}$ be a scalar-valued function constructed from a complete contraction of a (n, n)-tensor $T^{\xi_1 \cdots \xi_n}{}_{\eta_1 \ldots \eta_n}$, with the ordered index set $\{\eta_k\}$ and a bijective permutation $\{\xi_{\pi(k)}\}$ of the ordered index set $\{\xi_j\}$ such that $\eta_k \equiv \xi_{\pi(k)} \equiv \xi_j$. Then the following identity holds:*

$$
\begin{aligned}
&- \frac{\partial S}{\partial T^{\nu\,\xi_2 \ldots \xi_n}{}_{\eta_1 \ldots \eta_n}} T^{\mu\,\xi_2 \ldots \xi_n}{}_{\eta_1 \ldots \eta_n} - \cdots - \frac{\partial S}{\partial T^{\xi_1 \ldots \xi_{n-1}\,\nu}{}_{\eta_1 \ldots \eta_n}} T^{\xi_1 \ldots \xi_{n-1}\,\mu}{}_{\eta_1 \ldots \eta_n} \\
&+ \frac{\partial S}{\partial T^{\xi_1 \ldots \xi_n}{}_{\mu\,\eta_2 \ldots \eta_n}} T^{\xi_1 \ldots \xi_n}{}_{\nu\,\eta_2 \ldots \eta_n} + \cdots + \frac{\partial S}{\partial T^{\xi_1 \ldots \xi_n}{}_{\eta_1 \ldots \eta_{n-1}\,\mu}} T^{\xi_1 \ldots \xi_n}{}_{\eta_1 \ldots \eta_{n-1}\,\nu} \\
&\equiv 0. \qquad (3.1)
\end{aligned}
$$

D. Vasak et al., *Covariant Canonical Gauge Gravity*, FIAS Interdisciplinary Science Series, https://doi.org/10.1007/978-3-031-43717-5_3

Here in each derivative term a previously contracted index ξ_j resp. η_j of S is replaced by the now open indices ν resp. μ.

Proof The proof is based on the simple relation

$$\frac{\partial S}{\partial T^{\alpha}{}_{\mu}} T^{\alpha}{}_{\nu} = \frac{\partial T^{\beta}{}_{\beta}}{\partial T^{\alpha}{}_{\mu}} T^{\alpha}{}_{\nu} = \delta^{\beta}_{\alpha}\,\delta^{\mu}_{\beta}\,T^{\alpha}{}_{\nu} = T^{\mu}{}_{\nu} = \delta^{\beta}_{\nu}\,\delta^{\alpha}_{\beta}\,T^{\mu}{}_{\alpha} = \frac{\partial T^{\beta}{}_{\beta}}{\partial T^{\nu}{}_{\alpha}} T^{\mu}{}_{\alpha} = \frac{\partial S}{\partial T^{\nu}{}_{\alpha}} T^{\mu}{}_{\alpha}\,.$$

It is straightforward to see now that this relation applies for any pair of contracted indices in Eq. (3.1). Then for any pair ξ_j and η_k of contracted indices $\eta_k \equiv \xi_{\pi(k)} \equiv \xi_j$ of $S = T^{\xi_1\ldots\xi_n}{}_{\eta_1\ldots\eta_n}$ we find for their respective replacement by ν and μ:

$$\begin{aligned}
&- \frac{\partial S}{\partial T^{\xi_1\ldots\xi_{j-1}\,\nu\,\xi_{j+1}\ldots\xi_n}{}_{\eta_1\ldots\eta_{k-1}\,\eta_k\,\eta_{k+1}\ldots\eta_n}} T^{\xi_1\ldots\xi_{j-1}\,\mu\,\xi_{j+1}\ldots\xi_n}{}_{\eta_1\ldots\eta_{k-1}\,\eta_k\,\eta_{k+1}\ldots\eta_n} \\
&+ \frac{\partial S}{\partial T^{\xi_1\ldots\xi_{j-1}\,\xi_j\,\xi_{j+1}\ldots\xi_n}{}_{\eta_1\ldots\eta_{k-1}\,\mu\,\eta_{k+1}\ldots\eta_n}} T^{\xi_1\ldots\xi_{j-1}\,\xi_j\,\xi_{j+1}\ldots\xi_n}{}_{\eta_1\ldots\eta_{k-1}\,\nu\,\eta_{k+1}\ldots\eta_n} \\
&= -\delta^{\xi_j}_{\nu}\, T^{\xi_1\ldots\xi_{j-1}\,\mu\,\xi_{j+1}\ldots\xi_n}{}_{\eta_1\ldots\eta_{k-1}\,\eta_k\,\eta_{k+1}\ldots\eta_n} + \delta^{\mu}_{\eta_k}\, T^{\xi_1\ldots\xi_{j-1}\,\xi_j\,\xi_{j+1}\ldots\xi_n}{}_{\eta_1\ldots\eta_{k-1}\,\nu\,\eta_{k+1}\ldots\eta_n} \\
&= -\delta^{\xi_j}_{\nu}\, T^{\xi_1\ldots\xi_{j-1}\,\mu\,\xi_{j+1}\ldots\xi_n}{}_{\eta_1\ldots\eta_{k-1}\,\xi_j\,\eta_{k+1}\ldots\eta_n} + \delta^{\mu}_{\eta_k}\, T^{\xi_1\ldots\xi_{j-1}\,\eta_k\,\xi_{j+1}\ldots\xi_n}{}_{\eta_1\ldots\eta_{k-1}\,\nu\,\eta_{k+1}\ldots\eta_n} \\
&= -T^{\xi_1\ldots\xi_{j-1}\,\mu\,\xi_{j+1}\ldots\xi_n}{}_{\eta_1\ldots\eta_{k-1}\,\nu\,\eta_{k+1}\ldots\eta_n} + T^{\xi_1\ldots\xi_{j-1}\,\mu\,\xi_{j+1}\ldots\xi_n}{}_{\eta_1\ldots\eta_{k-1}\,\nu\,\eta_{k+1}\ldots\eta_n} \\
&\equiv 0.
\end{aligned}$$

This proves the assertion. □

The (n, n)-tensor $T^{\xi_1\ldots\xi_n}{}_{\eta_1\ldots\eta_n}$ may in particular be a tensor product of tensors of lower ranks. This is demonstrated in the following example:

✍ Example 1: Scalar from contraction of three tensors

Let S be a scalar emerging from the contraction of arbitrary tensors $A^{\alpha\beta}_{\xi}$, $B_{\alpha}{}^{\xi\eta}$, and C_{β}:

$$S = A^{\alpha\beta}_{\xi}\, B^{\xi\eta}_{\alpha}\, C_{\beta}\, C_{\eta}. \tag{3.2}$$

Then

$$\begin{aligned}
&- \frac{\partial S}{\partial A^{\nu\beta}{}_{\xi}} A^{\mu\beta}{}_{\xi} - \frac{\partial S}{\partial A^{\alpha\nu}{}_{\xi}} A^{\alpha\mu}{}_{\xi} + \frac{\partial S}{\partial A^{\alpha\beta}{}_{\mu}} A^{\alpha\beta}{}_{\nu} \\
&- \frac{\partial S}{\partial B^{\nu\eta}_{\alpha}} B^{\mu\eta}_{\alpha} - \frac{\partial S}{\partial B^{\xi\nu}_{\alpha}} B^{\xi\mu}_{\alpha} + \frac{\partial S}{\partial B^{\xi\eta}_{\mu}} B^{\xi\eta}_{\nu} + \frac{\partial S}{\partial C_{\mu}} C_{\nu} \equiv 0.
\end{aligned} \tag{3.3}$$

For the proof of the identity (3.3) the respective terms of the sum are worked out explicitly:

$$-\frac{\partial S}{\partial A^{\nu\beta}{}_{\xi}} A^{\mu\beta}{}_{\xi} = -\delta^{\alpha}_{\nu}\, A^{\mu\beta}{}_{\xi}\, B^{\xi\eta}_{\alpha}\, C_{\beta}\, C_{\eta} = -A^{\mu\beta}{}_{\xi}\, B^{\xi\eta}_{\nu}\, C_{\beta}\, C_{\eta}$$

$$
\begin{aligned}
-\frac{\partial S}{\partial A^{\alpha\nu}{}_{\xi}} A^{\alpha\mu}{}_{\xi} &= -\delta^{\beta}_{\nu} A^{\alpha\mu}{}_{\xi} B^{\xi\eta}_{\alpha} C_{\beta} C_{\eta} = -A^{\alpha\mu}{}_{\xi} B^{\xi\eta}_{\alpha} C_{\nu} C_{\eta} \\
\frac{\partial S}{\partial A^{\alpha\beta}{}_{\mu}} A^{\alpha\beta}{}_{\nu} &= \quad \delta^{\mu}_{\xi} A^{\alpha\beta}{}_{\nu} B^{\xi\eta}_{\alpha} C_{\beta} C_{\eta} = \quad A^{\alpha\beta}{}_{\nu} B^{\mu\eta}_{\alpha} C_{\beta} C_{\eta} \\
-\frac{\partial S}{\partial B^{\nu\eta}_{\alpha}} B^{\mu\eta}_{\alpha} &= -A^{\alpha\beta}{}_{\xi} \delta^{\xi}_{\nu} B^{\mu\eta}_{\alpha} C_{\beta} C_{\eta} = -A^{\alpha\beta}{}_{\nu} B^{\mu\eta}_{\alpha} C_{\beta} C_{\eta} \\
-\frac{\partial S}{\partial B_{\alpha}{}^{\xi\nu}} B_{\alpha}{}^{\xi\mu} &= -A^{\alpha\beta}{}_{\xi} \delta^{\eta}_{\nu} B^{\xi\mu}_{\alpha} C_{\beta} C_{\eta} = -A^{\alpha\beta}{}_{\xi} B^{\xi\mu}_{\alpha} C_{\beta} C_{\nu} \\
\frac{\partial S}{\partial B_{\mu}{}^{\xi\eta}} B_{\nu}{}^{\xi\eta} &= \quad A^{\alpha\beta}{}_{\xi} \delta^{\mu}_{\alpha} B^{\xi\eta}_{\nu} C_{\beta} C_{\eta} = \quad A^{\mu\beta}{}_{\xi} B^{\xi\eta}_{\nu} C_{\beta} C_{\eta} \\
\frac{\partial S}{\partial C_{\mu}} C_{\nu} &= \quad A^{\alpha\beta}{}_{\xi} B^{\xi\eta}_{\alpha} \delta^{\mu}_{\beta} C_{\nu} C_{\eta} + A^{\alpha\beta}{}_{\xi} B^{\xi\eta}_{\alpha} C_{\beta} \delta^{\mu}_{\eta} C_{\nu} \\
&= \quad A^{\alpha\mu}{}_{\xi} B^{\xi\eta}_{\alpha} C_{\nu} C_{\eta} + A^{\alpha\beta}{}_{\xi} B^{\xi\mu}_{\alpha} C_{\beta} C_{\nu}.
\end{aligned}
$$

The terms on the right-hand sides in each case occur twice with opposite signs—and thus sum up to zero.

Of course, this example can be generalized to the contraction of any number of tensors of any rank.

3.2 Scalar Functions Involving the Metric Tensor

With $S = S(g, T) \in \mathbb{R}$ a scalar-valued function constructed from the metric tensor $g_{\mu\nu}$ and an (n, m)-tensor $T^{\xi_1 \ldots \xi_n}{}_{\eta_1 \ldots \eta_m}$ of the physical fields, the following identity then holds for $(n - m)/2 \in \mathbb{Z}$:

$$
\begin{aligned}
&\frac{\partial S}{\partial g_{\mu\beta}} g_{\nu\beta} + \frac{\partial S}{\partial g_{\beta\mu}} g_{\beta\nu} \\
&\quad - \frac{\partial S}{\partial T^{\nu\,\xi_2 \ldots \xi_n}{}_{\eta_1 \ldots \eta_m}} T^{\mu\,\xi_2 \ldots \xi_n}{}_{\eta_1 \ldots \eta_m} - \cdots - \frac{\partial S}{\partial T^{\xi_1 \ldots \xi_{n-1}\,\nu}{}_{\eta_1 \ldots \eta_m}} T^{\xi_1 \ldots \xi_{n-1}\,\mu}{}_{\eta_1 \ldots \eta_m} \\
&\quad + \frac{\partial S}{\partial T^{\xi_1 \ldots \xi_n}{}_{\mu\,\eta_2 \ldots \eta_m}} T^{\xi_1 \ldots \xi_n}{}_{\nu\,\eta_2 \ldots \eta_m} + \cdots + \frac{\partial S}{\partial T^{\xi_1 \ldots \xi_n}{}_{\eta_1 \ldots \eta_{m-1}\,\mu}} T^{\xi_1 \ldots \xi_n}{}_{\eta_1 \ldots \eta_{m-1}\,\nu} \equiv 0.
\end{aligned}
\tag{3.4}
$$

The identity then provides the correlation of the *metric* and the *canonical* energy-momentum tensors of a given system.

Corollary 3.1 *Furthermore, the trace of Eq. (3.4) immediately gives the scalar identity:*

$$\frac{\partial S}{\partial g_{\alpha\beta}} g_{\alpha\beta} \equiv \frac{n-m}{2} \frac{\partial S}{\partial T^{\xi_1 \dots \xi_n}{}_{\eta_1 \dots \eta_m}} T^{\xi_1 \dots \xi_n}{}_{\eta_1 \dots \eta_m} . \tag{3.5}$$

3.3 Scalar Functions Involving Tensors with Multiple Index Classes

Corollary 3.2 *The theorem (3.1) holds also for scalars S constructed from generalized tensor objects—such as spinor-tensors—which are made of multiple index classes. Examples of such objects are Dirac matrices γ^μ, which are $(1, 1)$-spinor-$(1, 0)$-tensors, as well as vierbeins with one Lorentz and one spacetime index, respectively. One then encounters a specific identity for each particular index class, provided that all other indices are fully contracted.*

Proof For each fully contracted index group, Eq. (3.1) trivially vanishes. For the not fully contracted indices, Eq. (3.1) applies. □

3.4 Scalar Densities Built from Relative Tensors

Corollary 3.3 *Let a* relative scalar of weight *w—denoted by $\tilde{S} = S\left(\sqrt{-g}\right)^w \in \mathbb{R}$— be given as a function of the metric $g_{\mu\nu}$ and a tensor $T^{\xi_1 \dots \xi_n}{}_{\eta_1 \dots \eta_m}$ of rank (n, m). Then the following identity holds for $(n-m)/2 \in \mathbb{Z}$:*

$$\begin{aligned} &+ \frac{\partial \tilde{S}}{\partial g_{\mu\beta}} g_{\nu\beta} + \frac{\partial \tilde{S}}{\partial g_{\beta\mu}} g_{\beta\nu} \\ &- \frac{\partial \tilde{S}}{\partial T^{\nu\,\xi_2 \dots \xi_n}{}_{\eta_1 \dots \eta_m}} T^{\mu\,\xi_2 \dots \xi_n}{}_{\eta_1 \dots \eta_m} - \dots - \frac{\partial \tilde{S}}{\partial T^{\xi_1 \dots \xi_{n-1}\,\nu}{}_{\eta_1 \dots \eta_m}} T^{\xi_1 \dots \xi_{n-1}\,\mu}{}_{\eta_1 \dots \eta_m} \\ &+ \frac{\partial \tilde{S}}{\partial T^{\xi_1 \dots \xi_n}{}_{\mu\,\eta_2 \dots \eta_m}} T^{\xi_1 \dots \xi_n}{}_{\nu\,\eta_2 \dots \eta_m} + \dots + \frac{\partial \tilde{S}}{\partial T^{\xi_1 \dots \xi_n}{}_{\eta_1 \dots \eta_{m-1}\,\mu}} T^{\xi_1 \dots \xi_n}{}_{\eta_1 \dots \eta_{m-1}\,\nu} \\ &\equiv \delta^\mu_\nu\, w\, \tilde{S}. \end{aligned} \tag{3.6}$$

Proof Multiply Eq. (3.4) with $\left(\sqrt{-g}\right)^w$, add $\delta^\mu_\nu\, w\, \tilde{S}$ on both sides of the identity, and combine the appropriate terms on the left-hand side. With the general rule for the derivative of the determinant of the metric,

$$\frac{\partial \sqrt{-g}}{\partial g_{\mu\beta}} = \tfrac{1}{2} g^{\beta\mu} \sqrt{-g}, \tag{3.7}$$

one then gets:

$$\left(\frac{\partial S}{\partial g_{\mu\beta}} g_{\nu\beta} + \frac{\partial S}{\partial g_{\beta\mu}} g_{\beta\nu} + \delta^{\mu}_{\nu}\, w\, S\right)\left(\sqrt{-g}\right)^{w} = \frac{\partial \tilde{S}}{\partial g_{\mu\beta}} g_{\nu\beta} + \frac{\partial \tilde{S}}{\partial g_{\beta\mu}} g_{\beta\nu}.$$

Using the identity (3.4) gives Eq. (3.6). □

Equation (3.6) is obviously a representation of Euler's theorem on homogeneous functions in the realm of tensor calculus. The relative scalar $\tilde{S} = S\left(\sqrt{-g}\right)^{w}$ of weight w may in particular be the tensor product of some relative tensors of lower ranks and weights. The weight w of $\tilde{S}$ is then the *sum* of the weights of the relative tensors.

As a direct application, the above used relation for the derivative of the metric-determinant may also be recovered using Eq. (3.6).

✍ **Example 2:** Determinant of the metric tensor

The components of the covariant metric tensor, $g_{\mu\nu}(x)$, transform under the transition $x \mapsto X$ of the spacetime location as:

$$g_{\mu\nu}(X) = \frac{\partial x^{\alpha}}{\partial X^{\mu}}\, g_{\alpha\beta}(x) \frac{\partial x^{\beta}}{\partial X^{\nu}},$$

and hence its determinant $g \equiv \left|g_{\alpha\beta}(x)\right|$:

$$\left|g_{\mu\nu}(X)\right| = \left|g_{\alpha\beta}(x)\right| \left|\frac{\partial x}{\partial X}\right|^{2}.$$

The determinant g of the covariant metric tensor thus transforms as a *relative scalar* of weight $w = 2$. According to the general form of the identity for relative scalars of weight w from Eq. (3.6), we get for the relative scalar g due to the symmetry $g_{\beta\alpha} = g_{\alpha\beta}$:

$$\frac{\partial g}{\partial g_{\beta\mu}} g_{\beta\nu} + \frac{\partial g}{\partial g_{\mu\beta}} g_{\nu\beta} \equiv 2\,\delta^{\mu}_{\nu}\, g \quad \Rightarrow \quad \frac{\partial g}{\partial g_{\beta\mu}} g_{\beta\alpha} \equiv \delta^{\mu}_{\alpha}\, g. \tag{3.8}$$

Contracting (3.8) with the inverse metric $g^{\alpha\nu}$ yields the derivative of the determinant g of the covariant metric with respect to the component $g_{\nu\mu}$ of the metric:

$$\frac{\partial g}{\partial g_{\nu\mu}} \equiv g^{\mu\nu}\, g,$$

and thus for the negative square root of g we get Eq. (3.7) and also its dual version,

$$\frac{\partial \sqrt{-g}}{\partial g_{\nu\mu}} \equiv \tfrac{1}{2} g^{\mu\nu} \sqrt{-g}, \tag{3.9a}$$

$$\frac{\partial \sqrt{-g}}{\partial g^{\nu\mu}} \equiv -\tfrac{1}{2} g_{\mu\nu} \sqrt{-g}. \tag{3.9b}$$

3.5 Examples for Identities (3.1) Involving the Riemann Tensor

3.5.1 Riemann Tensor Squared

As a scalar, any Lagrangian $\mathcal{L}_R(R, g)$ built from the Riemann-Cartan tensor (2.43) and the metric satisfies the identity (3.1)

$$2\frac{\partial \mathcal{L}_R}{\partial g^{\nu\beta}} g^{\mu\beta} \equiv \frac{\partial \mathcal{L}_R}{\partial R^{\tau}{}_{\mu\beta\lambda}} R^{\tau}{}_{\nu\beta\lambda} - \frac{\partial \mathcal{L}_R}{\partial R^{\nu}{}_{\tau\beta\lambda}} R^{\mu}{}_{\tau\beta\lambda} + 2\frac{\partial \mathcal{L}_R}{\partial R^{\eta}{}_{\alpha\beta\mu}} R^{\eta}{}_{\alpha\beta\nu}. \tag{3.10}$$

The factors “2” emerge from the symmetry of the metric and the skew-symmetry of the Riemann-Cartan tensor in its last index pair. The identity is easily verified for a Lagrangian linear and quadratic in the Riemann tensor:

$$\mathcal{L}_R = R^{\eta}{}_{\alpha\beta\tau} \left[\tfrac{1}{4} R^{\alpha}{}_{\eta\xi\lambda} g^{\beta\xi} g^{\tau\lambda} - g_2 \left(\delta^{\beta}_{\eta}\, g^{\alpha\tau} - \delta^{\tau}_{\eta}\, g^{\alpha\beta}\right)\right].$$

The left-hand side of Eq. (3.10) evaluates to

$$2\frac{\partial \mathcal{L}_R}{\partial g^{\nu\beta}} g^{\mu\beta} = -R^{\eta\alpha\beta\mu}\, R_{\eta\alpha\beta\nu} + g_2 \left(R_{\nu}{}^{\mu} + R^{\mu}{}_{\nu}\right),$$

which indeed agrees with the terms obtained from the right-hand side:

$$\frac{\partial \mathcal{L}_R}{\partial R^{\tau}{}_{\mu\alpha\beta}} R^{\tau}{}_{\nu\alpha\beta} - \frac{\partial \mathcal{L}_R}{\partial R^{\nu}{}_{\tau\alpha\beta}} R^{\mu}{}_{\tau\alpha\beta} + 2\frac{\partial \mathcal{L}_R}{\partial R^{\tau}{}_{\alpha\beta\mu}} R^{\tau}{}_{\alpha\beta\nu} = -R^{\eta\alpha\beta\mu} R_{\eta\alpha\beta\nu} + g_2 \left(R_{\nu}{}^{\mu} + R^{\mu}{}_{\nu}\right).$$

3.5.2 Ricci Scalar

The Ricci scalar R is defined in Eq. (2.46) as the full contraction of the Riemann-Cartan tensor. Here the expression

$$R = R_{\eta\alpha\xi\lambda}\, g^{\eta\xi}\, g^{\alpha\lambda}. \tag{3.11}$$

builds the scalar Lagrangian $\mathcal{L}_R = R$. Then the generic identity (3.1) takes on the particular form

$$\frac{\partial R}{\partial g^{\nu\beta}} g^{\mu\beta} + \frac{\partial R}{\partial g^{\beta\nu}} g^{\beta\mu} - \frac{\partial R}{\partial R_{\mu\alpha\xi\lambda}} R_{\nu\alpha\xi\lambda} - \frac{\partial R}{\partial R_{\eta\mu\xi\lambda}} R_{\eta\nu\xi\lambda} - \frac{\partial R}{\partial R_{\eta\alpha\mu\lambda}} R_{\eta\alpha\nu\lambda} - \frac{\partial R}{\partial R_{\eta\alpha\xi\mu}} R_{\eta\alpha\xi\nu} \equiv 0. \tag{3.12}$$

Without making use of the symmetries of the Riemann tensor and the metric, this identity is actually fulfilled as

$$\frac{\partial R}{\partial g^{\nu\beta}} g^{\mu\beta} = R_{\eta\alpha\xi\lambda} \left(\delta^{\eta}_{\nu} \, \delta^{\xi}_{\beta} \, g^{\alpha\lambda} + g^{\eta\xi} \, \delta^{\alpha}_{\nu} \, \delta^{\lambda}_{\beta} \right) g^{\beta\mu}$$
$$= \left(R_{\nu\alpha\beta\lambda} \, g^{\alpha\lambda} + g^{\eta\xi} \, R_{\eta\nu\xi\beta} \right) g^{\beta\mu}$$
$$= R_{\nu\alpha}{}^{\mu\alpha} + R^{\xi}{}_{\nu\xi}{}^{\mu}.$$

Similarly

$$\frac{\partial R}{\partial g^{\beta\nu}} g^{\beta\mu} = R^{\mu}{}_{\alpha\nu}{}^{\alpha} + R^{\xi\mu}{}_{\xi\nu}.$$

The derivative terms of the Riemann tensor are

$$\frac{\partial R}{\partial R_{\mu\alpha\xi\lambda}} R_{\nu\alpha\xi\lambda} = \delta^{\mu}_{\eta} \, g^{\eta\xi} \, g^{\alpha\lambda} R_{\nu\alpha\xi\lambda} = R_{\nu\alpha}{}^{\mu\alpha}$$

and

$$\frac{\partial R}{\partial R_{\eta\mu\xi\lambda}} R_{\eta\nu\xi\lambda} = \delta^{\mu}_{\alpha} \, g^{\eta\xi} \, g^{\alpha\lambda} R_{\eta\nu\xi\lambda} = R^{\xi}{}_{\nu\xi}{}^{\mu}$$
$$\frac{\partial R}{\partial R_{\eta\alpha\mu\lambda}} R_{\eta\alpha\nu\lambda} = \delta^{\mu}_{\xi} \, g^{\eta\xi} \, g^{\alpha\lambda} R_{\eta\alpha\nu\lambda} = R^{\mu}{}_{\alpha\nu}{}^{\alpha}$$
$$\frac{\partial R}{\partial R_{\eta\alpha\xi\mu}} R_{\eta\alpha\xi\nu} = \delta^{\mu}_{\lambda} \, g^{\eta\xi} \, g^{\alpha\lambda} R_{\eta\alpha\xi\nu} = R^{\xi\mu}{}_{\xi\nu},$$

which obviously cancel the four terms emerging from the derivatives with respect to the metric.

Making now use of the skew-symmetries of the Riemann tensor in its first and second index pair and of the symmetry of the metric, Eq. (3.12) simplifies to

$$\frac{\partial R}{\partial g^{\nu\beta}} g^{\mu\beta} \equiv \frac{\partial R}{\partial R_{\mu\alpha\xi\lambda}} R_{\nu\alpha\xi\lambda} + \frac{\partial R}{\partial R_{\eta\alpha\xi\mu}} R_{\eta\alpha\xi\nu}. \tag{3.13}$$

For zero torsion, the Riemann tensor has the additional symmetry on exchange of both index pairs. Then

$$\frac{\partial R}{\partial g^{\nu\beta}} g^{\mu\beta} \equiv 2 \frac{\partial R}{\partial R_{\mu\alpha\xi\lambda}} R_{\nu\alpha\xi\lambda} \quad \Leftrightarrow \quad \frac{\partial R}{\partial g^{\nu\mu}} \equiv 2 \frac{\partial R}{\partial R^{\mu}{}_{\alpha\xi\lambda}} R_{\nu\alpha\xi\lambda}. \tag{3.14}$$

3.5.3 Ricci Tensor Squared

The scalar made of the (not necessarily symmetric) Ricci tensor $R_{\eta\alpha}$ is defined by the following contraction with the metric

$$\mathcal{L}_R = R_{\eta\alpha} R_{\xi\lambda}\, g^{\eta\xi}\, g^{\alpha\lambda}. \tag{3.15}$$

With Eq. (3.15), the general Eq. (3.1) now takes on the particular form

$$\frac{\partial\mathcal{L}_R}{\partial g^{\nu\beta}} g^{\mu\beta} + \frac{\partial\mathcal{L}_R}{\partial g^{\beta\nu}} g^{\beta\mu} - \frac{\partial\mathcal{L}_R}{\partial R_{\mu\beta}} R_{\nu\beta} - \frac{\partial\mathcal{L}_R}{\partial R_{\beta\mu}} R_{\beta\nu} \equiv 0. \tag{3.16}$$

Without making use of the symmetries of the Ricci tensor and the metric, this identity is actually fulfilled as

$$\begin{aligned}
\frac{\partial\mathcal{L}_R}{\partial g^{\nu\beta}} g^{\mu\beta} &= R_{\eta\alpha} R_{\xi\lambda} \left(\delta^{\eta}_{\nu}\, \delta^{\xi}_{\beta}\, g^{\alpha\lambda} + g^{\eta\xi}\, \delta^{\alpha}_{\nu}\, \delta^{\lambda}_{\beta}\right) g^{\beta\mu} \\
&= \left(R_{\nu\alpha} R_{\beta\lambda}\, g^{\alpha\lambda} + g^{\eta\xi}\, R_{\eta\nu} R_{\xi\beta}\right) g^{\beta\mu} \\
&= R_{\nu\beta} R^{\mu\beta} + R_{\beta\nu} R^{\beta\mu}.
\end{aligned}$$

Similarly

$$\frac{\partial\mathcal{L}_R}{\partial g^{\beta\nu}} g^{\beta\mu} = R_{\nu\beta} R^{\mu\beta} + R_{\beta\nu} R^{\beta\mu}.$$

The derivative terms of the Ricci tensor are

$$\frac{\partial\mathcal{L}_R}{\partial R_{\mu\beta}} R_{\nu\beta} = \left(\delta^{\mu}_{\eta}\delta^{\beta}_{\alpha} R_{\xi\lambda} + R_{\eta\alpha}\delta^{\mu}_{\xi}\delta^{\beta}_{\lambda}\right) g^{\eta\xi}\, g^{\alpha\lambda}\, R_{\nu\beta} = 2R_{\nu\beta}\, R^{\mu\beta}$$

and

$$\frac{\partial\mathcal{L}_R}{\partial R_{\beta\mu}} R_{\beta\nu} = \left(\delta^{\beta}_{\eta}\delta^{\mu}_{\alpha} R_{\xi\lambda} + R_{\eta\alpha}\delta^{\beta}_{\xi}\delta^{\mu}_{\lambda}\right) g^{\eta\xi}\, g^{\alpha\lambda}\, R_{\beta\nu} = 2R_{\beta\nu}\, R^{\beta\mu},$$

which obviously cancel the four terms emerging from the derivatives with respect to the metric.

For zero torsion, the Ricci tensor is symmetric. Then

$$\frac{\partial\mathcal{L}_R}{\partial g^{\nu\beta}} g^{\mu\beta} \equiv \frac{\partial\mathcal{L}_R}{\partial R_{\mu\beta}} R_{\nu\beta} \quad\Leftrightarrow\quad \frac{\partial\mathcal{L}_R}{\partial g^{\nu\mu}} \equiv \frac{\partial\mathcal{L}_R}{\partial R_{\alpha\beta}} R_{\nu\beta}\, g_{\alpha\mu}. \tag{3.17}$$

Reference

1. J. Struckmeier, A. van de Venn, D. Vasak, Identity for scalar-valued functions of tensors and its applications to energy-momentum tensors in classical field theories and gravity. Astronomische Nachrichten n/a.n/a (), e20220074. https://doi.org/10.1002/asna.20220074

Chapter 4
Gauge Theory of Gravity

In this chapter we provide a detailed description of the Covariant Canonical Gauge theory of Gravity (CCGG).

In Sect. 4.1 the formalism of canonical transformations in the realm of the De Donder-Weyl Hamiltonian description of classical field theories is introduced, the latter being founded by De Donder [1] and Weyl [2]. The real scalar field and its coupling to curved spacetime will be used as an initial example for working out the canonical transformation from Minkowski geometry to a locally curved spacetime. Our approach does not stipulate any a priori constraints beyond the requirement that initially a matter Lagrangian exists that is form-invariant under the global group $\mathrm{SO}(1,3)$, hence possesses the Lorentz symmetry. The request for the system's local invariance with respect to both, (passive) diffeomorphism group of chart transitions on the base manifold, and active Lorentz transformations of fields that are sections on the frame bundle with fibers (inertial frames attached to any point of that base manifold), is implemented via the construction of a generating function in the Hamiltonian picture. That generating function reflects the transformation properties of the involved fields under that selected symmetry group. The induced modifications of the original matter action yield, in the spirit of Weyl [3], Yang and Mills [4], Utiyama [5], Sciama [6], Kibble [7] and Hayashi [8], a gauge theory of dynamic spacetime. Spacetime is represented by an orthonormal frame bundle with the vierbein field soldering the global Minkowski metric on the tangent space with the metric on the base manifold. The Lorentz connection is then identified as the associated gauge field. The affine connection is found to be a dependent function of the vierbein and spin connection. Torsion of spacetime and non-metricity are formally not excluded. Nevertheless, metric compatibility appears as the natural choice for the structure of spacetime. Its resulting dynamics is thus driven by the existence of two independent fields, namely the vierbein and the Lorentz resp. the affine connection. Admitting independent dynamics for the metric, represented by the vierbein, and the connection, is referred to as *metric-affine* or *Palatini* formalism [7, 9].

In Sect. 4.2 the coupled canonical field equations are derived from the variation of the action that is now form-invariant with respect to the $\mathrm{SO}(1,3) \times \mathrm{Diff}(M)$

D. Vasak et al., *Covariant Canonical Gauge Gravity*, FIAS Interdisciplinary Science Series, https://doi.org/10.1007/978-3-031-43717-5_4

symmetry. The formulae derived are generic as they apply for any Hamiltonians resp. Lagrangians that describe the dynamics of the free (uncoupled) scalar, vector, and spinor fields, as well as the free gravitational field. The "free" Hamiltonians cannot be derived from first principles but have to be chosen such that they are consistent with the basic postulates and the experimentally verified phenomenology. Our theory starts off from a few basic physical principles and adheres to the rigorous formalism of the covariant canonical transformation theory to fix, in a straightforward manner, the coupling of spacetime to the various matter fields.

In order to set up the *physical* equations for the coupling of matter fields and gravity, we catch up on discussing the respective Hamiltonians for said fields in Sect. 4.3. The Klein-Gordon, Maxwell-Proca, and the regularized Dirac Lagrangians are selected, and their corresponding De Donder-Weyl Hamiltonians derived by means of *complete* (covariant) Legendre transformations. In addition, a particular choice for the Hamiltonian describing the free gravitational field—going beyond the version advocated in Ref. [10]—is addressed in Sect. 4.3.4. The discussion of the pertinent energy-momentum tensors that play a crucial role in the field dynamics is bundled in Sect. 4.4.

The coupled set of field equations of matter and dynamic spacetime are then discussed in Sect. 4.5. The particular choice of the gravity Hamiltonian, discussed in Sect. 4.3.4, is coupled to the Klein-Gordon, Maxwell-Proca, and Dirac fields. While for the scalar and vector fields the resulting equations of motion correspond to the "minimal coupling" recipe with torsion included, the dynamics of the spacetime field strengths tensors, curvature and torsion, gives novel field equations that show an intimate link between spin-momentum tensor of spinor matter and torsion of spacetime. Torsion is shown to be a propagating, totally anti-symmetric field. Its cosmological impact is discussed in Chap. 8.

The Dirac equation in curved spacetime is worked out in Sect. 4.3.3. The stipulated regularity of the Dirac Hamiltonian invokes a new length parameter $\ell = 1/M$ that, while spurious in the case of non-interacting spinors, becomes a physical parameter once interaction with spacetime or other gauge fields is turned on. We observe then that both, the momentum and the mass of the Dirac field, acquire anomalous , and novel spin-dependent contributions that couple to the torsion of spacetime. The curvature dependent mass term may have a considerable impact on the physics of dense matter in neutron stars and around black holes, and also on cosmology [11–14].

In Sect. 4.5.4 the so-called Consistency Equation is derived from that closed coupled set of equations for matter and gravitational fields, resulting in an extended version of an Einstein-type field equation which we refer to as *CCGG equation*. It turns out to involve both the symmetric strain-energy and stress-energy tensors, while the anti-symmetric portions of those tensors are crucial for the dynamics of torsion.

We conclude the Chapter in Sect. 4.6. In Sect. 4.6.1 a summary including a list of the key equations is given. A brief discussion follows in Sect. 4.6.2 on the ambiguities of our ansatz and on the consequences of including internal fermion-boson interactions stipulated by U(1) and SU(N) gauge invariance.

4.1 Gauge Theory of Gravitation in the Covariant Canonical Transformation Framework

4.1.1 Conventions and Notations

For convenience, and in order to support the advanced reader who has skipped Chap. xrefchap:RCspacespstime, we review here the key conventions and notations used below. Our conventions are those of Misner et al. [15] with natural units $\hbar = c = 1$ and metric signature $(+, -, -, -)$. In order to accommodate spinors in the formulation, we must introduce local inertial frames on the tangent space of the base manifold spanned by the unit vectors e_i. The Latin indices apply to the Lorentz basis and are called Lorentz indices in contrast to the Greek indices that denote the coordinates on an arbitrary chart of the base manifold. The $g_{\mu\nu}$ is then expressed as

$$g^{\mu\nu}(x) = e_i{}^{\mu}(x)\, e_j{}^{\nu}(x)\, \eta^{ij}, \qquad g_{\mu\nu}(x) = e_{\mu}{}^{i}(x)\, e_{\nu}{}^{j}(x)\, \eta_{ij}, \tag{4.1}$$

where the *vierbein field* (or tetrads) $e_i{}^{\mu}(x)$ and their duals, $e_{\mu}{}^{i}(x)$, represent the coordinates of vectors with respect to a global orthonormal basis, and η_{ij} denotes the globally constant Minkowski metric. Hence, from Eq. (2.8a), we have that

$$e_{\mu}{}^{i}\, e_j{}^{\mu} = \delta^i_j, \qquad e_{\nu}{}^{j}\, e_j{}^{\mu} = \delta^{\mu}_{\nu}$$

holds in a local (intertial) frame attached at each point p with coordinates x. If $\Lambda^I{}_i(x)$ defines a transformation within the local Lorentz frame with

$$\Lambda^I{}_i \Lambda^i{}_J = \delta^I_J, \qquad \det \Lambda^I{}_i = 1, \tag{4.2}$$

the vierbeins transform under a combined chart diffeomorphism and orthochronous Lorentz transformation, $\mathrm{SO}(1,3) \times \mathrm{Diff}(M)$, as

$$E_{\alpha}{}^{I}(X) = \Lambda^I{}_i(x)\, e_{\beta}{}^{i}(x)\, \frac{\partial x^{\beta}}{\partial X^{\alpha}}. \tag{4.3}$$

This means that the determinants of the vierbeins transform as relative scalars of weight $w = 1$, hence as scalar densities:

$$\left(\det E_{\alpha}{}^{I}\right)(X) = \left(\det e_{\beta}{}^{i}\right)(x) \left|\frac{\partial x}{\partial X}\right| = \varepsilon \left|\frac{\partial x}{\partial X}\right|. \tag{4.4}$$

In the vierbein formalism, the chart invariant volume form is obtained by multiplying d^4x with the determinant $\varepsilon := \det(e_{\alpha}{}^{i})$ of the dual vierbein $e_{\alpha}{}^{i}$:

$$\sqrt{-\det(g_{\alpha\beta})} = \det(e_{\alpha}{}^{i}) = \varepsilon.$$

Be $\mathcal{L}$ the Lagrangian of field theories in Minkowski space, i.e. an *absolute scalar*. Then in curved geometry the system's Lagrangian must be a scalar density, $\tilde{\mathcal{L}} = \mathcal{L}\sqrt{-g} = \mathcal{L}\varepsilon$, with

$$\tilde{\mathcal{L}}'(X) = \mathcal{L}'(X)\sqrt{-G} = \mathcal{L}(x)\sqrt{-G} = \mathcal{L}(x)\sqrt{-g}\left|\frac{\partial x}{\partial X}\right| = \tilde{\mathcal{L}}(x)\left|\frac{\partial x}{\partial X}\right|.$$

As a consequence, the integral (not the Lagrangian itself!) maintains its form under chart transition $x \mapsto X$:

$$\int_{V'} \tilde{\mathcal{L}}'(X)\,\mathrm{d}^4X = \int_V \tilde{\mathcal{L}}'(X)\left|\frac{\partial X}{\partial x}\right|\mathrm{d}^4x = \int_V \tilde{\mathcal{L}}(x)\,\mathrm{d}^4x. \tag{4.5}$$

Notice that in the following we use tilde to denote tensor densities, but beyond that we do not explicitly distinguish between absolute tensors and tensor densities unless it is required for clarity.

4.1.2 De Donder-Weyl Hamiltonian Formalism

In the De Donder-Weyl Hamiltonian formalism [1, 2], the $\pi^\mu(x)$ are the dual quantities of the derivatives $\partial\varphi/\partial x^\mu$ of a set of scalar fields φ, defined on the basis of a conventional Lagrangian $\mathcal{L}$ as

$$\pi^\mu(x) = \frac{\partial\mathcal{L}}{\partial\left(\frac{\partial\varphi}{\partial x^\mu}\right)}.$$

The corresponding definition in terms of a Lagrangian density $\tilde{\mathcal{L}} = \mathcal{L}\sqrt{-g}$ is then

$$\tilde{\pi}^\mu(x) = \pi^\mu(x)\sqrt{-g} = \frac{\partial\tilde{\mathcal{L}}}{\partial\left(\frac{\partial\varphi}{\partial x^\mu}\right)}, \qquad \tilde{\Pi}^\mu(X) = \Pi^\mu(X)\sqrt{-G} = \frac{\partial\tilde{\mathcal{L}}'}{\partial\left(\frac{\partial\Phi}{\partial X^\mu}\right)}. \tag{4.6}$$

Therefore, $\tilde{\pi}^\mu$ can be regarded as the dual of the derivative $\partial\varphi/\partial x^\mu$ with regard to the extended Lagrangian $\tilde{\mathcal{L}}$. While π^μ transforms as an absolute tensor, the related $\tilde{\pi}^\mu = \pi^\mu\sqrt{-g}$ transforms as a relative vector of weight $w = 1$, hence as a

$$\Pi^\mu(X) = \pi^\alpha(x)\frac{\partial X^\mu}{\partial x^\alpha} = \frac{\tilde{\Pi}^\mu(X)}{\sqrt{-G}} = \frac{\tilde{\pi}^\alpha(x)}{\sqrt{-g}}\frac{\partial X^\mu}{\partial x^\alpha},$$

hence by virtue of Eq. (2.37)

$$\tilde{\Pi}^{\mu}(X) = \tilde{\pi}^{\alpha}(x)\frac{\partial X^{\mu}}{\partial x^{\alpha}}\frac{\sqrt{-G}}{\sqrt{-g}} = \tilde{\pi}^{\alpha}(x)\frac{\partial X^{\mu}}{\partial x^{\alpha}}\left|\frac{\partial x}{\partial X}\right|. \tag{4.7}$$

Equation (4.7) is the general transformation rule for the vector density $\tilde{\pi}^{\mu}(x)$ under a chart transition.

The transition into the covariant Hamiltonian picture is achieved by applying the covariant Legendre transformation. Taking the scalar fields as example, this gives

$$\tilde{\mathcal{H}} := \tilde{\pi}^{\mu}\frac{\partial \varphi}{\partial x^{\mu}} - \tilde{\mathcal{L}}.$$

The DW Hamiltonian does not correspond to the energy of the system as it is the case in point mechanics. The energy in field theory is rather the 00-component of the canonical energy-momentum tensor (as discussed in Sect. 4.4),

$$T_{\mu}{}^{\nu} := \frac{\partial \mathcal{L}}{\partial\left(\frac{\partial \varphi}{\partial x^{\nu}}\right)}\frac{\partial \varphi}{\partial x^{\mu}} - \delta^{\nu}_{\mu}\,\mathcal{L}.$$

However, the trace of that tensor gives

$$T := T_{\mu}{}^{\mu} = \frac{\partial \mathcal{L}}{\partial\left(\frac{\partial \varphi}{\partial x^{\mu}}\right)}\frac{\partial \varphi}{\partial x^{\mu}} - 4\mathcal{L} = \pi_{\mu}\frac{\partial \varphi}{\partial x^{\mu}} - 4\mathcal{L} = \mathcal{H} - 3\mathcal{L},$$

which looks like the covariant version of the relation $T = \rho - 3p$ for a homogeneous fluid in its co-moving frame, where ρ is the energy density and p the pressure of that fluid.

4.1.3 The Role of the Vierbeins

In the realm of the Poincaré Gauge Theory the vierbeins, as transformation matrices between the inertial frames and the coordinate base of the tangent space, are considered to be gauge fields [7], or "translation potentials" [16, 17].

In the covariant De Donder-Weyl Hamiltonian approach that line of thought can be sketched using the example of a scalar field. In a Hamiltonian $\mathcal{H}(\phi, \pi^{i})$ the momentum field π^{i} is a vector at some point p, and its component index i is related to an inertial basis (endowed with the Minkowski metric η_{ij}) in the "field space". In that frame the field components transform under local (Lorentz) transformations according to

$$\pi^{I}(x) = \Lambda^{I}{}_{i}(x)\,\pi^{i}(x). \tag{4.8}$$

Here "local" refers to a chart $p \in U$ in which a local coordinate basis is defined such that the point p has the coordinates x^{μ}. Re-expressing or "gauging" the field components in that coordinate basis is an invertible linear transformation,

$$\pi^{\mu}(x) = e_{i}{}^{\mu}(x)\, \pi^{i}(x), \tag{4.9}$$

with the vierbein $e_{i}{}^{\mu}(x)$ as gauge field, and its dual $e^{i}{}_{\mu}(x)$ defined such that

$$e^{i}{}_{\mu}(x)\, e_{j}{}^{\mu}(x) = \delta^{i}_{j}$$

holds everywhere. Now under arbitrary chart transitions $x \mapsto X$ (diffeomorphisms), the newly defined components $\pi^{\mu}(x)$ must transform as tensors:

$$\pi^{\mu}(x) \mapsto \Pi^{\nu}(X) := \frac{\partial X^{\nu}}{\partial x^{\mu}}\, \pi^{\mu}(x). \tag{4.10}$$

Obviously, since the gauge transformation (4.9) must be independent of the choice of both, the field basis and the coordinate chart,

$$\Pi^{\mu}(X) = E_{I}{}^{\mu}(X)\, \Pi^{I}(X) \tag{4.11}$$

must hold as well. Inserting Eqs. (4.10), (4.9), and (4.8) into Eq. (4.11) gives

$$\frac{\partial X^{\nu}}{\partial x^{\mu}}\, e_{i}{}^{\mu}(x)\, \pi^{i}(x) = E_{I}{}^{\nu}(X)\, \Pi^{I}(X) = E_{I}{}^{\nu}(X)\, \Lambda^{I}{}_{i}(X)\, \Pi^{i}(X). \tag{4.12}$$

Since both X and x are the coordinates of the same point p, the components of the field vector depend on the inertial frame only, i.e. $\Pi^{i}(X) = \Pi^{i}(p) \equiv \pi^{i}(p) = \pi^{i}(x)$. Equation (4.12) must hold for any $\pi^{i} = \Pi^{i}$, such that

$$\frac{\partial X^{\nu}}{\partial x^{\mu}}\, e_{i}{}^{\mu}(x) = E_{I}{}^{\nu}(X)\, \Lambda^{I}{}_{i}(X),$$

and the gauge field vierbein must transform according to

$$e^{i}{}_{\mu}(x) \mapsto E^{I}{}_{\nu}(X) = \Lambda^{i}{}_{J}(x)\, e^{j}{}_{\mu}(x)\, \frac{\partial x^{\mu}}{\partial X^{\nu}}.$$

The re-gauged Hamiltonian now becomes

$$\mathcal{H}(\phi, \pi^{i}) \mapsto \mathcal{H}(\phi, \pi^{\mu}) \equiv \mathcal{H}(\phi, e_{i}{}^{\mu}\, \pi^{i}).$$

For ensuring in addition the invariance of the integration in the action, the Jacobian of that gauge transformation must be taken into account. Since $\det \Lambda^{i}{}_{J}(x) \equiv 1$ by definition, and

$$\left|\frac{\partial x}{\partial X}\right| \equiv \frac{\det E^{I}{}_{\nu}(X)}{\det e^{i}{}_{\mu}(x)},$$

the Hamiltonian scalar becomes a scalar density via

$$\mathcal{H}(\phi, \pi^\mu) \mapsto \tilde{\mathcal{H}}(\phi, \pi^\mu) \equiv \mathcal{H}(\phi, e_i{}^\mu \pi^i) \det e^i{}_\mu(x) = \varepsilon(x)\, \mathcal{H}(\phi, e_i{}^\mu \pi^i).$$

Putting in this way the vierbein and the connection under the same "gauge field" umbrella might appear quite appealing, but its physical significance is questionable, especially as it does not have any impact on neither Kibble's results nor on the following analysis. A gauge field in the sense of Weyl, Fock, Yang and Mills compensates for broken local symmetries and is intimately linked to (covariant) field derivatives. While the connection qualifies for this interpretation, it is here not the case for the vierbeins.[1] Therefore they are in the following treated as a given structure element of a frame bundle, the Lorentzian manifold. This is not equivalent to enforcing from the onset a curved spacetime, i.e. to plug in upfront any elements of gravity. It rather provides the "playground" for any kind of kinematics. In that philosophy, similar to Utiyama's [5], the essence of gravity is provided by the connection that, via the autoparallel equation, determines the trajectories of bodies under the influence of gravity. Whether spacetime is curved or not, i.e. whether gravity is present or not, is determined solely by a non-vanishing or vanishing connection in a holonomic cartesian coordinate frame.

The viewpoint that merely the connection can be regarded as the gauge field for raising a given global Lorentz symmetry to a local Lorentz-diffeomorphism-invariance is endorsed by the spinor representation of the gauge theory of gravity, presented in Chap. 5. In this formulation of the gauge formalism, the initially static matrices γ_μ representing the Dirac algebra are extended to spacetime-dependent matrices $\gamma_\mu(x)$. Thereby the system's initial Lorentz symmetry is lost and restored thereafter by introducing the spinor connection gauge field $\boldsymbol{\omega}_\nu(x)$. This field is intimately related with the spin connection $\omega^i{}_{j\nu}$, to be introduced in Eq. (4.27). Thus, in both descriptions the connection defines the only genuine gauge field.

4.1.4 Canonical Transformation Formalism for a Scalar Field in a Curved Spacetime

Based on the set of general transformations $\varphi(x) \mapsto \Phi(\varphi, X)$ of a scalar field with simultaneous transformations of the vierbeins $e_\alpha{}^i(x) \mapsto E_\beta{}^I(e, X)$ with arbitrary diffemorphisms $x \mapsto X(x)$, we consider in the following the particular subset of those transformations that maintain the form of the *variation* of the (4.5). This means, explicitly:

$$\delta \int_{V'} \tilde{\mathcal{L}}' \left(\Phi, \frac{\partial \Phi}{\partial X^\nu}, E_\mu{}^I, \frac{\partial E_\mu{}^I}{\partial X^\nu}, X \right) \mathrm{d}^4 X \stackrel{!}{=} \delta \int_V \tilde{\mathcal{L}} \left(\varphi, \frac{\partial \varphi}{\partial x^\nu}, e_\mu{}^i, \frac{\partial e_\mu{}^i}{\partial x^\nu}, x \right) \mathrm{d}^4 x. \tag{4.13}$$

[1] In Ref. [18] the gauge character of the vierbein is discussed, though, when embedding the Poincare group in the de Sitter group SO(1, 4).

The postulated variational invariance of the action integrals allows the *integrands* to differ by the divergence of an arbitrary vector function $\tilde{\mathcal{F}}_1^\nu\big|_x$ of the dynamic fields,

$$\tilde{\mathcal{L}}'\left(\Phi, \frac{\partial \Phi}{\partial X^\nu}, E_\mu{}^I, \frac{\partial E_\mu{}^I}{\partial X^\nu}, X\right)\left|\frac{\partial X}{\partial x}\right| \stackrel{!}{=} \tilde{\mathcal{L}}\left(\varphi, \frac{\partial \varphi}{\partial x^\nu}, e_\mu{}^i, \frac{\partial e_\mu{}^i}{\partial x^\nu}, x\right) - \frac{\partial \tilde{\mathcal{F}}_1^\nu}{\partial x^\nu}, \tag{4.14}$$

provided that the variation of the boundary term vanishes:

$$\delta \int_V \frac{\partial \tilde{\mathcal{F}}_1^\nu}{\partial x^\nu}\,\mathrm{d}^4x = \delta \oint_{\partial V} \tilde{\mathcal{F}}_1^\nu\,\mathrm{d}S_\nu = 0.$$

In the Lagrangian formulation the presence of the boundary term is thus required to maintain invariance of the system dynamics. In the covariant Hamiltonian canonical transformation theory that boundary term is promoted to a convenient vehicle for implementing any gauge formalism. It is thereby absorbed into a form-invariant Hamiltonian describing a closed system of interacting fields. Of course, the corresponding Lagrangian is then also form-invariant and void of boundary terms. This is shown in the following sections.

In order to implement the covariant canonical theory we have to work in the De Donder-Weyl Hamiltonian formulation. The integrand condition (4.14) is then equivalently expressed by means of a *complete* as:

$$\begin{aligned}
&\tilde{\pi}^\nu \frac{\partial \varphi}{\partial x^\nu} + \tilde{k}_i{}^{\mu\nu}\frac{\partial e_\mu{}^i}{\partial x^\nu} - \tilde{\mathcal{H}}\left(\varphi, \tilde{\pi}^\nu, e_\mu{}^i, \tilde{k}_i{}^{\mu\nu}, x\right) \\
&- \left[\tilde{\Pi}^\nu \frac{\partial \Phi}{\partial X^\nu} + \tilde{K}_I{}^{\mu\nu}\frac{\partial E_\mu{}^I}{\partial X^\nu} - \tilde{\mathcal{H}}'\left(\Phi, \tilde{\Pi}^\nu, E_\mu{}^I, \tilde{K}_I{}^{\mu\nu}, X\right)\right]\left|\frac{\partial X}{\partial x}\right| = \frac{\partial \tilde{\mathcal{F}}_1^\nu}{\partial x^\nu} \\
&= \frac{\partial \tilde{\mathcal{F}}_1^\nu}{\partial \varphi}\frac{\partial \varphi}{\partial x^\nu} + \frac{\partial \tilde{\mathcal{F}}_1^\beta}{\partial \Phi}\frac{\partial X^\nu}{\partial x^\beta}\frac{\partial \Phi}{\partial X^\nu} + \frac{\partial \tilde{\mathcal{F}}_1^\nu}{\partial e_\mu{}^i}\frac{\partial e_\mu{}^i}{\partial x^\nu} + \frac{\partial \tilde{\mathcal{F}}_1^\beta}{\partial E_\mu{}^I}\frac{\partial X^\nu}{\partial x^\beta}\frac{\partial E_\mu{}^I}{\partial X^\nu} + \left.\frac{\partial \tilde{\mathcal{F}}_1^\nu}{\partial x^\nu}\right|_{\text{expl}}.
\end{aligned} \tag{4.15}$$

Herein, $\tilde{k}_i{}^{\mu\nu}$ and $\tilde{K}_I{}^{\mu\nu}$ denote the canonical conjugates of the vierbeins $e_\mu{}^i$ and $E_\mu{}^I$, respectively:

$$\tilde{k}_i{}^{\mu\nu}(x) = \frac{\partial \tilde{\mathcal{L}}}{\partial \frac{\partial e_\mu{}^i}{\partial x^\nu}}, \qquad \tilde{K}_I{}^{\mu\nu}(X) = \frac{\partial \tilde{\mathcal{L}}'}{\partial \frac{\partial E_\mu{}^I}{\partial X^\nu}}.$$

Equation (4.15) is satisfied iff any matching partial derivatives cancel each other. For $\tilde{\mathcal{F}}_1^\nu\big|_x = \tilde{\mathcal{F}}_1^\nu\left(\varphi, \Phi, e_\mu{}^i, E_\mu{}^I, x\right)$ this gives the following for the fields:

$$\tilde{\pi}^{\nu}(x)=\frac{\partial\tilde{\mathcal{F}}_1^{\nu}}{\partial\varphi},\qquad \tilde{\Pi}^{\nu}(X)=-\frac{\partial\tilde{\mathcal{F}}_1^{\beta}}{\partial\Phi}\frac{\partial X^{\nu}}{\partial x^{\beta}}\left|\frac{\partial x}{\partial X}\right| \tag{4.16a}$$

$$\tilde{k}_i^{\ \mu\nu}(x)=\frac{\partial\tilde{\mathcal{F}}_1^{\nu}}{\partial e_{\mu}^{\ i}},\qquad \tilde{K}_I^{\ \mu\nu}(X)=-\frac{\partial\tilde{\mathcal{F}}_1^{\beta}}{\partial E_{\mu}^{\ I}}\frac{\partial X^{\nu}}{\partial x^{\beta}}\left|\frac{\partial x}{\partial X}\right|. \tag{4.16b}$$

The transformation rule for the involves a possible *explicit* dependence of $\tilde{\mathcal{F}}_1^{\nu}\big|_x$ on x:

$$\tilde{\mathcal{H}}'\left(\Phi,\tilde{\Pi}^{\nu},E_{\mu}^{\ I},\tilde{K}_I^{\ \mu\nu},X\right)\left|\frac{\partial X}{\partial x}\right|=\tilde{\mathcal{H}}\left(\varphi,\tilde{\pi}^{\nu},e_{\mu}^{\ i},\tilde{k}_i^{\ \mu\nu},x\right)+\left.\frac{\partial\tilde{\mathcal{F}}_1^{\nu}}{\partial x^{\nu}}\right|_{\text{expl}}. \tag{4.17}$$

As the vector density $\tilde{\mathcal{F}}_1^{\nu}\big|_x$ defines the transformation rules it is called the *Generating function* of type 1. There are four possible types of generating functions corresponding to four possible combinations of the original and transformed fields and their conjugate momenta. A canonical transformation $\tilde{\mathcal{F}}_3^{\nu}\big|_x$, referred to as a *generating function* of type 3, is defined as a function of the momenta $\tilde{\pi}^{\nu}$ and $\tilde{k}_i^{\ \mu\nu}$ in place of the fields φ and $e_{\mu}^{\ i}$ in $\tilde{\mathcal{F}}_1^{\nu}$. It is related to $\tilde{\mathcal{F}}_1^{\nu}$ by means of the Legendre transformation

$$\begin{aligned}\tilde{\mathcal{F}}_3^{\nu}\left(\tilde{\pi}^{\nu}(x),\Phi(X),\tilde{k}_i^{\ \mu\nu}(x),E_{\mu}^{\ I}(X),x\right)&=\tilde{\mathcal{F}}_1^{\nu}\left(\varphi(x),\Phi(X),e_{\mu}^{\ i}(x),E_{\mu}^{\ I}(X),x\right)\\&\quad-\tilde{\pi}^{\nu}(x)\,\varphi(x)-\tilde{k}_i^{\ \mu\nu}(x)\,e_{\mu}^{\ i}(x).\end{aligned}$$

The integrand condition (4.15) can now be equivalently expressed in terms of $\tilde{\mathcal{F}}_3^{\nu}\big|_x$:

$$\begin{aligned}\frac{\partial\tilde{\mathcal{F}}_1^{\nu}}{\partial x^{\nu}}&=\frac{\partial\tilde{\mathcal{F}}_3^{\nu}}{\partial x^{\nu}}+\frac{\partial\tilde{\pi}^{\nu}}{\partial x^{\nu}}\varphi+\cancel{\tilde{\pi}^{\nu}\frac{\partial\varphi}{\partial x^{\nu}}}+\frac{\partial\tilde{k}_i^{\ \mu\nu}}{\partial x^{\nu}}e_{\mu}^{\ i}+\cancel{\tilde{k}_i^{\ \mu\nu}\frac{\partial e_{\mu}^{\ i}}{\partial x^{\nu}}}\\&=\cancel{\tilde{\pi}^{\nu}\frac{\partial\varphi}{\partial x^{\nu}}}+\cancel{\tilde{k}_i^{\ \mu\nu}\frac{\partial e_{\mu}^{\ i}}{\partial x^{\nu}}}-\tilde{\mathcal{H}}\left(\varphi,\tilde{\pi}^{\nu},e_{\mu}^{\ i},\tilde{k}_i^{\ \mu\nu},x\right)\\&\quad-\left[\tilde{\Pi}^{\nu}\frac{\partial\Phi}{\partial X^{\nu}}+\tilde{K}_I^{\ \mu\nu}\frac{\partial E_{\mu}^{\ I}}{\partial X^{\nu}}-\tilde{\mathcal{H}}'\left(\Phi,\tilde{\Pi}^{\nu},E_{\mu}^{\ I},\tilde{K}_I^{\ \mu\nu},X\right)\right]\left|\frac{\partial X}{\partial x}\right|.\end{aligned}$$

This gives

$$\begin{aligned}&-\frac{\partial\tilde{\pi}^{\alpha}}{\partial x^{\alpha}}\varphi-\frac{\partial\tilde{k}_i^{\ \mu\alpha}}{\partial x^{\alpha}}e_{\mu}^{\ i}-\tilde{\mathcal{H}}\left(\varphi,\tilde{\pi}^{\nu},e_{\mu}^{\ i},\tilde{k}_i^{\ \mu\nu},x\right)\\&-\left[\tilde{\Pi}^{\nu}\frac{\partial\Phi}{\partial X^{\nu}}+\tilde{K}_I^{\ \mu\nu}\frac{\partial E_{\mu}^{\ I}}{\partial X^{\nu}}-\tilde{\mathcal{H}}'\left(\Phi,\tilde{\Pi}^{\nu},E_{\mu}^{\ I},\tilde{K}_I^{\ \mu\nu},X\right)\right]\left|\frac{\partial X}{\partial x}\right|\\&=\frac{\partial\tilde{\mathcal{F}}_3^{\alpha}}{\partial\Phi}\frac{\partial X^{\nu}}{\partial x^{\alpha}}\frac{\partial\Phi}{\partial X^{\nu}}+\frac{\partial\tilde{\mathcal{F}}_3^{\alpha}}{\partial E_{\mu}^{\ I}}\frac{\partial X^{\nu}}{\partial x^{\alpha}}\frac{\partial E_{\mu}^{\ I}}{\partial X^{\nu}}+\frac{\partial\tilde{\mathcal{F}}_3^{\nu}}{\partial\tilde{\pi}^{\alpha}}\frac{\partial\tilde{\pi}^{\alpha}}{\partial x^{\nu}}+\frac{\partial\tilde{\mathcal{F}}_3^{\nu}}{\partial\tilde{k}_i^{\ \mu\alpha}}\frac{\partial\tilde{k}_i^{\ \mu\alpha}}{\partial x^{\nu}}+\left.\frac{\partial\tilde{\mathcal{F}}_3^{\nu}}{\partial x^{\nu}}\right|_{\text{expl}}.\end{aligned} \tag{4.18}$$

Comparing the coefficients in Eq. (4.18), we obtain the for a of type $\tilde{\mathcal{F}}_3^\nu\big|_x$,

$$\delta_\alpha^\nu\,\varphi(x) = -\frac{\partial\tilde{\mathcal{F}}_3^\nu}{\partial\tilde{\pi}^\alpha} \qquad \tilde{\Pi}^\nu(X) = -\frac{\partial\tilde{\mathcal{F}}_3^\alpha}{\partial\Phi}\frac{\partial X^\nu}{\partial x^\alpha}\left|\frac{\partial x}{\partial X}\right| \tag{4.19a}$$

$$\delta_\alpha^\nu\,e_\mu{}^i(x) = -\frac{\partial\tilde{\mathcal{F}}_3^\nu}{\partial\tilde{k}_i{}^{\mu\alpha}} \qquad \tilde{K}_I{}^{\mu\nu}(X) = -\frac{\partial\tilde{\mathcal{F}}_3^\alpha}{\partial E_\mu{}^I}\frac{\partial X^\nu}{\partial x^\alpha}\left|\frac{\partial x}{\partial X}\right|, \tag{4.19b}$$

and the corresponding rule for the Hamiltonian:

$$\tilde{\mathcal{H}}'\left(\Phi,\tilde{\Pi}^\nu,E_\mu{}^I,\tilde{K}_I{}^{\mu\nu},X\right)\left|\frac{\partial X}{\partial x}\right| = \tilde{\mathcal{H}}\left(\varphi,\tilde{\pi}^\nu,e_\mu{}^i,\tilde{k}_i{}^{\mu\nu},x\right) + \left.\frac{\partial\tilde{\mathcal{F}}_3^\nu}{\partial x^\nu}\right|_{\text{expl}}. \tag{4.19c}$$

Notice that on the basis of the transformation rules (4.19b), the dependence of $e_\beta{}^i(x)$ on $E_\mu{}^I(X)$ follows as:

$$\delta_\alpha^\nu\,\frac{\partial e_\beta{}^i(x)}{\partial E_\mu{}^I(X)} = -\frac{\partial^2\tilde{\mathcal{F}}_3^\nu(x)}{\partial\tilde{k}_i{}^{\beta\alpha}(x)\partial E_\mu{}^I(X)} = \frac{\partial\tilde{K}_I{}^{\mu\lambda}(X)}{\partial\tilde{k}_i{}^{\beta\alpha}(x)}\frac{\partial x^\nu}{\partial X^\lambda}\left|\frac{\partial X}{\partial x}\right|.$$

This and similar symmetries between the original and transformed fields—as derived from a generating function—and original and transformed pertaining conjugate momenta is a feature of *canonical* transformations, in contrast to arbitrary transformations of fields and momenta.

4.1.5 Local Lorentz and Diffeomorphism Transformation and the Associated Gauge Field

In the next step the particular is defined for a canonical transformation of a scalar field φ under combined:

$$\tilde{\mathcal{F}}_3^\nu\left(\Phi,\tilde{\pi}^\nu,E^I{}_\mu,\tilde{k}_i{}^{\mu\nu},x\right) := -\tilde{\pi}^\nu\,\Phi - \tilde{k}_i{}^{\beta\nu}\,\Lambda^i{}_I\,E^I{}_\alpha\,\frac{\partial X^\alpha}{\partial x^\beta}. \tag{4.20}$$

Notice that for $\tilde{\mathcal{F}}_3^\nu\big|_x$ to depend on x only, the transformed (capital letters) must be expressed in form of the underlying symmetry transformation that is in general explicitly x-dependent. This means that the transformed Lagrangian density, $\tilde{\mathcal{L}}'$, becomes explicitly x-dependent in Eq. (4.14), even if the original Lagrangian $\tilde{\mathcal{L}}$ is not.

The particular transformation rules (4.19) follow as

$$\tilde{\Pi}^{\nu} \equiv -\frac{\partial \tilde{\mathcal{F}}_3^{\alpha}}{\partial \Phi}\frac{\partial X^{\nu}}{\partial x^{\alpha}}\left|\frac{\partial x}{\partial X}\right| = \tilde{\pi}^{\alpha}\frac{\partial X^{\nu}}{\partial x^{\alpha}}\left|\frac{\partial x}{\partial X}\right| \tag{4.21a}$$

$$\delta^{\nu}_{\alpha}\,\varphi \equiv -\frac{\partial \tilde{\mathcal{F}}_3^{\nu}}{\partial \tilde{\pi}^{\alpha}} = \delta^{\nu}_{\alpha}\,\Phi \tag{4.21b}$$

$$\tilde{K}_I^{\ \mu\nu} \equiv -\frac{\partial \tilde{\mathcal{F}}_3^{\alpha}}{\partial E^I_{\ \mu}}\frac{\partial X^{\nu}}{\partial x^{\alpha}}\left|\frac{\partial x}{\partial X}\right| = \tilde{k}_i^{\ \beta\alpha}\Lambda^i_{\ I}\frac{\partial X^{\mu}}{\partial x^{\beta}}\frac{\partial X^{\nu}}{\partial x^{\alpha}}\left|\frac{\partial x}{\partial X}\right| \tag{4.21c}$$

$$\delta^{\nu}_{\alpha}\,e_{\beta}^{\ i} \equiv -\frac{\partial \tilde{\mathcal{F}}_3^{\nu}}{\partial \tilde{k}_i^{\ \beta\alpha}} = \delta^{\nu}_{\alpha}\,\Lambda^i_{\ I}\,E_{\alpha}^{\ I}\frac{\partial X^{\alpha}}{\partial x^{\beta}}, \tag{4.21d}$$

which recover the given proper transformation rules for the fields and their conjugates.

The set of transformation rules is completed by the rule for the Hamiltonian density from Eq. (4.19c), which follows again from the explicit spacetime dependence of the generating function (4.20):

$$\begin{aligned}\left.\frac{\partial \tilde{\mathcal{F}}_3^{\nu}}{\partial x^{\nu}}\right|_{\text{expl}} &= -\tilde{k}_i^{\ \beta\nu}\frac{\partial}{\partial x^{\nu}}\left(\Lambda^i_{\ I}\frac{\partial X^{\alpha}}{\partial x^{\beta}}\right)E_{\alpha}^{\ I}\\ &= -\tilde{k}_i^{\ (\beta\nu)}\frac{\partial}{\partial x^{\nu}}\left(\Lambda^i_{\ I}\frac{\partial X^{\alpha}}{\partial x^{\beta}}\right)E_{\alpha}^{\ I} - \tilde{k}_i^{\ [\beta\nu]}\frac{\partial \Lambda^i_{\ I}}{\partial x^{\nu}}\frac{\partial X^{\alpha}}{\partial x^{\beta}}E_{\alpha}^{\ I}.\end{aligned} \tag{4.22}$$

In the last line, the right-hand side is split into the symmetric and the skew-symmetric contributions of $\tilde{k}_i^{\ \beta\nu}$ in β and ν, considering that the second derivative term of X^{α} does not contribute to the skew-symmetric portion of $\tilde{k}_i^{\ \nu\beta}$. In order to finally cast the transformation rule (4.22) into a symmetric form in the original and the transformed fields, we calculate the x^{ν}-derivative of Eq. (4.21d):

$$\frac{\partial}{\partial x^{\nu}}\left(\Lambda^i_{\ I}\frac{\partial X^{\alpha}}{\partial x^{\beta}}\right)E^I_{\ \alpha} + \Lambda^i_{\ I}\frac{\partial X^{\alpha}}{\partial x^{\beta}}\frac{\partial X^{\xi}}{\partial x^{\nu}}\frac{\partial E^I_{\ \alpha}}{\partial X^{\xi}} = \frac{\partial e_{\beta}^{\ i}}{\partial x^{\nu}}. \tag{4.23}$$

Inserting Eqs. (4.23) and (4.21a) for the symmetric contribution of $\tilde{k}_i^{\ \beta\nu}$ in β and ν yields:

$$\begin{aligned}\left.\frac{\partial \tilde{\mathcal{F}}_3^{\nu}}{\partial x^{\nu}}\right|_{\text{expl}} &= -\tilde{k}_i^{\ (\beta\nu)}\left(\frac{\partial e_{\beta}^{\ i}}{\partial x^{\nu}} - \Lambda^i_{\ I}\frac{\partial X^{\alpha}}{\partial x^{\beta}}\frac{\partial X^{\xi}}{\partial x^{\nu}}\frac{\partial E_{\alpha}^{\ I}}{\partial X^{\xi}}\right) - \tilde{k}_i^{\ [\beta\nu]}\frac{\partial \Lambda^i_{\ I}}{\partial x^{\nu}}\frac{\partial X^{\alpha}}{\partial x^{\beta}}E_{\alpha}^{\ I}\\ &= -\tilde{k}_i^{\ (\beta\nu)}\frac{\partial e_{\beta}^{\ i}}{\partial x^{\nu}} + \tilde{K}_I^{\ (\beta\nu)}\frac{\partial E_{\beta}^{\ I}}{\partial X^{\nu}}\left|\frac{\partial X}{\partial x}\right| + \tilde{k}_i^{\ [\beta\nu]}\Lambda^i_{\ I}\frac{\partial \Lambda^I_{\ J}}{\partial x^{\nu}}e_{\beta}^{\ j}\\ &= \tilde{\mathcal{H}}'\left|\frac{\partial X}{\partial x}\right| - \tilde{\mathcal{H}}.\end{aligned} \tag{4.24}$$

Obviously, this is the crucial condition for the requirement (4.13) of invariance of the system dynamics to hold. We observe now that the terms $-\tilde{k}_i^{\,(\beta\nu)}\frac{\partial e_\beta^{\;i}}{\partial x^\nu}$ and $\tilde{K}_I^{\,(\beta\nu)}\frac{\partial E_\beta^{\;I}}{\partial X^\nu}\left|\frac{\partial X}{\partial x}\right|$ in Eq. (4.24) can be combined with similar terms in Eq. (4.18) to give

$$
\begin{aligned}
&-\frac{\partial\tilde{\pi}^\alpha}{\partial x^\alpha}\varphi-\frac{\partial\tilde{k}_i^{\,[\mu\alpha]}}{\partial x^\alpha}e_\mu^{\;i}-\tilde{\mathcal{H}}\left(\varphi,\tilde{\pi}^\nu,e_\mu^{\;i},\tilde{k}_i^{\,\mu\nu},x\right)\\
&-\left[\tilde{\Pi}^\nu\frac{\partial\Phi}{\partial X^\nu}+\tilde{K}_I^{\,[\mu\nu]}\frac{\partial E_\mu^{\;I}}{\partial X^\nu}-\tilde{\mathcal{H}}'\left(\Phi,\tilde{\Pi}^\nu,E_\mu^{\;I},\tilde{K}_I^{\,\mu\nu},X\right)\right]\left|\frac{\partial X}{\partial x}\right|\\
&=\tilde{k}_i^{\,[\beta\nu]}\Lambda^i{}_I\frac{\partial\Lambda^i{}_J}{\partial x^\nu}e_\beta^{\;j}.
\end{aligned}
\tag{4.25}
$$

In this way a portion of the explicit x-dependence of $\tilde{\mathcal{F}}_3^\nu$ has been absorbed in a covariant form into the action. The remaining term on the right-hand side of Eq. (4.25) contains the spacetime-dependent Lorentz transformation coefficients $\Lambda^i{}_J(x)$. The only way to re-establish the invariance of the system dynamics is to introduce a "counter term" whose transformation rule absorbs the symmetry-breaking term proportional to $\partial\Lambda^i{}_J/\partial x^\nu$ in Eq. (4.25). That new term called a *Gauge Hamiltonian* must thus transform as

$$
\tilde{\mathcal{H}}'_{\mathrm{Gau}_1}\left|\frac{\partial X}{\partial x}\right|-\tilde{\mathcal{H}}_{\mathrm{Gau}_1}=\tilde{k}_i^{\,[\mu\nu]}\Lambda^i{}_I\frac{\partial\Lambda^i{}_J}{\partial x^\nu}e_\mu^{\;j},
\tag{4.26}
$$

and be constructed such that the dependence on the given transformation is eliminated. For that the structure of the *free* indices of $\Lambda^i{}_I\partial\Lambda^i{}_J/\partial x^\nu$ will be exactly matched by a newly introduced compensating *Gauge field* $\omega^i{}_{j\nu}$. Its simplest form is given by

$$
\tilde{\mathcal{H}}_{\mathrm{Gau}_1}=-\tilde{k}_i^{\,[\mu\nu]}\,\omega^i{}_{j\nu}\,e_\mu^{\;j},
\tag{4.27}
$$

and is required to be form-invariant in terms of the transformed gauge field $\Omega^i{}_{j\nu}$:

$$
\tilde{\mathcal{H}}'_{\mathrm{Gau}_1}=-\tilde{K}_I^{\,[\mu\nu]}\,\Omega^i{}_{j\nu}\,E^J{}_\mu.
\tag{4.28}
$$

The sign of the gauge field was chosen such that $\omega^i{}_{j\nu}$ can later be identified with the *spin connection*. We derive the ensuing transformation rule for the gauge field $\omega^i{}_{k\mu}$ by inserting the Hamiltonians (4.27) and (4.28) into Eq. (4.26). Beforehand, the Hamiltonian (4.28) is expressed in terms of the original fields according to the canonical transformation rules (4.21):

$$
\tilde{\mathcal{H}}'_{\mathrm{Gau}_1}=-\tilde{k}_i^{\,[\mu\nu]}\Lambda^i{}_I\,\Omega^I{}_{J\alpha}\,\Lambda^J{}_j\frac{\partial X^\alpha}{\partial x^\nu}e_\mu^{\;j}\left|\frac{\partial x}{\partial X}\right|.
$$

It follows that the gauge field ${\omega'}^{i}_{\ j\nu}$ transforms inhomogeneously as:

$$\boxed{{\omega'}^{i}_{\ j\nu} = {\Lambda'}^{i}_{\ I}\,{\Omega'}^{I}_{\ J\alpha}\,{\Lambda'}^{J}_{\ j}\,\frac{\partial X^{\alpha}}{\partial x^{\nu}} + {\Lambda'}^{i}_{\ I}\,\frac{\partial {\Lambda'}^{I}_{\ J}}{\partial x^{\nu}}.} \tag{4.29}$$

The inhomogeneous term in (4.29) is skew-symmetric in i, j according to Eq. (4.2):

$${\Lambda'}^{i}_{\ I}\frac{\partial {\Lambda'}^{I}_{\ J}}{\partial x^{\nu}} = -\frac{\partial {\Lambda'}^{i}_{\ I}}{\partial x^{\nu}}{\Lambda'}^{I}_{\ J}. \tag{4.30}$$

Consequently, for the gauge field $\omega_{n_{j\nu}} = \eta_{ni}\,{\omega'}^{i}_{\ j\nu}$, the condition $\omega_{n_{j\nu}} \neq \omega_{j_{n\nu}}$ must hold. But it is not a priori necessary that the spin connection is anti-symmetric. A symmetric portion $\Omega^{(IJ)}_{\quad\ \alpha}$ transforms as a tensor into a symmetric $\omega^{(ij)}_{\quad\ \alpha}$ without affecting the inhomogeneous term:

$$\omega^{ij}_{\ \ \alpha} = \Lambda_{i_I}\,\Omega^{(IJ)}_{\quad\ \alpha}\,\Lambda_{J_j} = \Lambda_{I_i}\,\Omega^{(IJ)}_{\quad\ \alpha}\,\Lambda_{j_J} = \Lambda_{j_J}\,\Omega^{(IJ)}_{\quad\ \alpha}\,\Lambda_{I_i} = \Lambda_{j_I}\,\Omega^{(IJ)}_{\quad\ \alpha}\,\Lambda_{J_i} = \omega^{ji}_{\ \ \alpha}.$$

Remarkably, by requiring that the variation of the action functional is invariant with respect to the given SO(1, 3)×Diff(M), the canonical transformation formalism leads to *form-invariance* of a "gauged" action functional itself:

$$S = \int_V \left[\tilde{\pi}^{\nu}\frac{\partial \varphi}{\partial x^{\nu}} + \tilde{k}_i^{\ [\mu\nu]}\left(\frac{\partial e_{\mu}^{\ i}}{\partial x^{\nu}} + {\omega'}^{i}_{\ j\nu}\,e_{\mu}^{\ j}\right) - \tilde{\mathcal{H}}_0\right]\mathrm{d}^4x. \tag{4.31}$$

Herein, the gauge field ${\omega'}^{i}_{\ j\nu}(x)$ enters as an *external* field whose dynamics is not described by the action (4.31). The form-invariant action functional of Eq. (4.31) fulfills the postulate historically referred to as the "Principle of General Relativity".

We observe that no *direct* coupling of the scalar field $\varphi(x)$ with the vierbein field $e_{\nu}^{\ j}$ and the gauge field ${\omega'}^{i}_{\ j\nu}$ emerges. Rather, the coupling occurs merely via the common dependence of $\tilde{\mathcal{H}}_0$ and $\tilde{\mathcal{H}}_{\mathrm{Gau}_1}$ on the vierbein field $e_{\nu}^{\ j}$. The reason is that $\tilde{\pi}^{\nu}\,\partial\varphi/\partial x^{\nu}$ constitutes a world scalar density and is, therefore, already form-invariant under the given symmetry. This changes, however, if we include non-zero spin fields into our analysis that transform non-trivially under the SO(1, 3)× Diff(M) group. This is worked out below.

As the gauge field ${\omega'}^{i}_{\ j\nu}$ is at this point an *external* object, the system described by the action (4.31) is not yet complete. The strategy to finally end up with a *closed* system of dynamic fields is the subject of the next step in the transformation theory.

4.1.6 Including the Canonical Transformation Rule of the Gauge Field $\omega^i{}_{j\mu}$

In the second step, the newly introduced gauge field $\omega^i{}_{j\mu}$, defined in the gauge Hamiltoninan (4.27), will be treated as an *internal dynamic* object. Then the action functional (4.31) must too include the conjugate momentum field $\tilde{q}_i{}^{j\mu\nu}$ defined as

$$\tilde{q}_i{}^{j\mu\nu}(x) := \frac{\partial\tilde{\mathcal{L}}}{\partial\left(\frac{\partial\omega^i{}_{j\mu}}{\partial x^\nu}\right)}.$$

That extends the action functional to

$$S = \int_V \tilde{\mathcal{L}}\,\mathrm{d}^4x \equiv \int_V \left[\tilde{\pi}^\nu \frac{\partial\varphi}{\partial x^\nu} + \tilde{k}_i{}^{[\mu\nu]}\frac{\partial e_\mu{}^i}{\partial x^\nu} + \tilde{q}_i{}^{j\mu\nu}\frac{\partial\omega^i{}_{j\mu}}{\partial x^\nu} - \tilde{\mathcal{H}}_{\mathrm{Gau}_1} - \tilde{\mathcal{H}}_0\right]\mathrm{d}^4x.$$

If we include in addition to scalar matter the real vector field, a_μ, and the complex spinor field, ψ, in the matter Lagrangian $\tilde{\mathcal{L}}_0$, the corresponding action functional is further extended to:

$$S = \int_V \mathrm{d}^4x \left(\tilde{\pi}^\nu \frac{\partial\varphi}{\partial x^\nu} + \tilde{p}^{\mu\nu}\frac{\partial a_\mu}{\partial x^\nu} + \tilde{\bar{\kappa}}^\nu \frac{\partial\psi}{\partial x^\nu} + \frac{\partial\bar{\psi}}{\partial x^\nu}\tilde{\kappa}^\nu \right.$$
$$\left. + \tilde{k}_i{}^{\mu\nu}\frac{\partial e_\mu{}^i}{\partial x^\nu} + \tilde{q}_i{}^{j\mu\nu}\frac{\partial\omega^i{}_{j\mu}}{\partial x^\nu} - \tilde{\mathcal{H}}_{\mathrm{Gau}_2} - \tilde{\mathcal{H}}_0\right). \tag{4.32}$$

Of course, the Hamiltonian density $\tilde{\mathcal{H}}_0$ of matter is the Legendre transform of free matter fields collected in $\tilde{\mathcal{L}}_0$. The form of the additional terms for the spin-1 and spin-1/2 fields follows from the canonical definitions of the respective momentum fields in analogy to the steps carried out in Sect. 4.1.4 for scalar fields:

$$\tilde{p}^{\mu\nu}(x) := \frac{\partial\tilde{\mathcal{L}}_0}{\partial\left(\frac{\partial a_\mu}{\partial x^\nu}\right)}, \qquad \tilde{\bar{\kappa}}^\nu(x) := \frac{\partial\tilde{\mathcal{L}}_0}{\partial\left(\frac{\partial\psi}{\partial x^\nu}\right)}, \qquad \tilde{\kappa}^\nu(x) := \frac{\partial\tilde{\mathcal{L}}_0}{\partial\left(\frac{\partial\bar{\psi}}{\partial x^\nu}\right)}.$$

In order to accommodate those additional fields in the gauging process, also the gauge Hamiltonian $\tilde{\mathcal{H}}_{\mathrm{Gau}_1}$ will be extended to $\tilde{\mathcal{H}}_{\mathrm{Gau}_2}$. Thus, in analogy to the procedure that led to the gauge Hamiltonian (4.27), the task is now to determine the gauge Hamiltonian $\tilde{\mathcal{H}}_{\mathrm{Gau}_2}$ that renders the action (4.32) form-invariant for a given Hamiltonian $\tilde{\mathcal{H}}_0$ describing the dynamics of scalar, vector, and spinor fields. In other words, $\tilde{\mathcal{H}}_{\mathrm{Gau}_2}$ must make the integrand of (4.32) into a *world scalar density*.

The starting point is now a generating function of type $\tilde{\mathcal{F}}_3$ that encompasses also the additional conjugate fields, i.e. some $\tilde{\mathcal{F}}_3 = \tilde{\mathcal{F}}_3^\nu(\Phi, \tilde{\pi}, A, \tilde{p}, \Psi, \tilde{\bar{\kappa}}, \bar{\Psi}, \tilde{\kappa}, E, \tilde{k}, \Omega, \tilde{q}, x)$. With this rather generic assumption the general transformation rules for an action of type (4.32) can be readily found. Details are worked out in the following explicit calculation.

✓ Explict Calculation

The integrand condition similar to Eq. (4.18) for the action (4.32) becomes

$$
\begin{aligned}
&\tilde{\pi}^{\nu}\frac{\partial\varphi}{\partial x^{\nu}}+\tilde{p}^{\mu\nu}\frac{\partial a_{\mu}}{\partial x^{\nu}}+\frac{\partial\bar{\psi}}{\partial x^{\nu}}\tilde{\kappa}^{\nu}+\tilde{\bar{\kappa}}^{\nu}\frac{\partial\psi}{\partial x^{\nu}}+\tilde{k}_{i}{}^{\mu\nu}\frac{\partial e_{\mu}{}^{i}}{\partial x^{\nu}}+\tilde{q}_{i}{}^{j\mu\nu}\frac{\partial\omega^{i}{}_{j\mu}}{\partial x^{\nu}}\\
&-\tilde{\mathcal{H}}\left(\varphi,\tilde{\pi}^{\nu},a_{\mu},\tilde{p}^{\mu\nu},\psi,\tilde{\bar{\kappa}}^{\nu},\bar{\psi},\tilde{\kappa}^{\nu},e_{\mu}{}^{i},\tilde{k}_{i}{}^{\mu\nu},\omega^{i}{}_{j\mu},\tilde{q}_{i}{}^{j\mu\nu},x\right)\\
&=\left(\tilde{\Pi}^{\nu}\frac{\partial\Phi}{\partial X^{\nu}}+\tilde{P}^{\mu\nu}\frac{\partial A_{\mu}}{\partial X^{\nu}}+\frac{\partial\bar{\Psi}}{\partial X^{\nu}}\tilde{\mathcal{K}}^{\nu}+\tilde{\bar{\mathcal{K}}}^{\nu}\frac{\partial\Psi}{\partial X^{\nu}}+\tilde{K}_{I}{}^{\mu\nu}\frac{\partial E^{I}{}_{\mu}}{\partial X^{\nu}}+\tilde{Q}_{I}{}^{J\mu\nu}\frac{\partial\Omega^{I}{}_{J\mu}}{\partial X^{\nu}}\right)\left|\frac{\partial X}{\partial x}\right|\\
&-\tilde{\mathcal{H}}'\left(\Phi,\tilde{\Pi}^{\nu},A_{\mu},\tilde{P}^{\mu\nu},\Psi,\tilde{\bar{\mathcal{K}}}^{\nu},\bar{\Psi},\tilde{\mathcal{K}}^{\nu},E^{I}{}_{\mu},\tilde{K}_{I}{}^{\mu\nu},\Omega^{I}{}_{J\mu},\tilde{Q}_{I}{}^{J\mu\nu},X\right)\left|\frac{\partial X}{\partial x}\right|\\
&+\frac{\partial}{\partial x^{\nu}}\left(\tilde{\mathcal{F}}_{3}^{\nu}+\tilde{\pi}^{\nu}\varphi+\tilde{p}^{\mu\nu}a_{\mu}+\bar{\psi}\,\tilde{\kappa}^{\nu}+\tilde{\bar{\kappa}}^{\nu}\psi+\tilde{k}_{i}{}^{\mu\nu}e_{\mu}{}^{i}+\tilde{q}_{i}{}^{j\mu\nu}\omega^{i}{}_{j\mu}\right).
\end{aligned}
\tag{4.33}
$$

Expanding the left-hand side of Eq. (4.33) yields:

$$
\begin{aligned}
&-\frac{\partial\tilde{\pi}^{\beta}}{\partial x^{\nu}}\delta^{\nu}_{\beta}\varphi-\frac{\partial\tilde{p}^{\mu\beta}}{\partial x^{\nu}}\delta^{\nu}_{\beta}a_{\mu}-\frac{\partial\tilde{\bar{\kappa}}^{\beta}}{\partial x^{\nu}}\delta^{\nu}_{\beta}\psi-\delta^{\nu}_{\beta}\bar{\psi}\frac{\partial\tilde{\kappa}^{\beta}}{\partial x^{\nu}}-\frac{\partial\tilde{k}_{i}{}^{\mu\beta}}{\partial x^{\nu}}\delta^{\nu}_{\beta}e_{\mu}{}^{i}-\frac{\partial\tilde{q}_{i}{}^{j\mu\beta}}{\partial x^{\nu}}\delta^{\nu}_{\beta}\omega^{i}{}_{j\mu}-\tilde{\mathcal{H}}\\
&=\left(\tilde{\Pi}^{\nu}\frac{\partial\Phi}{\partial X^{\nu}}+\tilde{P}^{\mu\nu}\frac{\partial A_{\mu}}{\partial X^{\nu}}+\frac{\partial\bar{\Psi}}{\partial X^{\nu}}\tilde{\mathcal{K}}^{\nu}+\tilde{\bar{\mathcal{K}}}^{\nu}\frac{\partial\Psi}{\partial X^{\nu}}+\tilde{K}_{I}{}^{\mu\nu}\frac{\partial E^{I}{}_{\mu}}{\partial X^{\nu}}+\tilde{Q}_{I}{}^{J\mu\nu}\frac{\partial\Omega^{I}{}_{J\mu}}{\partial X^{\nu}}\right)\left|\frac{\partial X}{\partial x}\right|\\
&+\frac{\partial\tilde{\mathcal{F}}_{3}^{\lambda}}{\partial\Phi}\frac{\partial X^{\nu}}{\partial x^{\lambda}}\frac{\partial\Phi}{\partial X^{\nu}}+\frac{\partial\tilde{\mathcal{F}}_{3}^{\nu}}{\partial\tilde{\pi}^{\beta}}\frac{\partial\tilde{\pi}^{\beta}}{\partial x^{\nu}}+\frac{\partial\tilde{\mathcal{F}}_{3}^{\lambda}}{\partial A_{\mu}}\frac{\partial X^{\nu}}{\partial x^{\lambda}}\frac{\partial A_{\mu}}{\partial X^{\nu}}+\frac{\partial\tilde{\mathcal{F}}_{3}^{\nu}}{\partial\tilde{p}^{\mu\beta}}\frac{\partial\tilde{p}^{\mu\beta}}{\partial x^{\nu}}\\
&+\frac{\partial\bar{\Psi}}{\partial X^{\nu}}\frac{\partial\tilde{\mathcal{F}}_{3}^{\lambda}}{\partial\bar{\Psi}}\frac{\partial X^{\nu}}{\partial x^{\lambda}}+\frac{\partial\tilde{\mathcal{F}}_{3}^{\nu}}{\partial\tilde{\kappa}^{\beta}}\frac{\partial\tilde{\kappa}^{\beta}}{\partial x^{\nu}}+\frac{\partial\tilde{\mathcal{F}}_{3}^{\lambda}}{\partial\Psi}\frac{\partial X^{\nu}}{\partial x^{\lambda}}\frac{\partial\Psi}{\partial X^{\nu}}+\frac{\partial\tilde{\bar{\kappa}}^{\beta}}{\partial x^{\nu}}\frac{\partial\tilde{\mathcal{F}}_{3}^{\nu}}{\partial\tilde{\bar{\kappa}}^{\beta}}\\
&+\frac{\partial\tilde{\mathcal{F}}_{3}^{\lambda}}{\partial E^{I}{}_{\mu}}\frac{\partial X^{\nu}}{\partial x^{\lambda}}\frac{\partial E^{I}{}_{\mu}}{\partial X^{\nu}}+\frac{\partial\tilde{\mathcal{F}}_{3}^{\nu}}{\partial\tilde{k}_{i}{}^{\mu\beta}}\frac{\partial\tilde{k}_{i}{}^{\mu\beta}}{\partial x^{\nu}}+\frac{\partial\tilde{\mathcal{F}}_{3}^{\lambda}}{\partial\Omega^{I}{}_{J\mu}}\frac{\partial X^{\nu}}{\partial x^{\lambda}}\frac{\partial\Omega^{I}{}_{J\mu}}{\partial X^{\nu}}\\
&+\frac{\partial\tilde{\mathcal{F}}_{3}^{\nu}}{\partial\tilde{q}_{i}{}^{j\mu\beta}}\frac{\partial\tilde{q}_{i}{}^{j\mu\beta}}{\partial x^{\nu}}+\left.\frac{\partial\tilde{\mathcal{F}}_{3}^{\nu}}{\partial x^{\nu}}\right|_{\text{expl}}-\tilde{\mathcal{H}}'\left|\frac{\partial X}{\partial x}\right|.
\end{aligned}
\tag{4.34}
$$

We again derive the canonical transformation rules for the fields by comparing the coefficients of Eq. (4.34):

$$
\tilde{\Pi}^{\nu}\equiv-\frac{\partial\tilde{\mathcal{F}}_{3}^{\lambda}}{\partial\Phi}\frac{\partial X^{\nu}}{\partial x^{\lambda}}\left|\frac{\partial x}{\partial X}\right|\qquad\delta^{\nu}_{\beta}\varphi\equiv-\frac{\partial\tilde{\mathcal{F}}_{3}^{\nu}}{\partial\tilde{\pi}^{\beta}}\tag{4.35a}
$$

$$
\tilde{P}^{\mu\nu}\equiv-\frac{\partial\tilde{\mathcal{F}}_{3}^{\lambda}}{\partial A_{\mu}}\frac{\partial X^{\nu}}{\partial x^{\lambda}}\left|\frac{\partial x}{\partial X}\right|\qquad\delta^{\nu}_{\beta}a_{\mu}\equiv-\frac{\partial\tilde{\mathcal{F}}_{3}^{\nu}}{\partial\tilde{p}^{\mu\beta}}\tag{4.35b}
$$

$$\tilde{\mathcal{K}}^{\nu} \equiv -\frac{\partial \tilde{\mathcal{F}}_3^{\lambda}}{\partial \bar{\Psi}} \frac{\partial X^{\nu}}{\partial x^{\lambda}} \left| \frac{\partial x}{\partial X} \right| \qquad \delta_{\beta}^{\nu} \bar{\psi} \equiv -\frac{\partial \tilde{\mathcal{F}}_3^{\nu}}{\partial \tilde{\kappa}^{\beta}} \tag{4.35c}$$

$$\tilde{\bar{\mathcal{K}}}^{\nu} \equiv -\frac{\partial \tilde{\mathcal{F}}_3^{\lambda}}{\partial \Psi} \frac{\partial X^{\nu}}{\partial x^{\lambda}} \left| \frac{\partial x}{\partial X} \right| \qquad \delta_{\beta}^{\nu} \psi \equiv -\frac{\partial \tilde{\mathcal{F}}_3^{\nu}}{\partial \tilde{\bar{\kappa}}^{\beta}} \tag{4.35d}$$

$$\tilde{K}_I^{\ \mu\nu} \equiv -\frac{\partial \tilde{\mathcal{F}}_3^{\lambda}}{\partial E^I_{\ \mu}} \frac{\partial X^{\nu}}{\partial x^{\lambda}} \left| \frac{\partial x}{\partial X} \right| \qquad \delta_{\beta}^{\nu} e_{\mu}^{\ i} \equiv -\frac{\partial \tilde{\mathcal{F}}_3^{\nu}}{\partial \tilde{k}_i^{\ \mu\beta}} \tag{4.35e}$$

$$\tilde{Q}_I^{\ J\mu\nu} \equiv -\frac{\partial \tilde{\mathcal{F}}_3^{\lambda}}{\partial \Omega^I_{\ J\mu}} \frac{\partial X^{\nu}}{\partial x^{\lambda}} \left| \frac{\partial x}{\partial X} \right| \qquad \delta_{\beta}^{\nu} \omega^i_{\ j\mu} \equiv -\frac{\partial \tilde{\mathcal{F}}_3^{\nu}}{\partial \tilde{q}_i^{\ j\mu\beta}} \tag{4.35f}$$

and, as usual, the rule for the Hamiltonian

$$\tilde{\mathcal{H}}' = \left(\tilde{\mathcal{H}} + \left. \frac{\partial \tilde{\mathcal{F}}_3^{\nu}}{\partial x^{\nu}} \right|_{\text{expl}} \right) \left| \frac{\partial x}{\partial X} \right| . \tag{4.35g}$$

In order to promote the connection $\omega^i_{\ j\alpha}$ to a dynamic field, the generating function $\tilde{\mathcal{F}}_3^{\nu}$ from Eq. (4.20) must also define the transformation law (4.29). Suppressing the spinor indices of ψ, $\tilde{\bar{\kappa}}$, $\bar{\psi}$, $\tilde{\kappa}$ and of the yet to be specified spinor transformation matrix, S (not to be confused with neither the symbol for torsion not with that for the action integral), representing the spinor representation of the Lorentz group elements, this gives:

$$\begin{aligned}
&\tilde{\mathcal{F}}_3^{\nu}\left(\Phi, \tilde{\pi}, A, \tilde{p}, \Psi, \tilde{\bar{\kappa}}, \bar{\Psi}, \tilde{\kappa}, E, \tilde{k}, \Omega, \tilde{q}, x\right) \\
&= -\tilde{\pi}^{\nu}\,\Phi - \tilde{p}^{\mu\nu}\,A_{\alpha}\frac{\partial X^{\alpha}}{\partial x^{\mu}} - \tilde{\bar{\kappa}}^{\nu}\,S^{-1}\,\Psi - \bar{\Psi}\,S\,\tilde{\kappa}^{\nu} - \tilde{k}_i^{\ \mu\nu}\Lambda^i_{\ I}\,E^I_{\ \alpha}\frac{\partial X^{\alpha}}{\partial x^{\mu}} \\
&\quad - \tilde{q}_i^{\ j\mu\nu}\left(\Lambda^i_{\ I}\,\Omega^I_{\ J\alpha}\,\Lambda^J_{\ j}\frac{\partial X^{\alpha}}{\partial x^{\mu}} + \Lambda^i_{\ I}\frac{\partial \Lambda^i_{\ J}}{\partial x^{\mu}}\right).
\end{aligned} \tag{4.36}$$

From the general set of canonical transformation rules (4.35), the complete set of specific rules for the generating function (4.36) are then:

$$\delta_{\beta}^{\nu}\varphi \equiv -\frac{\partial \tilde{\mathcal{F}}_3^{\nu}}{\partial \tilde{\pi}^{\beta}} = \delta_{\beta}^{\nu}\,\Phi \tag{4.37a}$$

$$\tilde{\Pi}^{\nu} \equiv -\frac{\partial \tilde{\mathcal{F}}_3^{\lambda}}{\partial \Phi}\frac{\partial X^{\nu}}{\partial x^{\lambda}}\left|\frac{\partial x}{\partial X}\right| = \tilde{\pi}^{\lambda}\,\frac{\partial X^{\nu}}{\partial x^{\lambda}}\left|\frac{\partial x}{\partial X}\right| \tag{4.37b}$$

$$\delta_{\beta}^{\nu}\,a_{\mu} \equiv -\frac{\partial \tilde{\mathcal{F}}_3^{\nu}}{\partial \tilde{p}^{\mu\beta}} = \delta_{\beta}^{\nu}\,A_{\alpha}\frac{\partial X^{\alpha}}{\partial x^{\mu}} \tag{4.37c}$$

$$\tilde{P}^{\mu\nu} \equiv -\frac{\partial \tilde{\mathcal{F}}_3^\lambda}{\partial A_\mu}\frac{\partial X^\nu}{\partial x^\lambda}\left|\frac{\partial x}{\partial X}\right| = \tilde{p}^{\xi\lambda}\frac{\partial X^\mu}{\partial x^\xi}\frac{\partial X^\nu}{\partial x^\lambda}\left|\frac{\partial x}{\partial X}\right| \tag{4.37d}$$

$$\delta^\nu_\beta \bar{\psi} \equiv -\frac{\partial \tilde{\mathcal{F}}_3^\nu}{\partial \tilde{\kappa}^\beta} = \delta^\nu_\beta \bar{\Psi}\, S \tag{4.37e}$$

$$\tilde{\mathcal{K}}^\nu \equiv -\frac{\partial \tilde{\mathcal{F}}_3^\lambda}{\partial \bar{\Psi}}\frac{\partial X^\nu}{\partial x^\lambda}\left|\frac{\partial x}{\partial X}\right| = S\,\tilde{\kappa}^\lambda\frac{\partial X^\nu}{\partial x^\lambda}\left|\frac{\partial x}{\partial X}\right| \tag{4.37f}$$

$$\delta^\nu_\beta \psi \equiv -\frac{\partial \tilde{\mathcal{F}}_3^\nu}{\partial \tilde{\bar{\kappa}}^\beta} = \delta^\nu_\beta S^{-1}\,\Psi \tag{4.37g}$$

$$\tilde{\bar{\mathcal{K}}}^\nu \equiv -\frac{\partial \tilde{\mathcal{F}}_3^\lambda}{\partial \Psi}\frac{\partial X^\nu}{\partial x^\lambda}\left|\frac{\partial x}{\partial X}\right| = \tilde{\bar{\kappa}}^\lambda S^{-1}\frac{\partial X^\nu}{\partial x^\lambda}\left|\frac{\partial x}{\partial X}\right| \tag{4.37h}$$

$$\delta^\nu_\beta\, e_\mu^{\ i} \equiv -\frac{\partial \tilde{\mathcal{F}}_3^\nu}{\partial \tilde{k}_i^{\ \mu\beta}} = \delta^\nu_\beta\,\Lambda^i_{\ I}\,E^I_{\ \alpha}\frac{\partial X^\alpha}{\partial x^\mu} \tag{4.37i}$$

$$\tilde{K}_I^{\ \mu\nu} \equiv -\frac{\partial \tilde{\mathcal{F}}_3^\lambda}{\partial E^I_{\ \mu}}\frac{\partial X^\nu}{\partial x^\lambda}\left|\frac{\partial x}{\partial X}\right| = \tilde{k}_i^{\ \xi\lambda}\,\Lambda^i_{\ I}\frac{\partial X^\mu}{\partial x^\xi}\frac{\partial X^\nu}{\partial x^\lambda}\left|\frac{\partial x}{\partial X}\right| \tag{4.37j}$$

and

$$\delta^\nu_\beta\,\omega^i_{\ j\mu} \equiv -\frac{\partial \tilde{\mathcal{F}}_3^\nu}{\partial \tilde{q}_i^{\ j\mu\beta}} = \delta^\nu_\beta\left(\Lambda^i_{\ I}\,\Omega^I_{\ J\alpha}\,\Lambda^J_{\ j}\frac{\partial X^\alpha}{\partial x^\mu} + \Lambda^i_{\ I}\frac{\partial \Lambda^i_{\ J}}{\partial x^\mu}\right) \tag{4.37k}$$

$$\tilde{Q}_I^{\ J\mu\nu} \equiv -\frac{\partial \tilde{\mathcal{F}}_3^\lambda}{\partial \Omega^I_{\ J\mu}}\frac{\partial X^\nu}{\partial x^\lambda}\left|\frac{\partial x}{\partial X}\right| = \tilde{q}_i^{\ j\xi\lambda}\,\Lambda^i_{\ I}\,\Lambda^J_{\ j}\frac{\partial X^\mu}{\partial x^\xi}\frac{\partial X^\nu}{\partial x^\lambda}\left|\frac{\partial x}{\partial X}\right|. \tag{4.37l}$$

Rule (4.37k), calculated explicitly in Sect. 4.1.7.2, reproduces the inhomogeneous transformation property of the gauge field $\omega^i_{\ j\mu}$, as required by Eq. (4.29), whereas rule (4.37l) determines the transformation property of the pertaining conjugate momentum field, $\tilde{q}_i^{\ j\mu\nu}$.

4.1.7 Derivation of the Gauge Hamiltonian

The key benefit of the canonical transformation framework is that it provides from the generating function the prescription for gauging the initial Hamiltonian density $\tilde{\mathcal{H}}_0$, hence to derive the gauge Hamiltonian $\tilde{\mathcal{H}}_{\mathrm{Gau}_2}$ such that the combined system $\tilde{\mathcal{H}}_0 + \tilde{\mathcal{H}}_{\mathrm{Gau}_2}$ becomes invariant with respect to the given transformation of the constituent fields. The gauge Hamiltonian is ultimately determined by the explicit x^μ-dependence of the generating function according to the general rule from Eq. (4.19c). For the actual generating function (4.36), the explicit x^ν-derivative of the parameters in the generating function is:

$$\left.\frac{\partial\tilde{\mathcal{F}}_3^\nu}{\partial x^\nu}\right|_{\mathrm{expl}} = -\,\tilde{p}^{(\mu\nu)}A_\alpha\,\frac{\partial^2 X^\alpha}{\partial x^\mu\partial x^\nu} - \tilde{\bar{\kappa}}^\nu\,\frac{\partial S^{-1}}{\partial x^\nu}\,\Psi - \bar{\Psi}\,\frac{\partial S}{\partial x^\nu}\,\tilde{\kappa}^\nu$$
$$-\,\tilde{k}_i^{\ \mu\nu}\frac{\partial}{\partial x^\nu}\left(\Lambda^i{}_I\,\frac{\partial X^\alpha}{\partial x^\mu}\right)E^I{}_\alpha$$
$$-\,\tilde{q}_i^{\ j\mu\nu}\left[\Omega^I{}_{J\alpha}\frac{\partial}{\partial x^\nu}\left(\Lambda^i{}_I\Lambda^J{}_j\,\frac{\partial X^\alpha}{\partial x^\mu}\right)+\frac{\partial}{\partial x^\nu}\left(\Lambda^i{}_I\frac{\partial\Lambda^I{}_J}{\partial x^\mu}\right)\right]. \tag{4.38}$$

The gauge Hamiltonian is then obtained from (4.38) by expressing all its parameters, namely $\Lambda^i{}_I$, S, $\partial X^\alpha/\partial x^\mu$ and their respective derivatives, in terms of the physical fields of the system according to the set of canonical transformation rules (4.37). The individual contributions from all involved fields are worked out in the next subsections.

4.1.7.1 Contribution of the Vector Field A_α to Eq. (4.38)

The scalar field $\varphi(x)$ does not contribute to the divergence of the generating function. Regarding the contribution of the vector field, we have two options, depending on the particular field properties. From the canonical transformation rule (4.37i), written in the form

$$\frac{\partial X^\alpha}{\partial x^\mu} = E_I{}^\alpha\,\Lambda^i{}_I\,e_\mu{}^i,$$

we can use its x^ν-derivative and Eq. (4.29) to compute the following transformation of the first (vector field) term in Eq. (4.38):

$$-\tilde{p}^{\mu\nu}A_\alpha\,\frac{\partial^2 X^\alpha}{\partial x^\mu\partial x^\nu} = -\tilde{p}^{\mu\nu}\,e_i{}^\beta\,a_\beta\left(\frac{\partial e_\mu{}^i}{\partial x^\nu}+\omega^i{}_{j\nu}\,e_\mu{}^j\right)$$
$$+\,\tilde{P}^{\mu\nu}\,E_i{}^\beta\,A_\beta\left(\frac{\partial E_\mu{}^I}{\partial X^\nu}+\Omega^i{}_{j\nu}E_\mu{}^J\right)\left|\frac{\partial X}{\partial x}\right|.$$

This provides a *direct* coupling of the vector field $e_i{}^\beta\,a_\beta$ with the connection $\omega^i{}_{j\nu}$ and leads to a *covariant* derivative term of a_β in the final, diffeomorphism-invariant action functional, as laid out in Ref. [10].

The second option is to express the second derivative of $X^\alpha(x)$ through the derivative of the canonical transformation rule (4.37c) of the vector field

$$a_\mu = A_\alpha\frac{\partial X^\alpha}{\partial x^\mu},$$

which yields

$$A_\alpha\frac{\partial^2 X^\alpha}{\partial x^\mu\partial x^\nu} = \frac{\partial a_\mu}{\partial x^\nu} - \frac{\partial A_\alpha}{\partial X^\tau}\frac{\partial X^\tau}{\partial x^\nu}\frac{\partial X^\alpha}{\partial x^\mu}.$$

Consequently, the term related to the vector field A_α in Eq. (4.38) transforms as:

$$-\tilde{p}^{\mu\nu} A_\alpha \frac{\partial^2 X^\alpha}{\partial x^\mu \partial x^\nu} = -\tilde{p}^{(\mu\nu)} \frac{\partial a_\mu}{\partial x^\nu} + \tilde{P}^{(\mu\nu)} \frac{\partial A_\mu}{\partial X^\nu} \left| \frac{\partial X}{\partial x} \right| .$$

As will be shown in the following, this leads to the conventional skew-symmetric field tensor $\tilde{p}^{\mu\nu}$ as the *exterior* derivative of a_β in the final, diffeomorphism-invariant action functional. As the covariant derivative is not invariant under U(1) resp. SU(N) symmetries, the first type of coupling is not allowed for those gauge theories. For this reason, we will stick here with the second option, the "Maxwell" form, involving the exterior derivative of a_μ. Nevertheless, the covariant derivative option may apply to gauge theories not featuring a U(1) resp. SU(N) symmetry but effective massive vector bosons.

In both cases, one encounters the required form of a transformation rule for the Hamiltonian. It no longer depends on parameters but instead on the respective physical fields in a symmetric representation in the original and—with the opposite sign—the transformed fields. This scheme will be be shown in the following to emerge for all terms of the sum of Eq. (4.38) and thereby to produce a valid gauge Hamiltonian enabling form-invariance of the total action.

4.1.7.2 Contribution of the Spinor Fields Ψ, $\bar{\Psi}$ to Eq. (4.38)

The transformation rule for the spinor representative S of the Lorentz transformation is derived in the explicit calculation below.

✓ Explict Calculation

The parameters of the transformation given by the spinor transformation matrix S such that $\Psi = S\psi$ depend on those of the Lorentz transformation $\Lambda_I^{\ i}$. With the Dirac matrices Γ^I and γ^i in the local Lorentz frames, the spinor representation of the Lorentz transformation follows from the requirement of form-invariance of the Dirac equation under Lorentz transformations:

$$\mathrm{i}\Gamma^I \frac{\partial \Psi}{\partial X^I} - m\Psi = 0 \quad \Leftrightarrow \quad \mathrm{i}\gamma^i \frac{\partial \psi}{\partial x^i} - m\psi = 0$$

exactly if

$$S^{-1}\Gamma^I S \Lambda_I^{\ i} = \gamma^i, \qquad S^{-1}\Gamma_I S = \gamma_i \Lambda^i_{\ I}. \tag{4.39}$$

For a *local* Lorentz transformation, the matrices S and $\Lambda^i_{\ I}$ are x-dependent. The derivative of Eq. (4.39) is

$$\Gamma_I \frac{\partial S}{\partial x^\mu} = \frac{\partial S}{\partial x^\mu} \gamma_i \Lambda^i{}_I + S \gamma_i \frac{\partial \Lambda^i{}_I}{\partial x^\mu}. \tag{4.40}$$

Contracting Eq. (4.40) with $S^{-1}\Gamma^I$ from the left yields

$$S^{-1}\Gamma^I \Gamma_I \frac{\partial S}{\partial x^\mu} = S^{-1}\Gamma^I \frac{\partial S}{\partial x^\mu} \gamma_i \Lambda^i{}_I + S^{-1}\Gamma^I S \gamma_i \frac{\partial \Lambda^i{}_I}{\partial x^\mu},$$

hence

$$4S^{-1} \frac{\partial S}{\partial x^\mu} = \gamma^j S^{-1} \frac{\partial S}{\partial x^\mu} \gamma_i \Lambda^i{}_I \Lambda^i{}_J + \Lambda^i{}_J \frac{\partial \Lambda^i{}_I}{\partial x^\mu} \gamma^j \gamma_i. \tag{4.41}$$

Contracting Eq. (4.40) with S^{-1} from the left and with $\Lambda^i{}_J \gamma^j$ from the right gives, on the other hand,

$$S^{-1} \Gamma_I \frac{\partial S}{\partial x^\mu} \Lambda^i{}_J \gamma^j = S^{-1} \frac{\partial S}{\partial x^\mu} \gamma_i \Lambda^i{}_I \Lambda^i{}_J \gamma^j + \gamma_i \frac{\partial \Lambda^i{}_I}{\partial x^\mu} \Lambda^i{}_J \gamma^j,$$

hence

$$4S^{-1} \frac{\partial S}{\partial x^\mu} = \gamma_i S^{-1} \frac{\partial S}{\partial x^\mu} \Lambda^i{}_I \Lambda^i{}_J \gamma^j - \Lambda^i{}_J \frac{\partial \Lambda^i{}_I}{\partial x^\mu} \gamma_i \gamma^j, \tag{4.42}$$

Equations (4.41) and (4.42) are now added to give

$$\begin{aligned} 8S^{-1} \frac{\partial S}{\partial x^\mu} &= \Lambda^i{}_J \frac{\partial \Lambda^i{}_I}{\partial x^\mu} \left(\gamma^j \gamma_i - \gamma_i \gamma^j \right) + S^{-1} \left(\Gamma^I \frac{\partial S}{\partial x^\mu} \gamma_j \Lambda^j{}_I + \Gamma_I \frac{\partial S}{\partial x^\mu} \gamma^j \Lambda^i{}_J \right) \\ &= -2\mathrm{i} \Lambda^i{}_I \frac{\partial \Lambda^i{}_J}{\partial x^\mu} \sigma_i{}^j + 2\gamma^i S^{-1} \frac{\partial S}{\partial x^\mu} \gamma_i. \end{aligned}$$

Herein, σ^{ij}, the commutator of the Dirac matrices γ^k, is the generator of the spinor representation $\mathrm{SL}(2, \mathbb{C})$ of the Lorentz group,

$$\sigma^{ij} = \frac{\mathrm{i}}{2} \left(\gamma^i \gamma^j - \gamma^j \gamma^i \right), \qquad \eta^{ij} \mathbb{1} = \frac{1}{2} \left(\gamma^i \gamma^j + \gamma^j \gamma^i \right). \tag{4.43}$$

η^{ij} is the Minkowski metric of the Lorentz frame, with $\mathbb{1}$ denoting the unit matrix in spinor space.

The last term vanishes by virtue of the Dirac algebra if S stands for the spinor representation S of the (infinitesimal) local Lorentz transformation $\Lambda^i{}_I = \delta_i^I + \epsilon^i{}_I$ with infinitesimal parameters $\epsilon_{ij}(x) = -\epsilon_{ji}(x)$ [19]:

$$S(x) = \exp\left(-\tfrac{i}{4}\epsilon_{ij}(x)\sigma^{ij}\right) \quad \Rightarrow \quad \gamma^k S^{-1}\frac{\partial S}{\partial x^\mu}\gamma_k = -\tfrac{i}{4}\frac{\partial \epsilon_{ij}}{\partial x^\mu}\underbrace{\gamma^k\sigma^{ij}\gamma_k}_{\equiv 0} \equiv 0.$$

We conclude that for *local* Lorentz transformations

$$\tfrac{i}{4}\Lambda^i{}_I\frac{\partial \Lambda^i{}_J}{\partial x^\mu}\sigma_i{}^j = -S^{-1}\frac{\partial S}{\partial x^\mu}. \tag{4.44}$$

With the commutators $\sigma^{ij} \equiv \tfrac{i}{2}\left(\gamma^i\gamma^j - \gamma^j\gamma^i\right)$ and $\Sigma^{IJ} \equiv \tfrac{i}{2}\left(\Gamma^I\Gamma^J - \Gamma^J\Gamma^I\right)$, we finally encounter the spinor representation of the transformation rule (4.37k) for the gauge field $\omega^i{}_{j\mu}$:

$$\tfrac{i}{4}\Omega_{IJ_\nu}\Sigma^{IJ} = \left(\tfrac{i}{4}\, S\,\omega_{ij_\mu}\,\sigma^{ij}\,S^{-1} + \frac{\partial S}{\partial x^\mu}S^{-1}\right)\frac{\partial x^\mu}{\partial X^\nu}. \tag{4.45}$$

Now by means of the transformation rule (4.45), the replacement of the coefficients associated with the spinor terms in the first line of Eq. (4.38) follows as:

$$\begin{aligned}
-\bar{\tilde{\kappa}}^\nu\frac{\partial S^{-1}}{\partial x^\nu}\Psi &= \bar{\tilde{\kappa}}^\nu S^{-1}\frac{\partial S}{\partial x^\nu}\psi \\
&= \tfrac{i}{4}\bar{\tilde{\kappa}}^\nu\left(S^{-1}\,\Omega_{IJ\alpha}\,\Sigma^{IJ}\,S\,\frac{\partial X^\alpha}{\partial x^\nu} - \omega_{ij\nu}\,\sigma^{ij}\right)\psi \\
&= \bar{\tilde{\mathcal{K}}}^\nu\,\tfrac{i}{4}\Omega_{IJ\nu}\,\Sigma^{IJ}\,\Psi\left|\frac{\partial X}{\partial x}\right| - \bar{\tilde{\kappa}}^\nu\,\tfrac{i}{4}\omega_{ij\nu}\,\sigma^{ij}\,\psi.
\end{aligned}$$

Similarly

$$\begin{aligned}
-\bar{\Psi}\,\frac{\partial S}{\partial x^\nu}\,\tilde{\kappa}^\nu &= -\bar{\psi}\,S^{-1}\,\frac{\partial S}{\partial x^\nu}\,\tilde{\kappa}^\nu \\
&= \bar{\psi}\,\tfrac{i}{4}\omega_{ij\nu}\,\sigma^{ij}\,\tilde{\kappa}^\nu - \bar{\Psi}\,\tfrac{i}{4}\Omega_{IJ\nu}\,\Sigma^{IJ}\tilde{\mathcal{K}}^\nu\left|\frac{\partial X}{\partial x}\right|.
\end{aligned}$$

Hence, the parameters of the spinor-related terms in Eq. (4.38) are absorbed in a covariant combination of the matter and gauge fields according to

$$\begin{aligned}
-\bar{\tilde{\kappa}}^\nu\,\frac{\partial S^{-1}}{\partial x^\nu}\,\Psi - \bar{\Psi}\,\frac{\partial S}{\partial x^\nu}\,\tilde{\kappa}^\nu &= \tfrac{i}{4}\left(\bar{\psi}\,\omega_{ij\nu}\,\sigma^{ij}\,\tilde{\kappa}^\nu - \bar{\tilde{\kappa}}^\nu\,\omega_{ij\nu}\,\sigma^{ij}\,\psi\right) \\
&\quad - \tfrac{i}{4}\left(\bar{\Psi}\,\Omega_{IJ\nu}\,\Sigma^{IJ}\,\tilde{\mathcal{K}}^\nu - \bar{\tilde{\mathcal{K}}}^\nu\,\Omega_{IJ\nu}\,\Sigma^{IJ}\,\Psi\right)\left|\frac{\partial X}{\partial x}\right|.
\end{aligned}$$

4.1.7.3 Contribution of the vierbein field $E^I{}_\alpha$ to Eq. (4.38)

The coefficients in the term proportional to $\tilde{k}_i{}^{\mu\nu}$ can similarly be expressed in terms of the physical fields $e_\mu{}^i$ and $\omega^i{}_{n\mu}$ as follows:

$$-\tilde{k}_i{}^{\mu\nu}\frac{\partial}{\partial x^\nu}\left(\Lambda^i{}_J\frac{\partial X^\alpha}{\partial x^\mu}\right)E_\alpha{}^J = -\tfrac{1}{2}\tilde{k}_i{}^{\mu\nu}\left(\frac{\partial e_\mu{}^i}{\partial x^\nu}+\frac{\partial e_\nu{}^i}{\partial x^\mu}-\omega^i{}_{j\nu}e_\mu{}^j+\omega^i{}_{j\mu}e_\nu{}^j\right)$$
$$+\tfrac{1}{2}\tilde{K}_I{}^{\mu\nu}\left(\frac{\partial E_\mu{}^I}{\partial X^\nu}+\frac{\partial E_\nu{}^i}{\partial X^\mu}-\Omega^i{}_{j\nu}E_\mu{}^J+\Omega^I{}_{J\mu}E_\nu{}^j\right)\left|\frac{\partial X}{\partial x}\right|.$$

The explicit calculation is worked out below.

✓ Explict Calculation

The coefficients related to the spacetime transformation in the transformation rule (4.38) for the Hamiltonian are to be converted into dependencies of the physical fields. This means for the term proportional to the momentum field $\tilde{k}_i{}^{\mu\nu}$:

$$-\tilde{k}_i{}^{\mu\nu}\frac{\partial}{\partial x^\nu}\left(\Lambda^i{}_J\frac{\partial X^\alpha}{\partial x^\mu}\right)E^J{}_\alpha = -\tilde{k}_i{}^{\mu\nu}\left(\frac{\partial\Lambda^i{}_J}{\partial x^\nu}\frac{\partial X^\alpha}{\partial x^\mu}+\Lambda^i{}_J\frac{\partial^2 X^\alpha}{\partial x^\mu\partial x^\nu}\right)E^J{}_\alpha$$

The derivative of $\Lambda^i{}_J$ is expressed in terms of the transformation rule (4.37k) for the gauge field $\omega^i{}_{j\nu}$

$$\frac{\partial\Lambda^i{}_J}{\partial x^\nu} = \Lambda^i{}_I\,\Omega^I{}_{J\xi}\frac{\partial X^\xi}{\partial x^\nu}-\omega^i{}_{j\nu}\,\Lambda^j{}_J,$$

whereas the second derivative of $X^\alpha(x)$ from is obtained from the canonical transformation rule (4.37i), written in the equivalent form

$$\frac{\partial X^\alpha}{\partial x^\mu} = E_I{}^\alpha\,\Lambda^i{}_I\,e_\mu{}^i$$

as

$$
\begin{aligned}
\frac{\partial^2 X^\alpha}{\partial x^\mu \partial x^\nu} &= \frac{\partial E_I{}^\alpha}{\partial x^\nu} \Lambda^i{}_I e_\mu{}^i + E_I{}^\alpha \Lambda^i{}_I \frac{\partial e_\mu{}^i}{\partial x^\nu} + E_I{}^\alpha \frac{\partial \Lambda^i{}_J}{\partial x^\nu} e_\mu{}^j \\
&= \frac{\partial E_I{}^\alpha}{\partial x^\nu} E_\eta{}^I \frac{\partial X^\eta}{\partial x^\mu} + \frac{\partial X^\alpha}{\partial x^\xi} e_i{}^\xi \frac{\partial e_\mu{}^i}{\partial x^\nu} + E_I{}^\alpha \left(\Lambda^i{}_I \omega^i{}_{j\nu} - \Omega^I{}_{J\xi} \Lambda^J{}_j \frac{\partial X^\xi}{\partial x^\nu} \right) e_\mu{}^j \\
&= -E_I{}^\alpha \frac{\partial E_\eta{}^I}{\partial X^\xi} \frac{\partial X^\xi}{\partial x^\nu} \frac{\partial X^\eta}{\partial x^\mu} + e_i{}^\xi \frac{\partial e_\mu{}^i}{\partial x^\nu} \frac{\partial X^\alpha}{\partial x^\xi} \\
&\quad + \frac{\partial X^\alpha}{\partial x^\xi} e_i{}^\xi \omega^i{}_{j\nu} e_\mu{}^j - E_I{}^\alpha \Omega^I{}_{J\xi} E_\eta{}^J \frac{\partial X^\eta}{\partial x^\mu} \frac{\partial X^\xi}{\partial x^\nu} \\
&= e_i{}^\xi \left(\frac{\partial e_\mu{}^i}{\partial x^\nu} + \omega^i{}_{j\nu} e_\mu{}^j \right) \frac{\partial X^\alpha}{\partial x^\xi} - E_I{}^\alpha \left(\frac{\partial E_\eta{}^I}{\partial X^\xi} + \Omega^I{}_{J\xi} E_\eta{}^J \right) \frac{\partial X^\eta}{\partial x^\mu} \frac{\partial X^\xi}{\partial x^\nu}.
\end{aligned} \tag{4.46}
$$

This yields

$$
\begin{aligned}
& -\tilde{k}_i{}^{\mu\nu} \frac{\partial}{\partial x^\nu} \left(\Lambda^i{}_J \frac{\partial X^\alpha}{\partial x^\mu} \right) E^J{}_\alpha \\
&= -\tilde{k}_i{}^{\mu\nu} \frac{\partial \Lambda^i{}_J}{\partial x^\nu} \frac{\partial X^\alpha}{\partial x^\mu} E^J{}_\alpha - \tilde{k}_i{}^{(\mu\nu)} \Lambda^i{}_J \frac{\partial^2 X^\alpha}{\partial x^\mu \partial x^\nu} E^J{}_\alpha \\
&= -\tilde{k}_i{}^{\mu\nu} \left(\Lambda^i{}_I \Omega^I{}_{J\xi} \frac{\partial X^\xi}{\partial x^\nu} - \omega^i{}_{j\nu} \Lambda^j{}_J \right) E^J{}_\alpha \frac{\partial X^\alpha}{\partial x^\mu} - \tilde{k}_i{}^{(\mu\nu)} \Lambda^i{}_J E^J{}_\alpha \frac{\partial^2 X^\alpha}{\partial x^\mu \partial x^\nu} \\
&= -\tilde{k}_i{}^{\mu\nu} \Lambda^i{}_I \Omega^I{}_{J\xi} E^J{}_\alpha \frac{\partial X^\alpha}{\partial x^\mu} \frac{\partial X^\xi}{\partial x^\nu} + \tilde{k}_i{}^{\mu\nu} \omega^i{}_{j\nu} \Lambda^j{}_J E^J{}_\alpha \frac{\partial X^\alpha}{\partial x^\mu} \\
&\quad + \tfrac{1}{2} \tilde{k}_i{}^{\mu\nu} \Lambda^i{}_I \frac{\partial X^\xi}{\partial x^\nu} \frac{\partial X^\eta}{\partial x^\mu} \left(\frac{\partial E^I{}_\eta}{\partial X^\xi} + \frac{\partial E^I{}_\xi}{\partial X^\eta} + \Omega^I{}_{J\xi} E^J{}_\eta + \Omega^I{}_{J\eta} E^J{}_\xi \right) \\
&\quad - \tfrac{1}{2} \tilde{k}_n{}^{\mu\nu} \Lambda^n{}_J E_\alpha{}^J \frac{\partial X^\alpha}{\partial x^\xi} e_i{}^\xi \left(\frac{\partial e_\mu{}^i}{\partial x^\nu} + \frac{\partial e_\nu{}^i}{\partial x^\mu} + \omega^i{}_{j\nu} e_\mu{}^j + \omega^i{}_{j\mu} e_\nu{}^j \right) \\
&= \tilde{k}_i{}^{\mu\nu} \omega^i{}_{j\nu} e_\mu{}^j - \tilde{K}_I{}^{\eta\xi} \Omega^I{}_{J\xi} E_\eta{}^J \left| \frac{\partial X}{\partial x} \right| \\
&\quad - \tfrac{1}{2} \tilde{k}_i{}^{\mu\nu} \left(\frac{\partial e_\mu{}^i}{\partial x^\nu} + \frac{\partial e_\nu{}^i}{\partial x^\mu} + \omega^i{}_{j\nu} e_\mu{}^j + \omega^i{}_{j\mu} e_\nu{}^j \right) \\
&\quad + \tfrac{1}{2} \tilde{K}_I{}^{\eta\xi} \left(\frac{\partial E_\eta{}^I}{\partial X^\xi} + \frac{\partial E_\xi{}^I}{\partial X^\eta} + \Omega^I{}_{J\xi} E_\eta{}^J + \Omega^I{}_{J\eta} E_\xi{}^J \right) \left| \frac{\partial X}{\partial x} \right|,
\end{aligned}
$$

hence finally

$$-\tilde{k}_i^{\ \mu\nu}\frac{\partial}{\partial x^\nu}\left(\Lambda^i{}_J\frac{\partial X^\alpha}{\partial x^\mu}\right)E^J{}_\alpha=-\tfrac{1}{2}\tilde{k}_i^{\ \mu\nu}\left(\frac{\partial e_\mu{}^i}{\partial x^\nu}+\frac{\partial e_\nu{}^i}{\partial x^\mu}-\omega^i{}_{j\nu}e_\mu{}^j+\omega^i{}_{j\mu}e_\nu{}^j\right)$$
$$+\tfrac{1}{2}\tilde{K}_I^{\ \mu\nu}\left(\frac{\partial E_\mu{}^I}{\partial X^\nu}+\frac{\partial E_\nu{}^i}{\partial X^\mu}-\Omega^i{}_{j\nu}E_\mu{}^J+\Omega^I{}_{J\mu}E_\nu{}^j\right)\left|\frac{\partial X}{\partial x}\right|.$$

4.1.7.4 Contribution of the Gauge Field $\Omega^I{}_{J\alpha}$ to Eq. (4.38)

As the last step, we express the coefficients in the term proportional to $\tilde{q}_i^{\ j\mu\nu}$ of Eq. (4.38) through the physical fields $\omega^i{}_{j\mu}$ according to the canonical transformation rules (4.37). The explicit calculation is again presented below, giving the final result:

$$-\tilde{q}_i^{\ j\mu\nu}\left[\Omega^I{}_{J\alpha}\frac{\partial}{\partial x^\nu}\left(\Lambda^i{}_I\Lambda^J{}_j\frac{\partial X^\alpha}{\partial x^\mu}\right)+\frac{\partial}{\partial x^\nu}\left(\Lambda^i{}_I\frac{\partial \Lambda^i{}_J}{\partial x^\mu}\right)\right]$$
$$=-\tfrac{1}{2}\tilde{q}_i^{\ j\mu\nu}\left(\frac{\partial \omega^i{}_{j\mu}}{\partial x^\nu}+\frac{\partial \omega^i{}_{j\nu}}{\partial x^\mu}+\omega^i{}_{n\mu}\,\omega^n{}_{j\nu}-\omega^i{}_{n\nu}\,\omega^n{}_{j\mu}\right)$$
$$+\tfrac{1}{2}\tilde{Q}_I^{\ J\mu\nu}\left(\frac{\partial \Omega^I{}_{J\mu}}{\partial X^\nu}+\frac{\partial \Omega^i{}_{j\nu}}{\partial X^\mu}+\Omega^I{}_{N\mu}\,\Omega^N{}_{J\nu}-\Omega^i{}_{n\nu}\,\Omega^N{}_{J\mu}\right)\left|\frac{\partial X}{\partial x}\right|. \tag{4.47}$$

✓ Explict Calculation

All coefficients in the term proportional to $\tilde{q}_i^{\ j\mu\nu}$ of Eq. (4.38) are now converted into dependencies on the physical fields. To this end, we first write this term in expanded form:

$$-\tilde{q}_i^{\ j\mu\nu}\left[\Omega^I{}_{J\alpha}\frac{\partial}{\partial x^\nu}\left(\Lambda^i{}_I\Lambda^J{}_j\frac{\partial X^\alpha}{\partial x^\mu}\right)+\frac{\partial}{\partial x^\nu}\left(\Lambda^i{}_I\frac{\partial \Lambda^i{}_J}{\partial x^\mu}\right)\right]$$
$$=-\tilde{q}_i^{\ j\mu\nu}\Omega^I{}_{J\alpha}\left(\frac{\partial \Lambda^i{}_I}{\partial x^\nu}\Lambda^J{}_j\frac{\partial X^\alpha}{\partial x^\mu}+\Lambda^i{}_I\frac{\partial \Lambda^J{}_j}{\partial x^\nu}\frac{\partial X^\alpha}{\partial x^\mu}+\Lambda^i{}_I\Lambda^J{}_j\frac{\partial^2 X^\alpha}{\partial x^\mu\partial x^\nu}\right)$$
$$-\tilde{q}_i^{\ j\mu\nu}\left(\frac{\partial \Lambda^i{}_I}{\partial x^\nu}\frac{\partial \Lambda^i{}_J}{\partial x^\mu}+\Lambda^i{}_I\frac{\partial^2 \Lambda^i{}_J}{\partial x^\mu\partial x^\nu}\right).$$

With the transformation rule (4.29) solved for $\Omega^{I}{}_{J\alpha}$

$$\Omega^{I}{}_{J\alpha} = \left(\Lambda^{I}{}_{n}\,\omega^{n}{}_{m\xi} - \frac{\partial \Lambda^{I}{}_{m}}{\partial x^{\xi}}\right)\Lambda^{m}{}_{J}\,\frac{\partial x^{\xi}}{\partial X^{\alpha}},$$

Equation (4.47) is expressed equivalently in terms of the original fields as

$$\begin{aligned}
&-\tilde{q}_{i}{}^{j\mu\nu}\Bigg[\left(\Lambda^{I}{}_{n}\omega^{n}{}_{m\xi} - \frac{\partial \Lambda^{I}{}_{m}}{\partial x^{\xi}}\right)\left(\frac{\partial \Lambda^{i}{}_{I}}{\partial x^{\nu}}\delta^{m}_{j}\delta^{\xi}_{\mu} + \Lambda^{i}{}_{I}\frac{\partial \Lambda^{J}{}_{j}}{\partial x^{\nu}}\Lambda^{m}{}_{J}\delta^{\xi}_{\mu} + \Lambda^{i}{}_{I}\delta^{m}_{j}\frac{\partial^{2} X^{\alpha}}{\partial x^{\mu}\partial x^{\nu}}\frac{\partial x^{\xi}}{\partial X^{\alpha}}\right)\\
&\qquad + \frac{\partial \Lambda^{i}{}_{I}}{\partial x^{\nu}}\frac{\partial \Lambda^{i}{}_{J}}{\partial x^{\mu}} + \Lambda^{i}{}_{I}\frac{\partial^{2} \Lambda^{i}{}_{J}}{\partial x^{\mu}\partial x^{\nu}}\Bigg]\\
&= -\tilde{q}_{i}{}^{j\mu\nu}\Bigg(\omega^{n}{}_{j\mu}\Lambda^{I}{}_{n}\frac{\partial \Lambda^{i}{}_{I}}{\partial x^{\nu}} + \omega^{i}{}_{n\mu}\Lambda^{n}{}_{J}\frac{\partial \Lambda^{J}{}_{j}}{\partial x^{\nu}} + \omega^{i}{}_{j\xi}\frac{\partial^{2} X^{\alpha}}{\partial x^{\mu}\partial x^{\nu}}\frac{\partial x^{\xi}}{\partial X^{\alpha}} - \cancel{\frac{\partial \Lambda^{i}{}_{J}}{\partial x^{\mu}}\frac{\partial \Lambda^{i}{}_{I}}{\partial x^{\nu}}}\\
&\qquad - \Lambda^{i}{}_{I}\frac{\partial \Lambda^{I}{}_{n}}{\partial x^{\mu}}\Lambda^{n}{}_{J}\frac{\partial \Lambda^{J}{}_{j}}{\partial x^{\nu}} - \Lambda^{i}{}_{I}\frac{\partial \Lambda^{i}{}_{J}}{\partial x^{\xi}}\frac{\partial^{2} X^{\alpha}}{\partial x^{\mu}\partial x^{\nu}}\frac{\partial x^{\xi}}{\partial X^{\alpha}} + \cancel{\frac{\partial \Lambda^{i}{}_{I}}{\partial x^{\nu}}\frac{\partial \Lambda^{i}{}_{J}}{\partial x^{\mu}}} + \Lambda^{i}{}_{I}\frac{\partial^{2} \Lambda^{i}{}_{J}}{\partial x^{\mu}\partial x^{\nu}}\Bigg)\\
&= -\tilde{q}_{i}{}^{j\mu\nu}\Bigg[\omega^{i}{}_{n\mu}\Lambda^{n}{}_{J}\frac{\partial \Lambda^{J}{}_{j}}{\partial x^{\nu}} - \omega^{n}{}_{j\mu}\Lambda^{i}{}_{I}\frac{\partial \Lambda^{I}{}_{n}}{\partial x^{\nu}} + \left(\omega^{i}{}_{j\xi} - \Lambda^{i}{}_{I}\frac{\partial \Lambda^{i}{}_{J}}{\partial x^{\xi}}\right)\frac{\partial^{2} X^{\alpha}}{\partial x^{\mu}\partial x^{\nu}}\frac{\partial x^{\xi}}{\partial X^{\alpha}}\\
&\qquad - \Lambda^{i}{}_{I}\frac{\partial \Lambda^{I}{}_{n}}{\partial x^{\mu}}\Lambda^{n}{}_{J}\frac{\partial \Lambda^{J}{}_{j}}{\partial x^{\nu}} + \Lambda^{i}{}_{I}\frac{\partial^{2} \Lambda^{i}{}_{J}}{\partial x^{\mu}\partial x^{\nu}}\Bigg].
\end{aligned}\tag{4.48}$$

Now, the transformation rule (4.29) is inserted in the form

$$\Lambda^{i}{}_{I}\frac{\partial \Lambda^{i}{}_{J}}{\partial x^{\nu}} = \omega^{i}{}_{j\nu} - \Lambda^{i}{}_{I}\,\Omega^{I}{}_{J\alpha}\Lambda^{J}{}_{j}\frac{\partial X^{\alpha}}{\partial x^{\nu}}$$

and its derivative

$$\begin{aligned}
\Lambda^{i}{}_{I}\frac{\partial^{2} \Lambda^{i}{}_{J}}{\partial x^{\mu}\partial x^{\nu}} &= \tfrac{1}{2}\Lambda^{i}{}_{I}\left(\frac{\partial \Lambda^{I}{}_{n}}{\partial x^{\nu}}\omega^{n}{}_{j\mu} + \frac{\partial \Lambda^{I}{}_{n}}{\partial x^{\mu}}\omega^{n}{}_{j\nu}\right) + \tfrac{1}{2}\left(\frac{\partial \omega^{i}{}_{j\mu}}{\partial x^{\nu}} + \frac{\partial \omega^{i}{}_{j\nu}}{\partial x^{\mu}}\right)\\
&\quad - \tfrac{1}{2}\Lambda^{i}{}_{I}\left(\frac{\partial \Omega^{I}{}_{J\xi}}{\partial X^{\alpha}} + \frac{\partial \Omega^{I}{}_{J\alpha}}{\partial X^{\xi}}\right)\Lambda^{J}{}_{j}\frac{\partial X^{\xi}}{\partial x^{\mu}}\frac{\partial X^{\alpha}}{\partial x^{\nu}}\\
&\quad - \tfrac{1}{2}\Lambda^{i}{}_{I}\Omega^{I}{}_{J\alpha}\left(\frac{\partial \Lambda^{J}{}_{j}}{\partial x^{\nu}}\frac{\partial X^{\alpha}}{\partial x^{\mu}} + \frac{\partial \Lambda^{J}{}_{j}}{\partial x^{\mu}}\frac{\partial X^{\alpha}}{\partial x^{\nu}}\right) - \Lambda^{i}{}_{I}\,\Omega^{I}{}_{J\alpha}\Lambda^{J}{}_{j}\frac{\partial^{2} X^{\alpha}}{\partial x^{\mu}\partial x^{\nu}}.
\end{aligned}$$

Putting it together yields

$$
\begin{aligned}
&-\tilde{q}_i{}^{j\mu\nu}\Bigg[\frac{1}{2}\left(\frac{\partial\omega^i{}_{j\mu}}{\partial x^\nu}+\frac{\partial\omega^i{}_{j\nu}}{\partial x^\mu}\right)-\frac{1}{2}\left(\frac{\partial\Omega^I{}_{K\beta}}{\partial X^\alpha}+\frac{\partial\Omega^I{}_{K\alpha}}{\partial X^\beta}\right)\Lambda^i{}_I\Lambda^K{}_j\frac{\partial X^\beta}{\partial x^\mu}\frac{\partial X^\alpha}{\partial x^\nu}\\
&\quad+\omega^i{}_{n\mu}\left(\omega^n{}_{j\nu}-\Lambda^n{}_I\,\Omega^I{}_{J\alpha}\Lambda^J{}_j\frac{\partial X^\alpha}{\partial x^\nu}\right)\\
&\quad-\frac{1}{2}\omega^n{}_{j\mu}\left(\omega^i{}_{n\nu}-\Lambda^i{}_I\,\Omega^I{}_{J\alpha}\Lambda^J{}_n\frac{\partial X^\alpha}{\partial x^\nu}\right)\\
&\quad+\frac{1}{2}\omega^n{}_{j\nu}\left(\omega^i{}_{n\mu}-\Lambda^i{}_I\,\Omega^I{}_{J\alpha}\Lambda^J{}_n\frac{\partial X^\alpha}{\partial x^\mu}\right)+\cancel{\Lambda^i{}_I\,\Omega^I{}_{J\alpha}\Lambda^J{}_j\frac{\partial^2X^\alpha}{\partial x^\mu\partial x^\nu}}\\
&\quad-\left(\omega^i{}_{n\mu}-\Lambda^i{}_I\,\Omega^I{}_{J\alpha}\Lambda^J{}_n\frac{\partial X^\alpha}{\partial x^\mu}\right)\left(\omega^n{}_{j\nu}-\Lambda^n{}_K\,\Omega^K{}_{L\beta}\Lambda^L{}_j\frac{\partial X^\beta}{\partial x^\nu}\right)\\
&\quad-\frac{1}{2}\Lambda^i{}_I\Omega^I{}_{J\alpha}\left(\Lambda^J{}_n\,\omega^n{}_{j\nu}-\Omega^J{}_{K\beta}\Lambda^K{}_j\frac{\partial X^\beta}{\partial x^\nu}\right)\frac{\partial X^\alpha}{\partial x^\mu}\\
&\quad-\frac{1}{2}\Lambda^i{}_I\Omega^I{}_{J\alpha}\left(\Lambda^J{}_n\,\omega^n{}_{j\mu}-\Omega^J{}_{K\beta}\Lambda^K{}_j\frac{\partial X^\beta}{\partial x^\mu}\right)\frac{\partial X^\alpha}{\partial x^\nu}\\
&\quad-\cancel{\Lambda^i{}_I\,\Omega^I{}_{J\alpha}\Lambda^J{}_j\frac{\partial^2X^\alpha}{\partial x^\mu\partial x^\nu}}\Bigg],
\end{aligned}
$$

which simplifies, after expanding

$$
\begin{aligned}
&-\tilde{q}_i{}^{j\mu\nu}\Bigg[\frac{1}{2}\left(\frac{\partial\omega^i{}_{j\mu}}{\partial x^\nu}+\frac{\partial\omega^i{}_{j\nu}}{\partial x^\mu}\right)-\frac{1}{2}\left(\frac{\partial\Omega^I{}_{K\beta}}{\partial X^\alpha}+\frac{\partial\Omega^I{}_{K\alpha}}{\partial X^\beta}\right)\Lambda^i{}_I\Lambda^K{}_j\frac{\partial X^\alpha}{\partial x^\nu}\frac{\partial X^\beta}{\partial x^\mu}\\
&\quad+\omega^i{}_{n\mu}\omega^n{}_{j\nu}-\omega^i{}_{n\mu}\Lambda^n{}_I\,\Omega^I{}_{J\alpha}\Lambda^J{}_j\frac{\partial X^\alpha}{\partial x^\nu}-\frac{1}{2}\omega^n{}_{j\mu}\omega^i{}_{n\nu}+\frac{1}{2}\omega^n{}_{j\mu}\Lambda^i{}_I\,\Omega^I{}_{J\alpha}\Lambda^J{}_n\frac{\partial X^\alpha}{\partial x^\nu}\\
&\quad+\frac{1}{2}\omega^n{}_{j\nu}\omega^i{}_{n\mu}-\frac{1}{2}\omega^n{}_{j\nu}\Lambda^i{}_I\,\Omega^I{}_{J\alpha}\Lambda^J{}_n\frac{\partial X^\alpha}{\partial x^\mu}-\omega^i{}_{n\mu}\,\omega^n{}_{j\nu}+\omega^i{}_{n\mu}\Lambda^n{}_I\,\Omega^I{}_{J\alpha}\Lambda^J{}_j\frac{\partial X^\alpha}{\partial x^\nu}\\
&\quad+\omega^n{}_{j\nu}\Lambda^i{}_I\,\Omega^I{}_{J\alpha}\Lambda^J{}_n\frac{\partial X^\alpha}{\partial x^\mu}-\Lambda^i{}_I\,\Omega^I{}_{J\alpha}\,\Omega^J{}_{K\beta}\Lambda^K{}_j\frac{\partial X^\alpha}{\partial x^\mu}\frac{\partial X^\beta}{\partial x^\nu}\\
&\quad-\frac{1}{2}\Lambda^i{}_I\Omega^I{}_{J\alpha}\Lambda^J{}_n\,\omega^n{}_{j\nu}\frac{\partial X^\alpha}{\partial x^\mu}+\frac{1}{2}\Lambda^i{}_I\Omega^I{}_{J\alpha}\Omega^J{}_{K\beta}\Lambda^K{}_j\frac{\partial X^\alpha}{\partial x^\mu}\frac{\partial X^\beta}{\partial x^\nu}\\
&\quad-\frac{1}{2}\Lambda^i{}_I\Omega^I{}_{J\alpha}\Lambda^J{}_n\,\omega^n{}_{j\mu}\frac{\partial X^\alpha}{\partial x^\nu}+\frac{1}{2}\Lambda^i{}_I\Omega^I{}_{J\alpha}\Omega^J{}_{K\beta}\Lambda^K{}_j\frac{\partial X^\alpha}{\partial x^\nu}\frac{\partial X^\beta}{\partial x^\mu}\Bigg]\\
&=-\frac{1}{2}\tilde{q}_i{}^{j\mu\nu}\Bigg[\frac{\partial\omega^i{}_{j\mu}}{\partial x^\nu}+\frac{\partial\omega^i{}_{j\nu}}{\partial x^\mu}+\omega^i{}_{n\mu}\,\omega^n{}_{j\nu}-\omega^i{}_{n\nu}\,\omega^n{}_{j\mu}\\
&\qquad-\left(\frac{\partial\Omega^I{}_{K\alpha}}{\partial X^\beta}+\frac{\partial\Omega^I{}_{K\beta}}{\partial X^\alpha}+\Omega^I{}_{J\alpha}\Omega^J{}_{K\beta}-\Omega^I{}_{J\beta}\Omega^J{}_{K\alpha}\right)\Lambda^i{}_I\Lambda^K{}_j\frac{\partial X^\alpha}{\partial x^\mu}\frac{\partial X^\beta}{\partial x^\nu}\Bigg]\\
&=-\frac{1}{2}\tilde{q}_i{}^{j\mu\nu}\left(\frac{\partial\omega^i{}_{j\mu}}{\partial x^\nu}+\frac{\partial\omega^i{}_{j\nu}}{\partial x^\mu}+\omega^i{}_{n\mu}\,\omega^n{}_{j\nu}-\omega^i{}_{n\nu}\,\omega^n{}_{j\mu}\right)\\
&\quad+\frac{1}{2}\tilde{Q}_I{}^{J\mu\nu}\left(\frac{\partial\Omega^I{}_{J\mu}}{\partial X^\nu}+\frac{\partial\Omega^i{}_{j\nu}}{\partial X^\mu}+\Omega^I{}_{K\mu}\Omega^K{}_{J\nu}-\Omega^I{}_{K\nu}\Omega^K{}_{J\mu}\right)\left|\frac{\partial X}{\partial x}\right|.
\end{aligned}
$$

4.1.7.5 Final Gauge Hamiltonian, Gauge-Invariant Action

Collecting all contributions to the explicit x^μ-dependence of the generating function $\tilde{\mathcal{F}}_3^\nu$, Eq. (4.38), gives

$$\begin{aligned}
\left.\frac{\partial \tilde{\mathcal{F}}_3^\nu}{\partial x^\nu}\right|_{\text{expl}} = &-\tfrac{1}{2}\tilde{p}^{\mu\nu}\left(\frac{\partial a_\mu}{\partial x^\nu}+\frac{\partial a_\nu}{\partial x^\mu}\right)-\tfrac{\mathrm{i}}{4}\left(\tilde{\bar{\kappa}}^\nu\,\omega_{ij\nu}\,\sigma^{ij}\,\psi-\bar{\psi}\,\omega_{ij\nu}\,\sigma^{ij}\,\tilde{\kappa}^\nu\right)\\
&-\tfrac{1}{2}\tilde{k}_i{}^{\mu\nu}\left(\frac{\partial e_\mu{}^i}{\partial x^\nu}+\frac{\partial e_\nu{}^i}{\partial x^\mu}+\omega^i{}_{j\mu}\,e_\nu{}^j-\omega^i{}_{j\nu}\,e_\mu{}^j\right)\\
&-\tfrac{1}{2}\tilde{q}_i{}^{j\mu\nu}\left(\frac{\partial \omega^i{}_{j\mu}}{\partial x^\nu}+\frac{\partial \omega^i{}_{j\nu}}{\partial x^\mu}+\omega^i{}_{n\mu}\,\omega^n{}_{j\nu}-\omega^i{}_{n\nu}\,\omega^n{}_{j\mu}\right)\\
&+\left|\frac{\partial X}{\partial x}\right|\left[\tfrac{1}{2}\tilde{P}^{\mu\nu}\left(\frac{\partial A_\mu}{\partial X^\nu}+\frac{\partial A_\nu}{\partial X^\mu}\right)\right.\\
&+\tfrac{\mathrm{i}}{4}\left(\tilde{\bar{\mathcal{K}}}^\nu\,\Omega_{IJ\nu}\,\Sigma^{IJ}\,\Psi-\bar{\Psi}\,\Omega_{IJ\nu}\,\Sigma^{IJ}\tilde{\mathcal{K}}^\nu\right)\\
&+\tfrac{1}{2}\tilde{K}_I{}^{\mu\nu}\left(\frac{\partial E^I{}_\mu}{\partial X^\nu}+\frac{\partial E^I{}_\nu}{\partial X^\mu}+\Omega^I{}_{J\mu}\,E^J{}_\nu-\Omega^i{}_{j\nu}\,E^J{}_\mu\right)\\
&\left.+\tfrac{1}{2}\tilde{Q}_I{}^{J\mu\nu}\left(\frac{\partial \Omega^I{}_{J\mu}}{\partial X^\nu}+\frac{\partial \Omega^i{}_{j\nu}}{\partial X^\mu}+\Omega^I{}_{N\mu}\,\Omega^N{}_{J\nu}-\Omega^i{}_{n\nu}\,\Omega^N{}_{J\mu}\right)\right].
\end{aligned} \tag{4.49}$$

The final gauge Hamiltonian $\tilde{\mathcal{H}}_{\text{Gau}_2}$ to be deployed in Eq. (4.35g) follows from Eq. (4.49) as:

$$\begin{aligned}
\tilde{\mathcal{H}}_{\text{Gau}_2} = &\tfrac{\mathrm{i}}{4}\tilde{\bar{\kappa}}^\beta\,\omega^i{}_{j\beta}\,\sigma_i{}^j\,\psi-\tfrac{\mathrm{i}}{4}\bar{\psi}\,\omega^i{}_{j\beta}\,\sigma_i{}^j\,\tilde{\kappa}^\beta\\
&+\tfrac{1}{2}\tilde{k}_i{}^{\alpha\beta}\left(\omega^i{}_{j\alpha}\,e_\beta{}^j-\omega^i{}_{j\beta}\,e_\alpha{}^j\right)+\tfrac{1}{2}\tilde{q}_i{}^{j\alpha\beta}\left(\omega^i{}_{n\alpha}\,\omega^n{}_{j\beta}-\omega^i{}_{n\beta}\,\omega^n{}_{j\alpha}\right).
\end{aligned} \tag{4.50}$$

The partial derivatives associated with $\tilde{p}^{\mu\nu}$, $\tilde{k}_i{}^{\mu\nu}$, and $\tilde{q}_i{}^{j\mu\nu}$ in (4.49) are merged with the corresponding derivatives contained in the initial action functional (4.32) to yield the following modified action functional

$$\begin{aligned}
S = \int_V \Bigg[&\tilde{\pi}^\nu\frac{\partial \varphi}{\partial x^\nu}+\tfrac{1}{2}\tilde{p}^{\mu\nu}\left(\frac{\partial a_\mu}{\partial x^\nu}-\frac{\partial a_\nu}{\partial x^\mu}\right)+\tilde{\bar{\kappa}}^\nu\frac{\partial \psi}{\partial x^\nu}+\frac{\partial \bar{\psi}}{\partial x^\nu}\tilde{\kappa}^\nu\\
&+\tfrac{1}{2}\tilde{k}_i{}^{\mu\nu}\left(\frac{\partial e_\mu{}^i}{\partial x^\nu}-\frac{\partial e_\nu{}^i}{\partial x^\mu}\right)+\tfrac{1}{2}\tilde{q}_i{}^{k\mu\nu}\left(\frac{\partial \omega^i{}_{k\mu}}{\partial x^\nu}-\frac{\partial \omega^i{}_{k\nu}}{\partial x^\mu}\right)-\tilde{\mathcal{H}}_0-\tilde{\mathcal{H}}_{\text{Gau}_2}\Bigg]\mathrm{d}^4x.
\end{aligned} \tag{4.51}$$

Finally, the "free gravity" Hamiltonian $\tilde{\mathcal{H}}_{\mathrm{Gr}}(\tilde{q}, \tilde{k}, e)$, that accounts for the dynamics of the free gravitational field, must be added to $\tilde{\mathcal{H}}_0$. The concrete form of that Hamiltonian will be discussed in Sect. 4.3.4.

The so constructed generic total system Hamiltonian,

$$\tilde{\mathcal{H}}_2 = \tilde{\mathcal{H}}_0 + \tilde{\mathcal{H}}_{\mathrm{Gr}} + \tilde{\mathcal{H}}_{\mathrm{Gau}_2}, \tag{4.52}$$

consists then of the Hamiltonian $\tilde{\mathcal{H}}_0$ of the free scalar, vector, and spinor source fields, the free gravity Hamiltonian $\tilde{\mathcal{H}}_{\mathrm{Gr}}$, and the gauge Hamiltonian $\tilde{\mathcal{H}}_{\mathrm{Gau}_2}$.

The corresponding action functional,

$$\begin{aligned} S = \int_V \mathrm{d}^4x \Bigg[& \tilde{\pi}^\nu \frac{\partial \varphi}{\partial x^\nu} + \tfrac{1}{2}\tilde{p}^{\mu\nu}\left(\frac{\partial a_\mu}{\partial x^\nu} - \frac{\partial a_\nu}{\partial x^\mu}\right) \\ & + \tilde{\bar{\kappa}}^\nu \left(\frac{\partial \psi}{\partial x^\nu} - \tfrac{\mathrm{i}}{4}\omega^i{}_{j\nu}\,\sigma_i{}^j\,\psi\right) + \left(\frac{\partial \bar{\psi}}{\partial x^\nu} + \tfrac{\mathrm{i}}{4}\bar{\psi}\,\omega^i{}_{j\nu}\,\sigma_i{}^j\right)\tilde{\kappa}^\nu \\ & + \tfrac{1}{2}\tilde{k}_i{}^{\mu\nu}\left(\frac{\partial e_\mu{}^i}{\partial x^\nu} - \frac{\partial e_\nu{}^i}{\partial x^\mu} + \omega^i{}_{j\nu}\,e_\mu{}^j - \omega^i{}_{j\mu}\,e_\nu{}^j\right) \\ & + \tfrac{1}{2}\tilde{q}_i{}^{j\mu\nu}\left(\frac{\partial \omega^i{}_{j\mu}}{\partial x^\nu} - \frac{\partial \omega^i{}_{j\nu}}{\partial x^\mu} + \omega^i{}_{n\nu}\,\omega^n{}_{j\mu} - \omega^i{}_{n\mu}\,\omega^n{}_{j\nu}\right) - \tilde{\mathcal{H}}_0 - \tilde{\mathcal{H}}_{\mathrm{Gr}} \Bigg], \end{aligned} \tag{4.53}$$

is now closed as it does not contain any external dependencies. It includes torsion of spacetime and is *not* (yet) restricted to metric compatibility. Obviously, only the skew-symmetric portions of $\tilde{k}_i{}^{\mu\nu}$ and $\tilde{q}_i{}^{j\mu\nu}$ in μ and ν contribute to the action S. We may thus generally assume these tensors to be skew-symmetric in those indices. Also, only the skew-symmetric portion of $\omega_{ij\nu}$ contributes to the action S in the spinor-related terms due to the skew-symmetry of σ^{ij}. Notice, too, that in this formulation no boundary terms are needed for ensuring form-invariance, in contrast to the Lagrangian formulation, e.g. of the Hilbert's ansatz [20].

4.1.8 Restriction to Metric Compatibility

In this section the complex expressions encountered in the final action integral (4.53) shall be linked to physical entities and thus allow to introduce a well defined simplified notations. Metric compatibility then pops up naturally by restricting the symmetry of the spin connection.

To start off we define the quantity $\gamma^\xi{}_{\mu\nu}$ in terms of the vierbein and the spin connection via

$$\gamma^\xi{}_{\mu\nu} := e_i{}^\xi \left(\frac{\partial e_\mu{}^i}{\partial x^\nu} + \omega^i{}_{j\nu}\,e_\mu{}^j\right). \tag{4.54}$$

This relation can easily be reversed to

$$\omega^j{}_{i\nu} = e_i{}^\alpha \left(\gamma^\xi{}_{\alpha\nu} e_\xi{}^j - \frac{\partial e_\alpha{}^j}{\partial x^\nu} \right) \tag{4.55}$$

or re-written as

$$\frac{\partial e_\mu{}^i}{\partial x^\nu} = e_\xi{}^i \gamma^\xi{}_{\mu\nu} - \omega^i{}_{j\nu} e_\mu{}^j, \tag{4.56}$$

and, similarly, for the inverse vierbein:

$$\frac{\partial e_i{}^\mu}{\partial x^\nu} = \omega^j{}_{i\nu} e_j{}^\mu - e_i{}^\xi \gamma^\mu{}_{\xi\nu}. \tag{4.57}$$

Applying now the transformation rule for the spin connections $\omega^i{}_{j\nu}$ from Eq. (4.29),

$$\omega^i{}_{j\nu}(x) = \Lambda^i{}_I \, \Omega^I{}_{J\alpha}(X) \, \Lambda^J{}_j \frac{\partial X^\alpha}{\partial x^\nu} + \Lambda^i{}_I \frac{\partial \Lambda^I{}_J}{\partial x^\nu}, \tag{4.58}$$

translates into the transformation rule for the $\gamma^\xi{}_{\mu\nu}$ as follows:

$$\gamma^\xi{}_{\mu\nu}(x) = \Gamma^\tau{}_{\beta\alpha}(X) \frac{\partial x^\xi}{\partial X^\tau} \frac{\partial X^\beta}{\partial x^\mu} \frac{\partial X^\alpha}{\partial x^\nu} + \frac{\partial^2 X^\alpha}{\partial x^\mu \partial x^\nu} \frac{\partial x^\xi}{\partial X^\alpha}. \tag{4.59}$$

This obviously represents the usual transformation law (2.34) for the *affine connection*, and the quantity $\gamma^\xi{}_{\mu\nu}$ can be *identified* with it. Resembling the Vierbein Equation (2.14) it is common to call Eq. (4.56) *Vierbein Postulate*, and deploy the covariant derivative introduced in (2.18),

$$e^i{}_{\mu;\nu} := \frac{\partial e_\mu{}^i}{\partial x^\nu} - e_\xi{}^i \gamma^\xi{}_{\mu\nu} + \omega^i_{k\nu} e_\mu{}^k = 0, \tag{4.60}$$

that acts on both kinds of indices, on the Lorentz indices using the spin connection, and on the coordinate indices using the affine connection. With the above definition of the affine connection, hence with the validity of the Vierbein Postulate, we conclude as discussed in Sect. 2.2, that for *Metric compatibility* the spin connection must be skew-symmetric in its Lorentz indices. This is consistent with the transformation rule for the spin connection as becomes evident when lowering the index i in Eq. (4.58). Conversely, the affine connection $\gamma^\xi{}_{\mu\nu}$ must have a symmetric portion due to the symmetry of the inhomogeneous term in μ, ν. Yet, it may as well have a skew-symmetric portion (aka torsion) if both, $\gamma^\xi{}_{\mu\nu}$ and $\Gamma^\tau{}_{\beta_\alpha}$ are not symmetric in their lower indices. The transformation laws (4.58) and (4.59) are thus equivalent as both do not require any symmetries of the connections $\omega^i{}_{j\nu}$ and $\gamma^\xi{}_{\mu\nu}$, hence both comprise the full range of $4^3 = 64$ independent coefficients in a four-dimensional spacetime.

Henceforth we request that the spin connection is anti-symmetric and allow the affine connection to be asymmetric, i.e. we stipulate metric compatibility and allow torsion as a structural element of spacetime.

The definitions

$$S^{i}{}_{\mu\nu} := \tfrac{1}{2}\left(\frac{\partial e_{\mu}{}^{i}}{\partial x^{\nu}} - \frac{\partial e_{\nu}{}^{i}}{\partial x^{\mu}} + \omega^{i}{}_{n\nu}\, e_{\mu}{}^{n} - \omega^{i}{}_{n\mu}\, e_{\nu}{}^{n}\right) \tag{4.61a}$$

$$R^{i}{}_{j\nu\mu} := \frac{\partial \omega^{i}{}_{j\mu}}{\partial x^{\nu}} - \frac{\partial \omega^{i}{}_{j\nu}}{\partial x^{\mu}} + \omega^{i}{}_{n\nu}\, \omega^{n}{}_{j\mu} - \omega^{i}{}_{n\mu}\, \omega^{n}{}_{j\nu} \tag{4.61b}$$

acquire a physical meaning once we plug in the definition (4.54) of the affine connection:

$$S^{i}{}_{\mu\nu} \equiv \tfrac{1}{2} e_{\xi}{}^{i}\left(\gamma^{\xi}{}_{\mu\nu} - \gamma^{\xi}{}_{\nu\mu}\right) = e_{\xi}{}^{i}\, S^{\xi}{}_{\mu\nu} \tag{4.62a}$$

$$R^{i}{}_{j\nu\mu} \equiv e_{\xi}{}^{i}\, e_{j}{}^{\eta} R^{\xi}{}_{\eta\nu\mu}. \tag{4.62b}$$

Obviously $S^{\xi}{}_{\mu\nu}$ thus denotes the Cartan torsion tensor, and $R^{\xi}{}_{\eta\nu\mu}$ is the Riemann-Cartan curvature tensor. The proof of the correlation of $R^{\xi}{}_{\eta\nu\mu}$ and $R^{i}{}_{j\nu\mu}$ is given in the following explicit calculation.

✓ Explict Calculation

By virtue of

$$\frac{\partial e_{\mu}{}^{i}}{\partial x^{\alpha}} = e_{\xi}{}^{i}\gamma^{\xi}{}_{\mu\alpha} - \omega^{i}{}_{j\alpha}\, e_{\mu}{}^{j},$$

the Riemann-Cartan tensor can be set up on the basis of the affine connection $\gamma^{\xi}{}_{\mu_\nu}$ as follows. The second derivatives of the vierbeins have the two equivalent representations

$$\frac{\partial^2 e_{\mu}{}^{i}}{\partial x^{\alpha}\partial x^{\beta}} = \frac{\partial e_{\xi}{}^{i}}{\partial x^{\beta}}\gamma^{\xi}{}_{\mu\alpha} + e_{\xi}{}^{i}\frac{\partial \gamma^{\xi}{}_{\mu\alpha}}{\partial x^{\beta}} - \frac{\partial \omega^{i}{}_{j\alpha}}{\partial x^{\beta}}\, e_{\mu}{}^{j} - \omega^{i}{}_{j\alpha}\frac{\partial e_{\mu}{}^{j}}{\partial x^{\beta}}$$

$$\frac{\partial^2 e_{\mu}{}^{i}}{\partial x^{\beta}\partial x^{\alpha}} = \frac{\partial e_{\xi}{}^{i}}{\partial x^{\alpha}}\gamma^{\xi}{}_{\mu\beta} + e_{\xi}{}^{i}\frac{\partial \gamma^{\xi}{}_{\mu\beta}}{\partial x^{\alpha}} - \frac{\partial \omega^{i}{}_{j\beta}}{\partial x^{\alpha}}\, e_{\mu}{}^{j} - \omega^{i}{}_{j\beta}\frac{\partial e_{\mu}{}^{j}}{\partial x^{\alpha}}.$$

The difference of both representations follows as:

$$
\begin{aligned}
0 = {} & \frac{\partial e_{\xi}{}^{i}}{\partial x^{\beta}} \gamma^{\xi}{}_{\mu\alpha} - \frac{\partial e_{\xi}{}^{i}}{\partial x^{\alpha}} \gamma^{\xi}{}_{\mu\beta} + \omega^{i}{}_{j\beta} \frac{\partial e_{\mu}{}^{j}}{\partial x^{\alpha}} - \omega^{i}{}_{j\alpha} \frac{\partial e_{\mu}{}^{j}}{\partial x^{\beta}} \\
& + e_{\xi}{}^{i} \left(\frac{\partial \gamma^{\xi}{}_{\mu\alpha}}{\partial x^{\beta}} - \frac{\partial \gamma^{\xi}{}_{\mu\beta}}{\partial x^{\alpha}} \right) - \left(\frac{\partial \omega^{i}{}_{j\alpha}}{\partial x^{\beta}} - \frac{\partial \omega^{i}{}_{j\beta}}{\partial x^{\alpha}} \right) e_{\mu}{}^{j}
\end{aligned}
$$

The above expression for the derivatives of the vierbeins can now be inserted:

$$
\begin{aligned}
0 = {} & \left(e_{\xi}{}^{i} \gamma^{\xi}{}_{\eta\beta} - \cancel{\omega^{i}{}_{j\beta} e_{\eta}{}^{j}} \right) \gamma^{\eta}{}_{\mu\alpha} - \left(e_{\xi}{}^{i} \gamma^{\xi}{}_{\eta\alpha} - \cancel{\omega^{i}{}_{j\alpha} e_{\eta}{}^{j}} \right) \gamma^{\eta}{}_{\mu\beta} \\
& + \omega^{i}{}_{j\beta} \left(\cancel{e_{\eta}{}^{j} \gamma^{\eta}{}_{\mu\alpha}} - \omega^{j}{}_{n\alpha} e_{\mu}{}^{n} \right) - \omega^{i}{}_{j\alpha} \left(\cancel{e_{\eta}{}^{j} \gamma^{\eta}{}_{\mu\beta}} - \omega^{j}{}_{n\beta} e_{\mu}{}^{n} \right) \\
& + e_{\xi}{}^{i} \left(\frac{\partial \gamma^{\xi}{}_{\mu\alpha}}{\partial x^{\beta}} - \frac{\partial \gamma^{\xi}{}_{\mu\beta}}{\partial x^{\alpha}} \right) - \left(\frac{\partial \omega^{i}{}_{j\alpha}}{\partial x^{\beta}} - \frac{\partial \omega^{i}{}_{j\beta}}{\partial x^{\alpha}} \right) e_{\mu}{}^{j},
\end{aligned}
$$

which yields, after reordering:

$$
\begin{aligned}
0 = {} & e_{\xi}{}^{i} \left(\frac{\partial \gamma^{\xi}{}_{\mu\alpha}}{\partial x^{\beta}} - \frac{\partial \gamma^{\xi}{}_{\mu\beta}}{\partial x^{\alpha}} + \gamma^{\xi}{}_{\eta\beta} \gamma^{\eta}{}_{\mu\alpha} - \gamma^{\xi}{}_{\eta\alpha} \gamma^{\eta}{}_{\mu\beta} \right) \\
& - \left(\frac{\partial \omega^{i}{}_{j\alpha}}{\partial x^{\beta}} - \frac{\partial \omega^{i}{}_{j\beta}}{\partial x^{\alpha}} + \omega^{i}{}_{n\beta} \omega^{n}{}_{j\alpha} - \omega^{i}{}_{n\alpha} \omega^{n}{}_{j\beta} \right) e_{\mu}{}^{j}.
\end{aligned}
$$

The sums in parentheses are the affine and the spin connection representations of the Riemann-Cartan tensor:

$$
\begin{aligned}
R^{\xi}{}_{\mu\beta\alpha} &= \frac{\partial \gamma^{\xi}{}_{\mu\alpha}}{\partial x^{\beta}} - \frac{\partial \gamma^{\xi}{}_{\mu\beta}}{\partial x^{\alpha}} + \gamma^{\xi}{}_{\eta\beta} \gamma^{\eta}{}_{\mu\alpha} - \gamma^{\xi}{}_{\eta\alpha} \gamma^{\eta}{}_{\mu\beta} \\
R^{i}{}_{j\beta\alpha} &= \frac{\partial \omega^{i}{}_{j\alpha}}{\partial x^{\beta}} - \frac{\partial \omega^{i}{}_{j\beta}}{\partial x^{\alpha}} + \omega^{i}{}_{n\beta} \omega^{n}{}_{j\alpha} - \omega^{i}{}_{n\alpha} \omega^{n}{}_{j\beta},
\end{aligned}
$$

hence

$$
R^{i}{}_{j\beta\alpha} = R^{\xi}{}_{\mu\beta\alpha} e_{\xi}{}^{i} e_{j}{}^{\mu} \quad \Leftrightarrow \quad R^{\xi}{}_{\mu\beta\alpha} = R^{i}{}_{j\beta\alpha} e_{i}{}^{\xi} e_{\mu}{}^{j}. \tag{4.62c}
$$

For further simplifying the notation in Eq. (4.53), we define the

$$
f_{\nu\mu} := \frac{\partial a_{\mu}}{\partial x^{\nu}} - \frac{\partial a_{\nu}}{\partial x^{\mu}}. \tag{4.62d}
$$

Notice, in addition, that a new *spinor covariant derivatives* have emerged for the spinor fields that we abbreviate as

$$\overrightarrow{\mathcal{D}}_{\mu} := \frac{\overrightarrow{\partial}}{\partial x^{\mu}} - \tfrac{\mathrm{i}}{4}\sigma^{lj}\,\omega_{lj\mu} \tag{4.63a}$$

$$\overleftarrow{\mathcal{D}}_{\mu} := \frac{\overleftarrow{\partial}}{\partial x^{\mu}} + \tfrac{\mathrm{i}}{4}\sigma^{lj}\,\omega_{lj\mu} \equiv \gamma^{0}\,\overrightarrow{\mathcal{D}}^{\dagger}_{\mu}\,\gamma^{0}. \tag{4.63b}$$

The Lorentz connection corrects the partial derivative as it is appropriate for gauge fields. Notice that while the spinor and coordinate indices in these tensor operators are open, the Lorentz indices are contracted out. With the spinor indices a, b made explicit this reads

$$\left(\overrightarrow{\mathcal{D}}_{\eta}\right)^{a}_{\ b} := \delta^{a}_{b}\,\frac{\overrightarrow{\partial}}{\partial x^{\eta}} - \tfrac{\mathrm{i}}{4}\left(\sigma^{lj}\right)^{a}_{\ b}\,\omega_{lj\eta}.$$

Then the action integral (4.53), invariant w.r.t. the transformation group SO(1, 3) $\times$Diff(M), can be re-written in the compact form

$$S = \int_{V} \mathrm{d}^{4}x \left\{ \tilde{\pi}^{\nu}\frac{\partial\varphi}{\partial x^{\nu}} + \tfrac{1}{2}\tilde{p}^{\mu\nu}\,f_{\nu\mu} + \tilde{\bar{\kappa}}^{\nu}\,\overrightarrow{\mathcal{D}}_{\nu}\,\psi + \bar{\psi}\,\overleftarrow{\mathcal{D}}_{\nu}\,\tilde{\kappa}^{\nu} \right.$$
$$\left. + \tilde{k}_{i}^{\ \mu\nu}\,S^{i}_{\ \mu\nu} + \tfrac{1}{2}\tilde{q}_{i}^{\ j\mu\nu}\,R^{i}_{\ j\nu\mu} - \tilde{\mathcal{H}}_{0} - \tilde{\mathcal{H}}_{\mathrm{Gr}} \right\}. \tag{4.64}$$

4.2 Canonical Field Equations

The variation of the action integral with respect to the conjugate pairs of fields is now based on the amended version Eq. (4.64), and is carried out in the next subsections.

4.2.1 Canonical Equations for the Scalar Field

As the scalar field φ does not couple to the gauge field $\omega^{i}_{\ j\nu}$ and thus does not contribute to $\mathcal{H}_{\mathrm{Gau}_2}$, the field equation take their usual form:

$$\frac{\partial\varphi}{\partial x^{\nu}} = \frac{\partial\tilde{\mathcal{H}}_{0}}{\partial\tilde{\pi}^{\nu}}, \qquad \frac{\partial\tilde{\pi}^{\nu}}{\partial x^{\nu}} = -\frac{\partial\tilde{\mathcal{H}}_{0}}{\partial\varphi}. \tag{4.65}$$

The coupling of scalar fields to a dynamic spacetime occurs merely via the common dependence of $\tilde{\mathcal{H}}_{0}$ and $\tilde{\mathcal{H}}_{\mathrm{Gr}}$ on the vierbein field $e_{\nu}^{\ i}$. This applies also to the Maxwell version of the Proca field, see next section. In contrast, the spinor field couples directly to the gauge field $\omega^{i}{}_{j\nu}$, as will be shown below.

4.2.2 Canonical Equations for the Vector Field

The equation for the derivative of the vector field a_μ emerges as:

$$\frac{\partial a_\mu}{\partial x^\nu} - \frac{\partial a_\nu}{\partial x^\mu} \equiv f_{\nu\mu} = 2\frac{\partial \tilde{\mathcal{H}}_2}{\partial \tilde{p}^{\mu\nu}} = 2\frac{\partial \tilde{\mathcal{H}}_0}{\partial \tilde{p}^{\mu\nu}}. \tag{4.66}$$

This shows that $\tilde{p}^{\mu\nu}$ must be skew-symmetric. The conjugate canonical equation for the divergence of the momentum field $\tilde{p}^{\mu\nu}$ is then:

$$\frac{\partial \tilde{p}^{[\mu\nu]}}{\partial x^\nu} = -\frac{\partial \tilde{\mathcal{H}}_2}{\partial a_\mu} = -\frac{\partial \tilde{\mathcal{H}}_0}{\partial a_\mu}, \tag{4.67}$$

which is actually a tensor equation as will be demonstrated explicitly below.

4.2.3 Canonical Equations for the Spinor Fields

For the spinors ψ and $\bar{\psi}$ the set of canonical equations is:

$$\frac{\partial \psi}{\partial x^\nu} = \frac{\partial \tilde{\mathcal{H}}_2}{\partial \tilde{\bar{\kappa}}^\nu} = \frac{\partial \tilde{\mathcal{H}}_0}{\partial \tilde{\bar{\kappa}}^\nu} + \tfrac{\mathrm{i}}{4}\,\omega^{i}{}_{j\nu}\,\sigma_{i}{}^{j}\,\psi \tag{4.68a}$$

$$\frac{\partial \tilde{\bar{\kappa}}^\nu}{\partial x^\nu} = -\frac{\partial \tilde{\mathcal{H}}_2}{\partial \psi} = -\frac{\partial \tilde{\mathcal{H}}_0}{\partial \psi} - \tfrac{\mathrm{i}}{4}\,\tilde{\bar{\kappa}}^\nu\,\omega^{i}{}_{j\nu}\,\sigma_{i}{}^{j} \tag{4.68b}$$

$$\frac{\partial \bar{\psi}}{\partial x^\nu} = \frac{\partial \tilde{\mathcal{H}}_2}{\partial \tilde{\kappa}^\nu} = \frac{\partial \tilde{\mathcal{H}}_0}{\partial \tilde{\kappa}^\nu} - \tfrac{\mathrm{i}}{4}\,\bar{\psi}\,\omega^{i}{}_{j\nu}\,\sigma_{i}{}^{j} \tag{4.68c}$$

$$\frac{\partial \tilde{\kappa}^\nu}{\partial x^\nu} = -\frac{\partial \tilde{\mathcal{H}}_2}{\partial \bar{\psi}} = -\frac{\partial \tilde{\mathcal{H}}_0}{\partial \bar{\psi}} + \tfrac{\mathrm{i}}{4}\,\omega^{i}{}_{j\nu}\,\sigma_{i}{}^{j}\,\tilde{\kappa}^\nu. \tag{4.68d}$$

hence in elaborate form with the spinor indices a, b displayed explicitly:

$$\frac{\partial \psi^a}{\partial x^\nu} = \frac{\partial \tilde{\mathcal{H}}_0}{\partial \tilde{\bar{\kappa}}^\nu_a} + \tfrac{\mathrm{i}}{4}\,\omega^{i}{}_{j\nu}\,\sigma^{a}{}_{bi}{}^{j}\,\psi^b$$

$$\frac{\partial \bar{\tilde{\kappa}}_a{}^{\nu}}{\partial x^{\nu}} = -\frac{\partial \tilde{\mathcal{H}}_0}{\partial \psi^a} - \tfrac{\mathrm{i}}{4}\,\omega^i{}_{j\nu}\,\bar{\tilde{\kappa}}_b{}^{\nu}\,\sigma^b{}_{ai}{}^j$$
$$\frac{\partial \bar{\psi}_a}{\partial x^{\nu}} = \frac{\partial \tilde{\mathcal{H}}_0}{\partial \tilde{\kappa}^{a\nu}} - \tfrac{\mathrm{i}}{4}\,\omega^i{}_{j\nu}\,\bar{\psi}_b\,\sigma^b{}_{ai}{}^j$$
$$\frac{\partial \tilde{\kappa}^{a\nu}}{\partial x^{\nu}} = -\frac{\partial \tilde{\mathcal{H}}_0}{\partial \bar{\psi}_a} + \tfrac{\mathrm{i}}{4}\,\omega^i{}_{j\nu}\,\sigma^a{}_{bi}{}^j\,\tilde{\kappa}^{b\nu}.$$

4.2.4 Canonical Equations for the Vierbein Field

For the canonical equation for the vierbein field $e_\mu{}^i$ we get

$$\frac{\partial e_\mu{}^i}{\partial x^\nu} - \frac{\partial e_\nu{}^i}{\partial x^\mu} = 2\frac{\partial \tilde{\mathcal{H}}_2}{\partial \tilde{k}_i{}^{\mu\nu}} = 2\frac{\partial \tilde{\mathcal{H}}_{\mathrm{Gr}}}{\partial \tilde{k}_i{}^{\mu\nu}} + \omega^i{}_{j\mu}\,e_\nu{}^j - \omega^i{}_{j\nu}\,e_\mu{}^j.$$

Regrouping gives then the simple relation

$$\frac{\partial \tilde{\mathcal{H}}_{\mathrm{Gr}}}{\partial \tilde{k}_i^{\mu\nu}} = \tfrac{1}{2}\left(\frac{\partial e_\mu{}^i}{\partial x^\nu} - \frac{\partial e_\nu{}^i}{\partial x^\mu} + \omega^i{}_{j\nu}\,e_\mu{}^j - \omega^i{}_{j\mu}\,e_\nu{}^j\right) = S^i{}_{\mu\nu}. \tag{4.69}$$

When deriving the action functional (4.32), the momentum field $\tilde{k}_i{}^{\mu\nu}$ turned out to be skew-symmetric in its last index pair. That is consistent with the action integral Eq. (4.64). Then the canonical equation for its divergence becomes

$$\begin{aligned}
\frac{\partial \tilde{k}_i{}^{[\mu\alpha]}}{\partial x^\alpha} &= -\frac{\partial \tilde{\mathcal{H}}_2}{\partial e_\mu{}^i} = -\frac{\partial \tilde{\mathcal{H}}_0}{\partial e_\mu{}^i} - \frac{\partial \tilde{\mathcal{H}}_{\mathrm{Gr}}}{\partial e_\mu{}^i} - \frac{\partial \tilde{\mathcal{H}}_{\mathrm{Gau}_2}}{\partial e_\mu{}^i}\\
&= -\frac{\partial \tilde{\mathcal{H}}_0}{\partial e_\mu{}^i} - \frac{\partial \tilde{\mathcal{H}}_{\mathrm{Gr}}}{\partial e_\mu{}^i} + \tfrac{1}{2}\left(\tilde{k}_j{}^{\mu\alpha} - \tilde{k}_j{}^{\alpha\mu}\right)\omega^j{}_{i\alpha}\\
&= -\frac{\partial \tilde{\mathcal{H}}_0}{\partial e_\mu{}^i} - \frac{\partial \tilde{\mathcal{H}}_{\mathrm{Gr}}}{\partial e_\mu{}^i} + \tilde{k}_j{}^{[\mu\alpha]}\,\omega^j{}_{i\alpha}.
\end{aligned} \tag{4.70}$$

Regrouping the terms of Eq. (4.70) yields:

$$\left(\frac{\partial \tilde{k}_i{}^{[\mu\alpha]}}{\partial x^\alpha} - \tilde{k}_j{}^{[\mu\alpha]}\,\omega^j{}_{i\alpha}\right)e_\nu{}^i = -\frac{\partial \tilde{\mathcal{H}}_{\mathrm{Gr}}}{\partial e_\mu{}^i}e_\nu{}^i - \frac{\partial \tilde{\mathcal{H}}_0}{\partial e_\mu{}^i}e_\nu{}^i. \tag{4.71}$$

The terms on the right-hand side of Eq. (4.71) are minus the Hamiltonian representation of the *metric* energy-momentum tensors of free gravity, $\tilde{\mathcal{H}}_{\mathrm{Gr}}$, and of free matter, $\tilde{\mathcal{H}}_0$. In terms of these, the field equations reduce to

$$\left(\frac{\partial \tilde{k}_i{}^{[\mu\alpha]}}{\partial x^\alpha} - \tilde{k}_j{}^{[\mu\alpha]}\,\omega^j{}_{i\alpha}\right) e_\nu{}^i = -\tilde{T}_{\mathrm{Gr}\,i}{}^\mu - \tilde{T}_{0\,i}{}^\mu. \tag{4.72}$$

With

$$\frac{1}{\varepsilon}\frac{\partial \tilde{k}_i{}^{[\mu\alpha]}}{\partial x^\alpha} = k_i{}^{[\mu\alpha]}{}_{;\alpha} + \omega^j{}_{i\alpha} k_j{}^{[\mu\alpha]} - \gamma^\mu{}_{\beta\alpha} k_i{}^{[\beta\alpha]} - \gamma^\alpha{}_{\beta\alpha} k_i{}^{[\mu\beta]} + \gamma^\alpha{}_{\alpha\beta} k_i{}^{[\mu\beta]} \tag{4.73}$$

this gives a tensor equation of motion for the momentum field $k_i{}^{[\mu\alpha]}$ the source of which is the total energy momentum of matter and gravity:

$$k_i{}^{[\mu\alpha]}{}_{;\alpha} - S^\mu{}_{\beta\alpha} k_i{}^{[\beta\alpha]} - 2S^\alpha{}_{\beta\alpha} k_i{}^{[\mu\beta]} = -T_{\mathrm{Gr}\,i}{}^\mu - T_{0\,i}{}^\mu. \tag{4.74}$$

4.2.5 Canonical Equations for the Connection Field

The canonical equation for the derivative of the gauge field $\omega^i{}_{j\mu}$ follows as

$$\frac{\partial \omega^i{}_{j\mu}}{\partial x^\nu} - \frac{\partial \omega^i{}_{j\nu}}{\partial x^\mu} = 2\frac{\partial \tilde{\mathcal{H}}_2}{\partial \tilde{q}_i{}^{j\mu\nu}} = 2\frac{\partial \tilde{\mathcal{H}}_{\mathrm{Gr}}}{\partial \tilde{q}_i{}^{j\mu\nu}} + \omega^i{}_{n\mu}\,\omega^n{}_{j\nu} - \omega^i{}_{n\nu}\,\omega^n{}_{j\mu}. \tag{4.75}$$

which shows that $\tilde{q}_i{}^{j\mu\nu}$ is skew-symmetric in its last index pair. With the definition (4.62b) this gives:

$$\frac{\partial \tilde{\mathcal{H}}_{\mathrm{Gr}}}{\partial \tilde{q}_i{}^{j\mu\nu}} = \tfrac{1}{2}\left(\frac{\partial \omega^i{}_{j\mu}}{\partial x^\nu} - \frac{\partial \omega^i{}_{j\nu}}{\partial x^\mu} + \omega^i{}_{n\nu}\,\omega^n{}_{j\mu} - \omega^i{}_{n\mu}\,\omega^n{}_{j\nu}\right) = \tfrac{1}{2} R^i{}_{j\nu\mu}. \tag{4.76}$$

The canonical equation for the divergence of $\tilde{q}_i{}^{j\mu\nu}$ follows as

$$\begin{aligned}\frac{\partial \tilde{q}_i{}^{j\mu\nu}}{\partial x^\nu} &= -\frac{\partial \tilde{\mathcal{H}}_2}{\partial \omega^i{}_{j\mu}} = -\frac{\partial \tilde{\mathcal{H}}_{\mathrm{Gau}_2}}{\partial \omega^i{}_{j\mu}} \\ &= -\tfrac{\mathrm{i}}{4}\bar{\tilde{\kappa}}^\mu\,\sigma_i{}^j\,\psi + \tfrac{\mathrm{i}}{4}\bar{\psi}\,\sigma_i{}^j\,\tilde{\kappa}^\mu + \tilde{k}_i{}^{\beta\mu} e_\beta{}^j + \tilde{q}_i{}^{n\beta\mu}\omega^j{}_{n\beta} - \tilde{q}_n{}^{j\beta\mu}\omega^n{}_{i\beta}.\end{aligned} \tag{4.77}$$

Again, realizing that

$$\begin{aligned}\frac{1}{\varepsilon}\frac{\partial \tilde{q}_i{}^{j\mu\nu}}{\partial x^\nu} = {}& q_i{}^{j\mu\nu}{}_{;\nu} - \omega^n{}_{i\nu} q_n{}^{j\nu\mu} + \omega^j{}_{n\nu} q_i{}^{n\nu\mu} \\ &- \gamma^\mu{}_{\lambda\nu} q_i{}^{n\lambda\nu} - \gamma^\nu{}_{\lambda\nu} q_i{}^{n\mu\lambda} - \gamma^\nu{}_{\nu\lambda} q_i{}^{n\mu\lambda},\end{aligned}$$

we can re-write Eq. (4.77) as

$$q_i^{\ j\mu\nu}{}_{;\nu} - S^{\mu}{}_{\lambda\nu}\, q_i^{\ j\lambda\nu} - 2S^{\nu}{}_{\lambda\nu}\, q_i^{\ j\mu\lambda} = k_i^{\ \beta\mu}\, e_{\beta}^{\ j} + \Sigma_i^{\ j\mu}. \tag{4.78}$$

4.2.6 Summary of the Coupled Set of Field Equations

The dynamics of matter fields in dynamic spacetime geometry is described by the action functional (4.53) with the Hamiltonian $\tilde{\mathcal{H}}_0$, that has been deducted following the rigorous formalism of the canonical transformation theory. By variation of that action the complete closed set of twelve coupled field equations for the system of scalar, vector, and spinor fields in curved spacetime, expressed by the vierbein $e_{\mu}^{\ j}$ and connection fields $\omega^i{}_{j\nu}$, is derived. The matter fields obey the following eight equation:

$$\frac{\partial\varphi}{\partial x^{\nu}} = \frac{\partial\tilde{\mathcal{H}}_0}{\partial\tilde{\pi}^{\nu}} \tag{4.79a}$$

$$\frac{\partial\tilde{\pi}^{\nu}}{\partial x^{\nu}} = -\frac{\partial\tilde{\mathcal{H}}_0}{\partial\varphi} \tag{4.79b}$$

$$f_{\nu\mu} = 2\,\frac{\partial\tilde{\mathcal{H}}_0}{\partial\tilde{p}^{\,\mu\nu}} \tag{4.79c}$$

$$\frac{\partial\tilde{p}^{[\mu\nu]}}{\partial x^{\nu}} = -\frac{\partial\tilde{\mathcal{H}}_0}{\partial a_{\mu}} \tag{4.79d}$$

$$\overrightarrow{\mathcal{D}}_{\nu}\,\psi = \frac{\partial\tilde{\mathcal{H}}_0}{\partial\bar{\tilde{\kappa}}^{\nu}} \tag{4.79e}$$

$$\bar{\tilde{\kappa}}^{\nu}\,\overleftarrow{\mathcal{D}}_{\nu} = -\frac{\partial\tilde{\mathcal{H}}_0}{\partial\psi} \tag{4.79f}$$

$$\bar{\psi}\,\overleftarrow{\mathcal{D}}_{\nu} = \frac{\partial\tilde{\mathcal{H}}_0}{\partial\tilde{\kappa}^{\nu}} \tag{4.79g}$$

$$\overrightarrow{\mathcal{D}}_{\nu}\,\tilde{\kappa}^{\nu} = -\frac{\partial\tilde{\mathcal{H}}_0}{\partial\bar{\psi}}. \tag{4.79h}$$

Four further equation govern the dynamics of the vierbeins, the connection fields, and their respective conjugates:

$$S^{i}{}_{\mu\nu} = \frac{\partial\tilde{\mathcal{H}}_{\mathrm{Gr}}}{\partial\tilde{k}_i^{\ \mu\nu}} \tag{4.79i}$$

$$R^{i}{}_{j\nu\mu} = 2\frac{\partial\tilde{\mathcal{H}}_{\mathrm{Gr}}}{\partial\tilde{q}_i^{\ j\mu\nu}} \tag{4.79j}$$

$$k_i{}^{[\mu\nu]}{}_{;\nu} - S^{\mu}{}_{\beta\nu}\, k_i{}^{[\beta\nu]} - 2S^{\nu}{}_{\beta\nu}\, k_i{}^{[\mu\beta]} = -T_{\mathrm{Gr}\,i}{}^{\mu} - T_{0\,i}{}^{\mu} \tag{4.79k}$$

$$q_i{}^{j\mu\nu}{}_{;\nu} - S^{\mu}{}_{\beta\nu}\, q_i{}^{j\beta\nu} - 2S^{\nu}{}_{\beta\nu}\, q_i{}^{j\mu\beta} = k_i{}^{\beta\mu} e_{\beta}{}^{j} - \Sigma_i{}^{j\mu}. \tag{4.79l}$$

4.2.7 *Consistency Equation*

The fields in the canonical equations display symmetries that can be utilized to derive the so called *Consistency equation* which relates matter and spacetime expressions.

We start by taking a derivative of equation (4.77) with respect to x^{β}. The left-hand side vanishes due to the skew-symmetry of $\tilde{q}_i{}^{j[\beta\alpha]}$ in its last index pair:

$$\begin{aligned} 0 = {} & \frac{\partial}{\partial x^{\beta}} \left(\tfrac{\mathrm{i}}{4} \bar{\psi}\, \sigma_i{}^{j}\, \tilde{\kappa}^{\beta} - \tfrac{\mathrm{i}}{4} \bar{\tilde{\kappa}}^{\beta}\, \sigma_i{}^{j}\, \psi + \tilde{k}_i{}^{[\alpha\beta]} e_{\alpha}{}^{j} \right) \\ & + \tilde{q}_i{}^{n[\alpha\beta]} \frac{\partial \omega^{j}{}_{n\alpha}}{\partial x^{\beta}} - \frac{\partial \omega^{n}{}_{i\alpha}}{\partial x^{\beta}} \tilde{q}_n{}^{j[\alpha\beta]} + \frac{\partial \tilde{q}_i{}^{n\beta\alpha}}{\partial x^{\alpha}} \omega^{j}{}_{n\beta} - \omega^{n}{}_{i\beta} \frac{\partial \tilde{q}_n{}^{j\beta\alpha}}{\partial x^{\alpha}}. \end{aligned} \tag{4.80}$$

Inserting the above cited field Eq. (4.77) for the first derivatives of $\tilde{q}$ gives:

$$\begin{aligned} 0 = {} & \frac{\partial}{\partial x^{\beta}} \left(\tfrac{\mathrm{i}}{4} \bar{\psi} \sigma_i{}^{j} \tilde{\kappa}^{\beta} - \tfrac{\mathrm{i}}{4} \bar{\tilde{\kappa}}^{\beta} \sigma_i{}^{j} \psi + \tilde{k}_i{}^{[\alpha\beta]} e_{\alpha}{}^{j} \right) + \tilde{q}_i{}^{n[\alpha\beta]} \frac{\partial \omega^{j}{}_{n\alpha}}{\partial x^{\beta}} - \frac{\partial \omega^{n}{}_{i\alpha}}{\partial x^{\beta}} \tilde{q}_n{}^{j[\alpha\beta]} \\ & + \left(\tfrac{\mathrm{i}}{4} \bar{\psi} \sigma_i{}^{n} \tilde{\kappa}^{\beta} - \tfrac{\mathrm{i}}{4} \bar{\tilde{\kappa}}^{\beta} \sigma_i{}^{n} \psi + \tilde{k}_i{}^{[\alpha\beta]} e_{\alpha}{}^{n} + \tilde{q}_i{}^{m[\alpha\beta]} \omega^{n}{}_{m\alpha} - \cancel{\tilde{q}_m{}^{n[\alpha\beta]} \omega^{m}{}_{i\alpha}} \right) \omega^{j}{}_{n\beta} \\ & - \omega^{n}{}_{i\beta} \left(\tfrac{\mathrm{i}}{4} \bar{\psi} \sigma_n{}^{j} \tilde{\kappa}^{\beta} - \tfrac{\mathrm{i}}{4} \bar{\tilde{\kappa}}^{\beta} \sigma_n{}^{j} \psi + \tilde{k}_n{}^{[\alpha\beta]} e_{\alpha}{}^{j} + \cancel{\tilde{q}_n{}^{m[\alpha\beta]} \omega^{j}{}_{m\alpha}} - \tilde{q}_m{}^{j[\alpha\beta]} \omega^{m}{}_{n\alpha} \right). \end{aligned} \tag{4.81}$$

Combining this with the canonical equations for the spacetime fields, we obtain after some algebra detailed in the explicit calculation below, a generic Einstein-type field equation:

$$\begin{aligned} 0 = {} & \frac{\mathrm{i}}{4} \left(\frac{\partial \tilde{\mathcal{H}}_0}{\partial \psi} \sigma_i{}^{j} \psi - \bar{\tilde{\kappa}}^{\alpha} \sigma_i{}^{j} \frac{\partial \tilde{\mathcal{H}}_0}{\partial \bar{\tilde{\kappa}}^{\alpha}} + \frac{\partial \tilde{\mathcal{H}}_0}{\partial \tilde{\kappa}^{\alpha}} \sigma_i{}^{j} \tilde{\kappa}^{\alpha} - \bar{\psi}\, \sigma_i{}^{j} \frac{\partial \tilde{\mathcal{H}}_0}{\partial \bar{\psi}} \right) - \frac{\partial \tilde{\mathcal{H}}_0}{\partial e_{\alpha}{}^{i}} e_{\alpha}{}^{j} \\ & - \frac{\partial \tilde{\mathcal{H}}_{\mathrm{Gr}}}{\partial e_{\alpha}{}^{i}} e_{\alpha}{}^{j} + \tilde{k}_i{}^{[\alpha\beta]} \frac{\partial \tilde{\mathcal{H}}_{\mathrm{Gr}}}{\partial \tilde{k}_j{}^{\alpha\beta}} + \tilde{q}_i{}^{m[\alpha\beta]} \frac{\partial \tilde{\mathcal{H}}_{\mathrm{Gr}}}{\partial \tilde{q}_j{}^{m\alpha\beta}} - \tilde{q}_m{}^{j[\alpha\beta]} \frac{\partial \tilde{\mathcal{H}}_{\mathrm{Gr}}}{\partial \tilde{q}_m{}^{i\alpha\beta}}. \end{aligned} \tag{4.82}$$

It relates the derivatives of the Hamiltonian $\tilde{\mathcal{H}}_0$ of matter fields with the derivatives emerging from the model Hamiltonian $\tilde{\mathcal{H}}_{\mathrm{Gr}}$ for gravity. It holds for any given Hamiltonian densities $\tilde{\mathcal{H}}_0$ and $\tilde{\mathcal{H}}_{\mathrm{Gr}}$. The particular equation for the scalar, vector, spinor, and gravitational fields can be set up only after specifying those Hamiltonians which will be done in the following sections.

✓ Explict Calculation

The eight terms in Eq. (4.81) involving the spinor fields add up to:

$$\begin{aligned}
&\tfrac{i}{4}\Bigg(\frac{\partial\bar{\tilde{\kappa}}^{\alpha}}{\partial x^{\alpha}}\sigma^{ij}\psi+\bar{\tilde{\kappa}}^{\alpha}\sigma^{ij}\frac{\partial\psi}{\partial x^{\alpha}}-\frac{\partial\bar{\psi}}{\partial x^{\alpha}}\sigma^{ij}\tilde{\kappa}^{\alpha}-\bar{\psi}\sigma^{ij}\frac{\partial\tilde{\kappa}^{\alpha}}{\partial x^{\alpha}}\\
&\qquad-\bar{\tilde{\kappa}}^{\alpha}\sigma^{jn}\psi\,\omega^{i}{}_{n\alpha}+\bar{\psi}\sigma^{jn}\tilde{\kappa}^{\alpha}\omega^{i}{}_{n\alpha}+\bar{\tilde{\kappa}}^{\alpha}\sigma^{in}\psi\,\omega^{j}{}_{n\alpha}-\bar{\psi}\sigma^{in}\tilde{\kappa}^{\alpha}\omega^{j}{}_{n\alpha}\Bigg)\\
&=\tfrac{i}{4}\left(-\frac{\partial\tilde{\mathcal{H}}_0}{\partial\psi}-\tfrac{i}{4}\bar{\tilde{\kappa}}^{\alpha}\,\omega_{nm\alpha}\,\sigma^{nm}\right)\sigma^{ij}\psi+\tfrac{i}{4}\bar{\tilde{\kappa}}^{\alpha}\sigma^{ij}\left(\frac{\partial\tilde{\mathcal{H}}_0}{\partial\bar{\tilde{\kappa}}^{\alpha}}+\tfrac{i}{4}\,\omega_{nm\alpha}\,\sigma^{nm}\,\psi\right)\\
&\quad+\tfrac{i}{4}\left(-\frac{\partial\tilde{\mathcal{H}}_0}{\partial\tilde{\kappa}^{\alpha}}+\tfrac{i}{4}\,\bar{\psi}\,\omega_{nm\alpha}\,\sigma^{nm}\right)\sigma^{ij}\tilde{\kappa}^{\alpha}+\tfrac{i}{4}\bar{\psi}\,\sigma^{ij}\left(\frac{\partial\tilde{\mathcal{H}}_0}{\partial\bar{\psi}}-\tfrac{i}{4}\,\omega_{nm\alpha}\,\sigma^{nm}\,\tilde{\kappa}^{\alpha}\right)\\
&\quad+\tfrac{i}{4}\Big[\bar{\psi}\,\sigma^{jn}\,\tilde{\kappa}^{\alpha}\omega^{i}{}_{n\alpha}-\bar{\tilde{\kappa}}^{\alpha}\,\sigma^{jn}\,\psi\,\omega^{i}{}_{n\alpha}+\bar{\tilde{\kappa}}^{\alpha}\,\sigma^{in}\,\psi\,\omega^{j}{}_{n\alpha}-\bar{\psi}\,\sigma^{in}\,\tilde{\kappa}^{\alpha}\omega^{j}{}_{n\alpha}\Big]\\
&=-\tfrac{i}{4}\frac{\partial\tilde{\mathcal{H}}_0}{\partial\psi}\sigma^{ij}\psi+\tfrac{i}{4}\bar{\tilde{\kappa}}^{\alpha}\sigma^{ij}\frac{\partial\tilde{\mathcal{H}}_0}{\partial\bar{\tilde{\kappa}}^{\alpha}}-\tfrac{i}{4}\frac{\partial\tilde{\mathcal{H}}_0}{\partial\tilde{\kappa}^{\alpha}}\sigma^{ij}\tilde{\kappa}^{\alpha}+\tfrac{i}{4}\bar{\psi}\,\sigma^{ij}\frac{\partial\tilde{\mathcal{H}}_0}{\partial\bar{\psi}}\\
&\quad+\tfrac{i}{4}\Big[\tfrac{i}{4}\bar{\tilde{\kappa}}^{\alpha}\left(\sigma^{ij}\sigma^{nm}-\sigma^{nm}\sigma^{ij}\right)\psi-\tfrac{i}{4}\bar{\psi}\left(\sigma^{ij}\sigma^{nm}-\sigma^{nm}\sigma^{ij}\right)\tilde{\kappa}^{\alpha}\Big]\omega_{nm\alpha}\\
&\quad+\tfrac{i}{4}\Big[\bar{\tilde{\kappa}}^{\alpha}\left(\sigma^{in}\,\omega^{j}{}_{n\alpha}-\sigma^{jn}\,\omega^{i}{}_{n\alpha}\right)\psi-\bar{\psi}\left(\sigma^{in}\,\omega^{j}{}_{n\alpha}-\sigma^{jn}\,\omega^{i}{}_{n\alpha}\right)\tilde{\kappa}^{\alpha}\Big].
\end{aligned}\tag{4.83}$$

By virtue of the algebra of the Dirac matrices, the commutator of σ-matrices amounts to

$$\sigma^{ij}\,\sigma^{nm}-\sigma^{nm}\,\sigma^{ij}\equiv-2i\left(\eta^{im}\,\sigma^{nj}-\eta^{nj}\,\sigma^{im}+\eta^{mj}\,\sigma^{in}-\eta^{in}\,\sigma^{mj}\right),\tag{4.84}$$

hence, contracted with $\frac{i}{4}\omega_{nm\alpha}$:

$$\tfrac{i}{4}\omega_{nm\alpha}\left(\sigma^{ij}\sigma^{nm}-\sigma^{nm}\sigma^{ij}\right)=\tfrac{1}{2}\left(-\omega^{i}{}_{n\alpha}\sigma^{nj}-\omega^{j}{}_{m\alpha}\sigma^{im}-\omega^{j}{}_{n\alpha}\sigma^{in}-\omega^{i}{}_{m\alpha}\sigma^{mj}\right)\tag{4.85}$$

$$=-\sigma^{in}\,\omega^{j}{}_{n\alpha}+\sigma^{jn}\,\omega^{i}{}_{n\alpha}.\tag{4.86}$$

The last two lines of Eq. (4.83) thus cancel. The consistency equation (4.80) thus reduces to:

$$
\begin{aligned}
0 = \frac{\mathrm{i}}{4}\left(\frac{\partial\tilde{\mathcal{H}}_0}{\partial\psi}\sigma_i{}^j\psi - \bar{\tilde{\kappa}}^\alpha\sigma_i{}^j\frac{\partial\tilde{\mathcal{H}}_0}{\partial\bar{\tilde{\kappa}}^\alpha} + \frac{\partial\tilde{\mathcal{H}}_0}{\partial\tilde{\kappa}^\alpha}\sigma_i{}^j\tilde{\kappa}^\alpha - \bar{\psi}\sigma_i{}^j\frac{\partial\tilde{\mathcal{H}}_0}{\partial\bar{\psi}}\right)\\
+\left(\frac{\partial\tilde{k}_i{}^{[\alpha\beta]}}{\partial x^\beta} - \omega^n{}_{i\beta}\tilde{k}_n{}^{[\alpha\beta]}\right)e_\alpha{}^j + \tilde{k}_i{}^{[\alpha\beta]}\left(\frac{\partial e_\alpha{}^j}{\partial x^\beta} + \omega^j{}_{n\beta}e_\alpha{}^n\right)\\
+\tilde{q}_i{}^{m[\alpha\beta]}\left(\frac{\partial\omega^j{}_{m\alpha}}{\partial x^\beta} + \omega^j{}_{n\beta}\omega^n{}_{m\alpha}\right) - \tilde{q}_m{}^{j[\alpha\beta]}\left(\frac{\partial\omega^m{}_{i\alpha}}{\partial x^\beta} - \omega^m{}_{n\alpha}\omega^n{}_{i\beta}\right).
\end{aligned}
$$

Finally, the canonical equations (4.79i), (4.79j), and (4.79k) are inserted to get Eq. (4.82).

4.3 Free Field Hamiltonians in Curved Spacetime

The derivations presented in the previous sections were based solely on the transformation properties of the spin-0, spin-1, and spin-1/2 fields and the derived transformation properties of the gauge fields. While the transformation framework fixed the interaction between the matter and spacetime fields, the specific form of the pertinent "kinetic" terms, the free, non-interacting Lagrangians $\tilde{\mathcal{L}}$ or equivalently Hamiltonians $\tilde{\mathcal{H}}$ have so far been kept open. In this section that gap shall be closed and the details of the matter fields, namely the free Klein-Gordon, Maxwell-Proca and Dirac systems specified. We thereby transfer the well known Lorentz covariant field equations of those matter fields with help of the vierbeins from the inertial frames, with the standard Minkowski metric, into a curvilinear geometry with an arbitrary metric. Then the particular Hamiltonians of mutually non-interacting scalar, vector, and spinor matter fields are derived, constituting the total free matter Hamiltonian density:

$$
\tilde{\mathcal{H}}_0 = \tilde{\mathcal{H}}_{\mathrm{KG}} + \tilde{\mathcal{H}}_{\mathrm{P}} + \tilde{\mathcal{H}}_{\mathrm{D}}.
$$

(Introducing mutual matter-matter interactions, e.g. Yang-Mills or Higgs models, are delegated to upcoming work. We are also aware that the Proca field is not a fundamental field but rather an element of effective field theories. Nevertheless, since vector fields will be key to abelian and non-abelian gauge theories we have included it in its Maxwell form in our analyzes.) In addition, we present a particular, phenomenologically justified model Hamiltonian for the free gravitational field.

4.3.1 Scalar Fields

As the first term in the yet unspecified matter Hamiltonian $\tilde{\mathcal{H}}_0$ we select for the scalar field its real massive Klein-Gordon version. That gives a simple first impression of how the curvilinear nature of a a manifold is transferred to the dynamics of physical fields that live in local inertial frames. The standard Lorentz invariant free Klein-Gordon Lagrangian in the inertial frame is given by

$$\mathcal{L}^0_{\text{KG}}(\phi, \partial\phi) = \tfrac{1}{2}\frac{\partial\varphi}{\partial x^i}\eta^{ij}\frac{\partial\varphi}{\partial x^j} - \tfrac{1}{2}m^2\varphi^2. \tag{4.87}$$

It is a function of fields and their derivatives, analogous to generalized coordinates and velocities in point dynamics. The Lagrangian must have the dimension $[\mathcal{L}_{\text{KG}}] = L^{-4}$, so that we find for the scalar field $[\varphi] = L^{-1}$. As $\partial\varphi/\partial x^i$ is a vector in the tangent space of the frame bundle written in the inertial frame base, it can be re-written in the coordinate base as

$$\frac{\partial\varphi}{\partial x^i} = \frac{\partial\varphi}{\partial x^\alpha}e_i{}^\alpha, \qquad \frac{\partial\varphi}{\partial x^\alpha} = e_\alpha{}^i\frac{\partial\varphi}{\partial x^i}.$$

In order to express this Klein-Gordon Lagrangian in the coordinate frame, we use the relations (4.1) and the orthonormality identities (2.8a) of the vierbein fields. The coordinate frame Klein-Gordon Lagrangian is then

$$\begin{aligned}\tilde{\mathcal{L}}^0_{\text{KG}}\left(\varphi, \partial_\alpha\varphi, e_k{}^\alpha\right) &= \tfrac{1}{2}\varepsilon\left(\frac{\partial\varphi}{\partial x^\alpha}e_k{}^\alpha\,\eta^{kj}\,e_j{}^\beta\frac{\partial\varphi}{\partial x^\beta} - m^2\varphi^2\right)\\ &= \tfrac{1}{2}\varepsilon\left(\frac{\partial\varphi}{\partial x^\alpha}g^{\alpha\beta}\frac{\partial\varphi}{\partial x^\beta} - m^2\varphi^2\right)\end{aligned} \tag{4.88}$$

The factor ε makes the absolute scalar Lagrangian $\mathcal{L}^0_{\text{KG}}$ from Eq. (4.88) into a scalar density, in order to maintains the correct integral measure as discussed in Sect. 4.1.1.

To derive the corresponding Hamiltonian, we apply in the following the generic definition given in Sect. 4.1.4 for the canonical conjugate "momentum" field:

$$\tilde{\pi}^\nu = \frac{\partial\tilde{\mathcal{L}}^0_{\text{KG}}}{\partial\left(\frac{\partial\varphi}{\partial x^\nu}\right)} = \varepsilon e_k{}^\nu\eta^{kj}e_j{}^\beta\frac{\partial\varphi}{\partial x^\beta}, \tag{4.89}$$

which has the dimensions $[\tilde{\pi}^\nu] = L^{-2}$. In analogy to point mechanics one might call the momentum field canonically conjugate to the free field a *kinetic* momentum field, and that conjugate to the integrating Lagrangian including all gauge terms a *canonical* momentum field.

The Hamiltonian can now be obtained by inverting the above relation and inserting it into the definition of the Hamiltonian:

$$\tilde{\mathcal{H}}_{\mathrm{KG}}(\phi, \tilde{\pi}^{\nu}, e_{\mu}{}^{i}) = \tilde{\pi}^{\nu}\frac{\partial\varphi}{\partial x^{\nu}} - \tilde{\mathcal{L}}^{0}_{\mathrm{KG}}(\phi, \partial\phi, e_{\mu}{}^{i})$$
$$= \frac{1}{2\varepsilon}\tilde{\pi}^{\alpha}e_{\alpha}{}^{n}\eta_{nm}e_{\beta}{}^{m}\tilde{\pi}^{\beta} + \frac{\varepsilon}{2}m^2\varphi^2. \tag{4.90}$$

This Hamiltonian can now simply be inserted into the Eqs. (4.79a) and (4.79b) to obtain the field equations for a scalar field in curved spacetime. This step will be performed in Sect. 4.5.

Notice in passing that the above calculations and those following below also hold if we replace the mass term $\frac{1}{2}m^2\varphi^2$ by a more general, possibly explicitly spacetime-dependent potential term $V(\varphi, x)$.

4.3.2 Vector Fields

Next, we want to deduce the Hamiltonian for the vector field. The particle model employed is the Maxwell-type Proca system for a massive spin-1 field. Maxwell type means that the field-strength tensor is the exterior derivative of the vector potential and thus a 2-form:

$$f_{ij} = \frac{\partial a_j}{\partial x^i} - \frac{\partial a_i}{\partial x^j}.$$

The Maxwell-Proca Lagrangian in the local inertial frame is given by

$$\mathcal{L}^{0}_{\mathrm{P}} = -\tfrac{1}{4}f_{ij}\eta^{ik}\eta^{jl}f_{kl} + \tfrac{1}{2}m^2a_i\eta^{ij}a_j. \tag{4.91}$$

The field-strength tensor can be re-written in the general coordinate base with help of the proper vierbeins,

$$f_{ij} = e_i{}^{\alpha}e_j{}^{\beta}f_{\alpha\beta}, \qquad a_i = e_i{}^{\alpha}a_{\alpha}. \tag{4.92}$$

Multiplying then $\mathcal{L}^{0}_{\mathrm{P}}$ by the volume factor ε gives the coordinate-frame Lagrangian density,

$$\tilde{\mathcal{L}}^{0}_{\mathrm{P}} = \left(-\tfrac{1}{4}\eta^{nk}\eta^{jl}\,e_n{}^{\tau}\,e_k{}^{\sigma}\,e_j{}^{\lambda}\,e_l{}^{\beta}\,f_{\tau\lambda}\,f_{\sigma\beta} + \tfrac{1}{2}m^2\eta^{jl}\,e_j{}^{\alpha}\,e_l{}^{\beta}\,a_{\alpha}\,a_{\beta}\right)\varepsilon. \tag{4.93}$$

For swapping into the Hamiltonian picture, we need to compute the pertinent conjugate momentum field:

$$\tilde{p}^{\mu\nu} = \frac{\partial\tilde{\mathcal{L}}^{0}_{\mathrm{P}}}{\partial\left(\frac{\partial a_{\mu}}{\partial x^{\nu}}\right)} = \varepsilon\eta^{nk}\eta^{jl}e_n{}^{\mu}e_k{}^{\sigma}e_j{}^{\nu}e_l{}^{\beta}f_{\sigma\beta} \equiv \tilde{f}^{\mu\nu} \tag{4.94}$$

The skew-symmetry of $\tilde{f}^{\mu\nu}$ thus induces the skew-symmetry of $\tilde{p}^{\mu\nu}$. The Maxwell-Proca Hamiltonian now follows as

$$\begin{aligned}\tilde{\mathcal{H}}_{\mathrm{P}}(a_\mu, \tilde{p}^{\mu\nu}, e_\mu{}^i) &= \tilde{p}^{\mu\nu} \frac{\partial a_\mu}{\partial x^\nu} - \tilde{\mathcal{L}}^0_{\mathrm{P}} \left(a_\mu, \partial_\nu a_\mu, e_\mu{}^i\right) \\ &= -\frac{1}{4\varepsilon} \eta_{ik}\, \eta_{jl}\, e_\xi{}^i\, e_\alpha{}^k\, e_\lambda{}^j\, e_\beta{}^l\, \tilde{p}^{\xi\lambda}\, \tilde{p}^{\alpha\beta} - \tfrac{1}{2} m^2 \eta^{ij}\, e_i{}^\alpha\, e_j{}^\beta\, a_\alpha\, a_\beta\, \varepsilon \\ &= -\frac{1}{4\varepsilon} g_{\xi\alpha}\, g_{\lambda\beta} \tilde{p}^{\xi\lambda}\, \tilde{p}^{\alpha\beta} - \tfrac{1}{2} m^2 g^{\alpha\beta}\, a_\alpha\, a_\beta\, \varepsilon. \end{aligned} \tag{4.95}$$

4.3.3 Spinor Fields

The dynamics of uncoupled particles and antiparticles with spin-1/2 and mass m is in the inertial frame described by the Dirac equation and its conjugate,

$$\mathrm{i}\gamma^k \frac{\partial \psi}{\partial x^k} - m\,\psi = 0 \tag{4.96a}$$

$$\frac{\partial \bar{\psi}}{\partial x^k}\, \mathrm{i}\gamma^k + m\bar{\psi} = 0, \tag{4.96b}$$

with $\gamma^k, k = 0, \ldots, 3$ denoting the 4×4 Dirac matrices in the Minkowski spacetime. ψ and $\bar{\psi}$ are the independent fields, where ψ is a four component Dirac spinor and $\bar{\psi} \equiv \psi^\dagger \gamma^0$ the adjoint spinor of ψ.

The symmetric form of the usual free Dirac Lagrangian $\mathcal{L}^0_{\mathrm{D}}\left(\psi, \partial_k \psi, \bar{\psi}, \partial_k \bar{\psi}\right)$ which entails Eqs. (4.96) by means of the Euler-Lagrange equations

$$\frac{\partial}{\partial x^k} \frac{\partial \mathcal{L}^0_{\mathrm{D}}}{\partial(\partial_k \psi)} - \frac{\partial \mathcal{L}^0_{\mathrm{D}}}{\partial \psi} = 0, \qquad \frac{\partial}{\partial x^k} \frac{\partial \mathcal{L}^0_{\mathrm{D}}}{\partial(\partial_k \bar{\psi})} - \frac{\partial \mathcal{L}^0_{\mathrm{D}}}{\partial \bar{\psi}} = 0 \tag{4.97}$$

is given by

$$\mathcal{L}^0_{\mathrm{D}} = \frac{\mathrm{i}}{2} \left(\bar{\psi} \gamma^k \frac{\partial \psi}{\partial x^k} - \frac{\partial \bar{\psi}}{\partial x^k} \gamma^k\, \psi \right) - m\, \bar{\psi} \psi. \tag{4.98}$$

As this Lagrangian is merely *linear* in the derivatives of the fields, it is degenerate and cannot directly be Legendre-transformed into an equivalent Dirac Hamiltonian $\mathcal{H}_{\mathrm{D}}$. In order for the Legendre transformation to exist, this Lagrangian must be amended by adding the divergence of the vector

$$F^j = \frac{\mathrm{i}}{6M} \left(\bar{\psi}\, \sigma^{jk} \frac{\partial \psi}{\partial x^k} + \frac{\partial \bar{\psi}}{\partial x^k} \sigma^{kj}\, \psi \right) \quad \Rightarrow \quad \frac{\partial F^j}{\partial x^j} = \frac{\mathrm{i}}{3M} \frac{\partial \bar{\psi}}{\partial x^k} \sigma^{kj} \frac{\partial \psi}{\partial x^j}.$$

M is a new coupling constant with mass dimension.

The regularized Dirac Lagrangian proposed by Gasiorowicz [21] is thus given by:

$$\mathcal{L}_{\mathrm{D}}^{0}=\frac{\mathrm{i}}{2}\left(\bar{\psi}\,\gamma^{k}\frac{\partial\psi}{\partial x^{k}}-\frac{\partial\bar{\psi}}{\partial x^{k}}\gamma^{k}\,\psi\right)-m\,\bar{\psi}\psi+\frac{\mathrm{i}}{3M}\frac{\partial\bar{\psi}}{\partial x^{k}}\sigma^{kj}\frac{\partial\psi}{\partial x^{j}}. \tag{4.99}$$

Making use of the identities

$$\gamma_{k}\,\sigma^{kj}\equiv\sigma^{jk}\,\gamma_{k}\equiv 3\mathrm{i}\,\gamma^{j},\qquad \gamma_{k}\,\sigma^{kj}\gamma_{j}\equiv 12\mathrm{i}\,\mathbb{1}, \tag{4.100}$$

it is equivalently expressed as the product

$$\mathcal{L}_{\mathrm{D}}^{0}=\left(\frac{\partial\bar{\psi}}{\partial x^{k}}-\frac{\mathrm{i}M}{2}\bar{\psi}\,\gamma_{k}\right)\frac{\mathrm{i}\sigma^{kj}}{3M}\left(\frac{\partial\psi}{\partial x^{j}}+\frac{\mathrm{i}M}{2}\gamma_{j}\psi\right)-(m-M)\,\bar{\psi}\psi. \tag{4.101}$$

This form remarkably suggests a "minimal" coupling of the spinor to the γ-matrices with coupling constant $M/2$. The Euler-Lagrange equations for (4.101) yield the standard Dirac equations (4.96), as amending the Lagrangian (4.98) by the divergence term $\partial F^{k}/\partial x^{k}$ does not modify the resulting field equations. The newly introduced constant M is thus merely spurious for the non-interacting Dirac field. However, as will be shown below, once interactions of the Dirac field are added, the constant M becomes a fundamental physical quantity, introducing a new "emergent" length, $\ell = 1/M$.

The inertial frame vectors $\partial\psi/\partial x^{k}$ and $\partial\bar{\psi}/\partial x^{k}$ in Eq. (4.101) can again be expressed in the coordinate frame by means of the vierbeins, which provides the Dirac Lagrangian density $\tilde{\mathcal{L}}_{\mathrm{D}}^{0}(\psi,\partial\psi,\bar{\psi},\partial\bar{\psi},{e_{k}}^{\alpha})$ in the form

$$\tilde{\mathcal{L}}_{\mathrm{D}}^{0}=\left[\left(\frac{\partial\bar{\psi}}{\partial x^{\alpha}}{e_{k}}^{\alpha}-\frac{\mathrm{i}M}{2}\bar{\psi}\gamma_{k}\right)\frac{\mathrm{i}\sigma^{kj}}{3M}\left({e_{j}}^{\beta}\frac{\partial\psi}{\partial x^{\beta}}+\frac{\mathrm{i}M}{2}\gamma_{j}\psi\right)-(m-M)\bar{\psi}\psi\right]\varepsilon. \tag{4.102}$$

This yields the conjugate momenta:

$$\tilde{\kappa}^{\alpha}=\frac{\partial\tilde{\mathcal{L}}_{\mathrm{D}}^{0}}{\partial\left(\frac{\partial\bar{\psi}}{\partial x^{\alpha}}\right)}=\left(-\tfrac{\mathrm{i}}{2}\,{e_{k}}^{\alpha}\,\gamma^{k}\,\psi+\frac{\mathrm{i}}{3M}{e_{k}}^{\alpha}\sigma^{kj}{e_{j}}^{\beta}\frac{\partial\psi}{\partial x^{\beta}}\right)\varepsilon, \tag{4.103}$$

$$\tilde{\bar{\kappa}}^{\alpha}=\frac{\partial\tilde{\mathcal{L}}_{\mathrm{D}}^{0}}{\partial\left(\frac{\partial\psi}{\partial x^{\alpha}}\right)}=\left(\tfrac{\mathrm{i}}{2}\bar{\psi}\,\gamma^{j}\,{e_{j}}^{\alpha}+\frac{\mathrm{i}}{3M}\frac{\partial\bar{\psi}}{\partial x^{\beta}}{e_{k}}^{\beta}\sigma^{kj}{e_{j}}^{\alpha}\right)\varepsilon. \tag{4.104}$$

Since the Lagrangian is quadratic in the "velocities", the conjugate momenta have a linear "velocity" dependence. Those linear relations can be inverted, and the corresponding covariant Hamiltonian [22, 23] constructed via a regular Legendre transformation:

$$
\begin{aligned}
\tilde{\mathcal{H}}_{\mathrm{D}} &= \tilde{\bar{\kappa}}^{\alpha}\frac{\partial\psi}{\partial x^{\alpha}} + \frac{\partial\bar{\psi}}{\partial x^{\alpha}}\tilde{\kappa}^{\alpha} - \tilde{\mathcal{L}}_{\mathrm{D}}^{0}(\psi, \partial\psi, \bar{\psi}, \partial\bar{\psi}, e_{k}{}^{\alpha}) \\
&= \frac{\mathrm{i}M}{2}\left(\bar{\psi}\,\gamma_{k}\,e_{\alpha}{}^{k}\,\tilde{\kappa}^{\alpha} - \tilde{\bar{\kappa}}^{\alpha}\,e_{\alpha}{}^{k}\frac{6\tau_{kj}}{\varepsilon}e_{\beta}{}^{j}\,\tilde{\kappa}^{\beta} - \tilde{\bar{\kappa}}^{\alpha}\,e_{\alpha}{}^{k}\,\gamma_{k}\,\psi\right) + (m-M)\,\bar{\psi}\psi\,\varepsilon, \\
&= \left(\tilde{\bar{\kappa}}^{\alpha} - \frac{\mathrm{i}}{2}\varepsilon\bar{\psi}\gamma^{i}e_{i}{}^{\alpha}\right)\frac{3Me_{\alpha}{}^{k}\tau_{kj}e_{\beta}{}^{j}}{\mathrm{i}\varepsilon}\left(\tilde{\kappa}^{\beta} + \frac{\mathrm{i}}{2}e_{l}{}^{\beta}\gamma^{l}\psi\varepsilon\right) + m\bar{\psi}\psi\varepsilon. \qquad (4.105)
\end{aligned}
$$

τ_{kj} is the inverse of the matrix σ^{jk}, the latter defined in Eq. (4.43):

$$
\tau_{kj} = \frac{\mathrm{i}}{6}\left(\gamma_{k}\gamma_{j} + 3\gamma_{j}\gamma_{k}\right), \qquad \tau_{ik}\,\sigma^{kj} = \delta_{i}^{j}\,\mathbb{1}.
$$

This definition entails the identities

$$
\gamma^{k}\tau_{kj} = \frac{1}{3\mathrm{i}}\gamma_{j}, \qquad \tau_{kj}\gamma^{j} = \frac{1}{3\mathrm{i}}\gamma_{k}. \qquad (4.106)
$$

4.3.4 Gravitational Field

At last, we have to postulate the dynamics of the free gravitational field. Its Hamiltonian $\tilde{\mathcal{H}}_{\mathrm{Gr}}$ is defined as [10]:

$$
\begin{aligned}
\tilde{\mathcal{H}}_{\mathrm{Gr}} &= \frac{1}{4g_{1}\varepsilon}\tilde{q}_{l}{}^{m\alpha\beta}\tilde{q}_{m}{}^{l\xi\lambda}\,\eta_{kn}\,\eta_{ij}\,e_{\alpha}{}^{k}\,e_{\xi}{}^{n}\,e_{\beta}{}^{i}\,e_{\lambda}{}^{j} - g_{2}\,\tilde{q}_{l}{}^{m\alpha\beta}\,e_{\alpha}{}^{l}\,e_{\beta}{}^{n}\,\eta_{mn} \\
&\quad + \frac{1}{2g_{3}\varepsilon}\tilde{k}_{l}{}^{\alpha\beta}\tilde{k}_{m}{}^{\xi\lambda}\,\eta^{lm}\eta_{kn}\,\eta_{ij}\,e_{\alpha}{}^{k}\,e_{\xi}{}^{n}\,e_{\beta}{}^{i}\,e_{\lambda}{}^{j}. \qquad (4.107)
\end{aligned}
$$

It is quadratic in the conjugate momentum $\tilde{q}_{l}{}^{m\alpha\beta}$ of the spin connection $\omega^{l}{}_{m\alpha}$ with g_{1} the pertaining dimensionless coupling constant. That quadratic $\tilde{q}$-term then describes a "deformation" of the actual theory from classical General Relativity represented by the linear term, and its relative contribution is governed by the coupling constant g_{1}, the *deformation parameter*. The term *linear* in $\tilde{q}$ is associated with the coupling constant g_{2} of dimension Length^{-2} and corresponds to the linear dependence on the Riemann-Cartan tensor in the Lagrangian description, as given by the Hilbert-Einstein Lagrangian of classical General Relativity.

It is important to stress here that in any Hamiltonian description at least a *quadratic* momentum dependence of the Hamiltonian is necessary in order to ensure regularity, i.e. the reversibility of Legendre transformations between the Hamiltonian and Lagrangian pictures, and hence the existence of the action integral. That quadratic term must necessarily be a full momentum tensor concomitant [24] for the Legendre transformation to retain all information on the gauge field in the equivalent Lagrangian description. Other contractions of the quadratic tensor product in

Eq. (4.107) would inhibit a unique relation between the "momentum" field $\tilde{q}_{\alpha}{}^{\sigma\tau\lambda}$ and the dual ("velocity") field $R^{\alpha}{}_{\sigma\tau\lambda}$.

The contribution of $\tilde{k}_{l}{}^{\alpha\beta}$ describes the free (uncoupled) dynamics of the dimensionless vierbein field $e_{\alpha}{}^{l}$ and represents, as usual, the *dual* of the "velocity" $\partial e_{\alpha}{}^{l}/\partial x^{\beta}$. We chose by Occam's razor the simplest, quadratic term in $\tilde{k}_{l}{}^{\alpha\beta}$. Since $\tilde{k}$ is of dimension Length^{-3}, the pertaining coupling constant g_3 in the denominator must have the dimension Length^{-2}.

In order to understand now the physical content of the conjugate momenta $k^{j}{}_{\mu\nu}$ and $q^{j}{}_{i\mu\nu}$, we compute the equations (4.79i) and (4.79j), respectively. From

$$S^{j}{}_{\mu\nu} = \frac{\partial\tilde{\mathcal{H}}_{\text{Gr}}}{\partial\tilde{k}_{j}{}^{\mu\nu}} = \frac{1}{g_3}\,k^{j}{}_{\mu\nu}, \tag{4.108}$$

we conclude that $k^{i}{}_{\mu\nu}$ can be identified with the torsion tensor $S^{\xi}{}_{\mu\nu} = S^{j}{}_{\mu\nu}\,e_{j}{}^{\xi}$, and confirm, by check against the derivation of Eq. (4.31), that it is skew-symmetric in μ,ν:

$$k^{j}{}_{\mu\nu} \equiv k^{j}{}_{[\mu\nu]} = g_3\,e_{\xi}{}^{j}\,S^{\xi}{}_{\mu\nu}. \tag{4.109}$$

Equation (4.79j) gives, on the other hand:

$$-R^{j}{}_{i\mu\nu} = 2\frac{\partial\tilde{\mathcal{H}}_{\text{Gr}}}{\partial\tilde{q}_{j}{}^{i\mu\nu}} = \frac{1}{g_1}q_{i}{}^{j\xi\lambda}\,g_{\mu\xi}\,g_{\nu\lambda} - g_2\,e_{\mu}{}^{j}\,e_{i}{}^{\lambda}\,g_{\nu\lambda} + g_2\,e_{\nu}{}^{j}\,e_{i}{}^{\lambda}\,g_{\mu\lambda}.$$

Resolving for the momentum field gives

$$\begin{aligned} q^{j}{}_{i\mu\nu} &= g_1\,R^{j}{}_{i\mu\nu} + g_1\,g_2\left(e_{\nu}{}^{j}\,g_{\mu\lambda} - e_{\mu}{}^{j}\,g_{\nu\lambda}\right)e_{i}{}^{\lambda} \\ &= g_1\left(R^{j}{}_{i\mu\nu} - \bar{R}^{j}{}_{i\mu\nu}\right), \end{aligned} \tag{4.110}$$

wherein $\bar{R}^{j}{}_{i\mu\nu}$ denotes the Riemann tensor of the maximally symmetric spacetime:

$$\bar{R}^{j}{}_{i\mu\nu} = g_2\left(e_{\mu}{}^{j}\,g_{\nu\lambda} - e_{\nu}{}^{j}\,g_{\mu\lambda}\right)e_{i}{}^{\lambda}. \tag{4.111}$$

The momentum tensor $q^{j}{}_{i\mu\nu}$ thus describes the *deformation* of the actual curvature $R^{j}{}_{i\mu\nu}$ with respect to the de Sitter ($g_2 > 0$) or Anti-de Sitter ($g_2 < 0$) "ground state" [25, 26], given by $\bar{R}^{j}{}_{i\mu\nu}$. Calling the parameter g_1 "deformation parameter" has thus a twofold meaning—deformation of curvature, and deformation of the theory with respect to the linear Einstein ansatz. Moreover, being in the denominator in the Hamiltonian, and multiplying the "velocity" represented by the curvature tensor, it resembles the role of mass in point mechanics. Here it thus assigns to geometry a property similar to *inertia* w.r.t. curvature deformations [25]. Spacetime is thus not following matter rigidly as in General Relativity, but acquires a dynamic resistance ability on its own.

For the sake of completeness we list here also the pertaining Lagrangian density, now in contrast to the previous sections, as the Legendre-transform of the free gravity Hamiltonian $\tilde{\mathcal{H}}_{\mathrm{Gr}}$ in the final action functional (4.53):

$$\tilde{\mathcal{L}}_{\mathrm{Gr}} = \tilde{k}_i^{\;\mu\nu} S^i_{\;\mu\nu} + \tfrac{1}{2}\,\tilde{q}_i^{\;j\mu\nu} R^i_{\;j\mu\nu} - \tilde{\mathcal{H}}_{\mathrm{Gr}}. \tag{4.112}$$

Recall that $S^i_{\;\mu\nu}$ and $R^i_{\;j\mu\nu}$, defined by Eqs. (4.61a) and (4.61b), are determined by the Hamiltonian $\tilde{\mathcal{H}}_{\mathrm{Gr}}$ via the canonical equations (4.79i) and (4.79j). Hence

$$\tilde{\mathcal{L}}_{\mathrm{Gr}} = \tilde{k}_i^{\;\mu\nu} \frac{\partial \tilde{\mathcal{H}}_{\mathrm{Gr}}}{\partial \tilde{k}_i^{\;\mu\nu}} + \tilde{q}_i^{\;j\mu\nu} \frac{\partial \tilde{\mathcal{H}}_{\mathrm{Gr}}}{\partial \tilde{q}_i^{\;j\mu\nu}} - \tilde{\mathcal{H}}_{\mathrm{Gr}}. \tag{4.113}$$

Inserting $\tilde{\mathcal{H}}_{\mathrm{Gr}}$ from Eq. (4.107) into Eq. (4.113) gives

$$\begin{aligned}\tilde{\mathcal{L}}_{\mathrm{Gr}} &= \frac{1}{4g_1\varepsilon}\tilde{q}_l^{\;m\alpha\beta}\tilde{q}_m^{\;l\xi\lambda}\,\eta_{kn}\,\eta_{ij}\,e^k_{\;\alpha}\,e^n_{\;\xi}\,e^i_{\;\beta}\,e^j_{\;\lambda}\\ &+ \frac{1}{2g_3\varepsilon}\tilde{k}_l^{\;\alpha\beta}\tilde{k}_m^{\;\xi\lambda}\,\eta^{lm}\eta_{kn}\,\eta_{ij}\,e^k_{\;\alpha}\,e^n_{\;\xi}\,e^i_{\;\beta}\,e^j_{\;\lambda}.\end{aligned}$$

This reproduces the quadratic momentum terms of the Hamiltonian, whereas the term linear in $\tilde{q}$ cancels. Of course, any Lagrangian density proper must be expressed in terms of the derivatives of the respective fields rather than by the canonical momenta of the Hamiltonian description. To this end, the canonical momenta $\tilde{k}_i^{\;\mu\nu}$ and $\tilde{q}_i^{\;j\mu\nu}$ must be expressed in terms of the actual Lagrangian dynamic quantities given by the specific canonical equations (4.109) and (4.110), which leads to

$$\begin{aligned}\tilde{\mathcal{L}}_{\mathrm{Gr}} &= \frac{g_1\varepsilon}{4}\left(R_l^{\;m\alpha\beta} - \bar{R}_l^{\;m\alpha\beta}\right)\left(R_m^{\;l\xi\lambda} - \bar{R}_m^{\;l\xi\lambda}\right)\eta_{kn}\,\eta_{ij}\,e^k_{\;\alpha}\,e^n_{\;\xi}\,e^i_{\;\beta}\,e^j_{\;\lambda}\\ &+ \frac{g_3\varepsilon}{2}S_\sigma^{\;\alpha\beta}\,S_\rho^{\;\xi\lambda}\,\eta^{lm}\eta_{kn}\,\eta_{ij}\,e_l^{\;\sigma}\,e_m^{\;\rho}\,e^k_{\;\alpha}\,e^n_{\;\xi}\,e^i_{\;\beta}\,e^j_{\;\lambda}.\end{aligned}$$

Inserting Eq. (4.111) for $\bar{R}$ yields

$$\begin{aligned}\tilde{\mathcal{L}}_{\mathrm{Gr}} &= g_1\varepsilon\left(\tfrac{1}{4}R^l_{\;m\xi\lambda}R^m_{\;l\alpha\beta}\,\eta^{kn}\,\eta^{ij}\,e_k^{\;\alpha}\,e_n^{\;\xi}\,e_i^{\;\beta}\,e_j^{\;\lambda} + g_2\,R^{lj}_{\;\;\lambda\beta}\,e_j^{\;\beta}\,e_l^{\;\lambda} - 6g_2^2\right)\\ &+ \frac{g_3\varepsilon}{2}S_\sigma^{\;\alpha\beta}\,S_\rho^{\;\xi\lambda}\,\eta^{lm}\eta_{kn}\,\eta_{ij}\,e_l^{\;\sigma}\,e_m^{\;\rho}\,e^k_{\;\alpha}\,e^n_{\;\xi}\,e^i_{\;\beta}\,e^j_{\;\lambda}.\end{aligned}$$

The corresponding Lagrangian $\mathcal{L}_{\mathrm{Gr}}$ is expressed equivalently in terms of the metric and the Ricci scalar $R \equiv R^{lj}_{\;\;\lambda\beta}\,e_j^{\;\beta}\,e_l^{\;\lambda}$ as:

$$\mathcal{L}_{\mathrm{Gr}} = \frac{g_1}{4} R^{l}{}_{m\xi\lambda} R^{m}{}_{l\alpha\beta}\, g^{\alpha\xi}\, g^{\beta\lambda} + g_1\, g_2\, R - 6 g_1\, g_2^2 \\ + \frac{g_3}{2} S_{\sigma}{}^{\alpha\beta}\, S_{\rho}{}^{\xi\lambda}\, g^{\sigma\rho}\, g_{\alpha\xi}\, g_{\beta\lambda}. \tag{4.114}$$

One thus encounters the Hilbert Lagrangian with an additional (cosmological) constant, plus the quadratic Riemann tensor term that is exactly the one proposed by Einstein in his personal letter to H. Weyl [25]. (For a discussion of the physical meaning of the constants see 4.5.4.)

4.3.5 Hamiltonian of the Overall System

At this point, it is instructive to once again write down the action integral (4.64) of the entire now specific system as the basis for the following analyzes. Splitting up the matter Lagrangians, which now include the gravitational interaction and hence do not carry the superscript 0, and the corresponding Hamiltonian expressions by the different matter fields,

$$\int_V \mathrm{d}^4x \left\{\tilde{\mathcal{L}}_{\mathrm{matter}} + \tilde{\mathcal{L}}_{\mathrm{Gr}}\right\} = \int_V \mathrm{d}^4x \left\{\tilde{\mathcal{L}}_{\mathrm{KG}} + \tilde{\mathcal{L}}_{\mathrm{P}} + \tilde{\mathcal{L}}_{\mathrm{D}} + \tilde{\mathcal{L}}_{\mathrm{Gr}}\right\} \tag{4.115}$$

$$= \int_V \mathrm{d}^4x \Bigg\{ \underbrace{\left(\tilde{\pi}^{\nu} \frac{\partial\varphi}{\partial x^{\nu}} - \tilde{\mathcal{H}}_{\mathrm{KG}}\right)}_{=:\tilde{\mathcal{L}}_{\mathrm{KG}}} + \underbrace{\left(\tfrac{1}{2}\tilde{p}^{\mu\nu} f_{\nu\mu} - \tilde{\mathcal{H}}_{\mathrm{P}}\right)}_{=:\tilde{\mathcal{L}}_{\mathrm{P}}} + \underbrace{\left(\bar{\tilde{\kappa}}^{\nu} \overrightarrow{\mathcal{D}} \psi + \bar{\psi}\, \overleftarrow{\mathcal{D}} \tilde{\kappa}^{\nu} - \tilde{\mathcal{H}}_{\mathrm{D}}\right)}_{=:\tilde{\mathcal{L}}_{\mathrm{D}}} \\ + \underbrace{\left(\tilde{k}_{i}{}^{\mu\nu} S^{i}{}_{\mu\nu} + \tfrac{1}{2}\tilde{q}_{i}{}^{j\mu\nu} R^{i}{}_{j\mu\nu} - \tilde{\mathcal{H}}_{\mathrm{Gr}}\right)}_{=:\tilde{\mathcal{L}}_{\mathrm{Gr}}} \Bigg\},$$

we get by collecting all pieces discussed above expressed in terms of the proper dynamic fields:

$$S = \int_V \mathrm{d}^4x \Bigg\{ \tilde{\pi}^{\nu} \frac{\partial\varphi}{\partial x^{\nu}} - \frac{1}{2\varepsilon}\tilde{\pi}^{\alpha} g_{\alpha\beta}\tilde{\pi}^{\beta} + \frac{\varepsilon}{2} m^2 \varphi^2 \tag{4.116}$$

$$+ \tfrac{1}{2}\tilde{p}^{\mu\nu}\left(\frac{\partial a_{\mu}}{\partial x^{\nu}} - \frac{\partial a_{\nu}}{\partial x^{\mu}}\right) + \frac{1}{4\varepsilon} g_{\xi\alpha} g_{\lambda\beta} \tilde{p}^{\xi\lambda}\tilde{p}^{\alpha\beta} + \tfrac{1}{2} m^2 g^{\alpha\beta} a_{\alpha} a_{\beta}\varepsilon$$

$$+ \bar{\tilde{\kappa}}^{\nu}\left(\frac{\partial\psi}{\partial x^{\nu}} - \tfrac{\mathrm{i}}{4}\omega^{i}{}_{j\nu}\,\sigma_{i}{}^{j}\,\psi\right) + \left(\frac{\partial\bar{\psi}}{\partial x^{\nu}} + \tfrac{\mathrm{i}}{4}\bar{\psi}\,\omega^{i}{}_{j\nu}\,\sigma_{i}{}^{j}\right)\tilde{\kappa}^{\nu}$$

$$- \frac{3M}{\mathrm{i}\varepsilon}\left(\bar{\tilde{\kappa}}^{\alpha} - \frac{\mathrm{i}}{2}\varepsilon\bar{\psi}\gamma^{i} e_{i}{}^{\alpha}\right) e_{\alpha}{}^{k}\tau_{kj} e_{\beta}{}^{j}\left(\tilde{\kappa}^{\beta} + \frac{\mathrm{i}}{2} e_{l}{}^{\beta}\gamma^{l}\psi\varepsilon\right) - m\bar{\psi}\psi$$

$$+ \tfrac{1}{2}\tilde{k}_{i}{}^{\mu\nu}\left(\frac{\partial e_{\mu}{}^{i}}{\partial x^{\nu}} - \frac{\partial e_{\nu}{}^{i}}{\partial x^{\mu}} + \omega^{i}{}_{j\nu}\, e_{\mu}{}^{j} - \omega^{i}{}_{j\mu}\, e_{\nu}{}^{j}\right) - \frac{1}{2g_3\varepsilon}\tilde{k}_{l}{}^{\alpha\beta}\tilde{k}_{m}{}^{\xi\lambda}\,\eta^{lm}\, g_{\alpha\xi} g_{\beta\lambda}$$

$$+\tfrac{1}{2}\tilde{q}_i^{\ j\mu\nu}\left(\frac{\partial\omega^i{}_{j\mu}}{\partial x^\nu}-\frac{\partial\omega^i{}_{j\nu}}{\partial x^\mu}+\omega^i{}_{n\nu}\,\omega^n{}_{j\mu}-\omega^i{}_{n\mu}\,\omega^n{}_{j\nu}\right)$$
$$-\frac{1}{4g_1\varepsilon}\tilde{q}_l^{\ m\alpha\beta}\tilde{q}_m^{\ l\xi\lambda}g_{\alpha\xi}g_{\beta\lambda}+g_2\,\tilde{q}_l^{\ m\alpha\beta}e_\alpha^{\ l}e_\beta^{\ n}\eta_{mn}\bigg\}.$$

4.4 Metric Energy-Momentum Tensors

The energy-momentum tensor of matter, aka stress-energy tensor, has been the key physical entity in Einstein's derivation of General Relativity, since it determines the curvature of spacetime. In the CCGG framework the energy-momentum tensor density is naturally identified with the "metric" version, $g^{\nu\alpha}e_\alpha^{\ i}\partial\tilde{\mathcal{H}}/\partial e_\mu^{\ i}$. And that definition applies to both, matter and spacetime. Its dynamic impact depends on the concrete specification of the involved matter fields and on the underlying gravity model. For the matter portion of the total Lagrangian density in the action integral (4.115),

$$\tilde{\mathcal{L}}_{\text{matter}} \equiv \tilde{\mathcal{L}}_{\text{KG}} + \tilde{\mathcal{L}}_{\text{P}} + \tilde{\mathcal{L}}_{\text{D}}, \tag{4.117}$$

the terms of the Legendre transform to the Hamiltonian density do not depend on the vierbein field. Then

$$\tilde{T}^{\mu\nu} := g^{\nu\alpha}e_\alpha^{\ i}\frac{\partial}{\partial e_\mu^{\ i}}\left(\tilde{\mathcal{H}}_{\text{KG}}+\tilde{\mathcal{H}}_{\text{P}}+\tilde{\mathcal{H}}_{\text{D}}\right) \equiv -g^{\nu\alpha}e_\alpha^{\ i}\frac{\partial\tilde{\mathcal{L}}_{\text{matter}}}{\partial e_\mu^{\ i}}, \tag{4.118}$$

which is the standard textbook definition of the energy-momentum tensor. The specific expressions for each matter field and gravity are derived below.

4.4.1 Energy-Momentum Tensor of the Scalar Field

For the free Klein-Gordon field the metric energy-momentum tensor is computed using Eq. (2.8) from the Hamiltonian (4.90). This gives

$$\begin{aligned}
\tilde{T}^{\nu\mu}_{\text{KG}} &= g^{\nu\alpha}e_\alpha^{\ i}\frac{\partial\tilde{\mathcal{H}}_{\text{KG}}}{\partial e_\mu^{\ i}} \\
&= \frac{1}{\varepsilon}g^{\nu\alpha}\tilde{\pi}^\mu e_\alpha^{\ i}\eta_{im}e_\beta^{\ m}\tilde{\pi}^\beta - g^{\nu\mu}\left(\frac{1}{2\varepsilon}\tilde{\pi}^\gamma e_\gamma^{\ n}\eta_{nm}e_\beta^{\ m}\tilde{\pi}^\beta-\frac{\varepsilon}{2}m^2\varphi^2\right) \\
&= \frac{1}{\varepsilon}\tilde{\pi}^\nu\tilde{\pi}^\mu - g^{\nu\mu}\left(\frac{1}{2\varepsilon}\tilde{\pi}^\gamma g_{\gamma\beta}\tilde{\pi}^\beta-\frac{\varepsilon}{2}m^2\varphi^2\right).
\end{aligned} \tag{4.119}$$

In terms of the field and its derivative we appply Eq. (4.79a),

$$\tilde{\pi}^{\mu} = g^{\mu\nu}\,\varepsilon\,\partial_{\nu}\varphi,$$

to get the manifestly symmetric energy-momentum tensor density

$$\tilde{T}^{\nu\mu}_{\mathrm{KG}} = \varepsilon\left[g^{\nu\alpha}g^{\mu\beta}\partial_{\alpha}\varphi\partial_{\beta}\varphi - g^{\nu\mu}\tfrac{1}{2}\left(g^{\alpha\beta}\partial_{\alpha}\varphi\partial_{\beta}\varphi - m^2\varphi^2\right)\right]. \tag{4.120}$$

Obviously, dividing this equation by ε gives the corresponding energy-momentum tensor:

$$T^{\nu\mu}_{\mathrm{KG}} = g^{\nu\alpha}g^{\mu\beta}\partial_{\alpha}\varphi\partial_{\beta}\varphi - g^{\nu\mu}\tfrac{1}{2}\left(g^{\alpha\beta}\partial_{\alpha}\varphi\partial_{\beta}\varphi - m^2\varphi^2\right). \tag{4.121}$$

4.4.2 Energy-Momentum Tensor of the Vector Field

The energy-momentum tensor of the vector field is obtained in analogous steps from the Maxwell-Proca Hamiltonian (4.95). Taking into account the derivative of ${e_n}^{\tau}$ with respect to the inverse vierbein ${e_{\xi}}^{i}$, Eq. (2.8e), we obtain again a symmetric expression:

$$\begin{aligned}
\tilde{T}^{\nu\mu}_{\mathrm{P}} &= g^{\nu\alpha}{e_{\alpha}}^{i}\frac{\partial\tilde{\mathcal{H}}_{\mathrm{P}}}{\partial {e_{\mu}}^{i}} \\
&= -\frac{1}{\varepsilon}g^{\nu\alpha}\eta_{nk}\eta_{jl}{e_{\alpha}}^{n}{e_{\gamma}}^{k}{e_{\lambda}}^{j}{e_{\beta}}^{l}\tilde{p}^{\mu\lambda}\tilde{p}^{\gamma\beta} + g^{\nu\alpha}m^2\eta^{ij}{e_i}^{\mu}{e_j}^{\beta}a_{\alpha}a_{\beta}\varepsilon \\
&\quad - g^{\nu\mu}\left(-\frac{1}{4}\eta_{ik}\eta_{jl}{e_{\xi}}^{i}{e_{\alpha}}^{k}{e_{\lambda}}^{j}{e_{\beta}}^{l}\tilde{p}^{\xi\lambda}\tilde{p}^{\alpha\beta} + \tfrac{1}{2}m^2\eta^{ij}{e_i}^{\alpha}{e_j}^{\beta}a_{\alpha}a_{\beta}\varepsilon\right) \\
&= -\frac{1}{\varepsilon}\tilde{p}^{\nu\lambda}g_{\lambda\beta}\tilde{p}^{\mu\beta} + m^2g^{\nu\lambda}a_{\lambda}g^{\mu\alpha}a_{\alpha}\varepsilon \\
&\quad - g^{\nu\mu}\left(-\frac{1}{4\varepsilon}g_{\xi\alpha}g_{\lambda\beta}\tilde{p}^{\xi\lambda}\tilde{p}^{\alpha\beta} + \frac{1}{2}m^2g^{\alpha\beta}a_{\alpha}a_{\beta}\varepsilon\right).
\end{aligned} \tag{4.122}$$

Expressing this in terms of the field and its derivatives means replacing $\tilde{p}^{\mu\nu}$ by $\varepsilon f^{\mu\nu}$. The absolute energy-momentum tensor $T^{\nu\mu}_{\mathrm{P}}$ is obtained by division by ε:

$$T^{\nu\mu}_{\mathrm{P}} = -f^{\nu\lambda}g_{\lambda\beta}f^{\mu\beta} + m^2g^{\nu\lambda}a_{\lambda}g^{\mu\alpha}a_{\alpha} + g^{\nu\mu}\left(\tfrac{1}{4}g_{\xi\alpha}g_{\lambda\beta}f^{\xi\lambda}f^{\alpha\beta} - \tfrac{1}{2}m^2g^{\alpha\beta}a_{\alpha}a_{\beta}\right). \tag{4.123}$$

4.4.3 Energy-Momentum Tensor of the Spinor Field

Just like before, the metric energy-momentum tensor density of the Dirac field follows from the derivative of $\tilde{\mathcal{H}}_{\mathrm{D}}$, Eq. (4.105), with respect to the inverse vierbein:

$$
\begin{aligned}
\tilde{T}_{\mathrm{D}}^{\nu\mu} &= g^{\nu\alpha} e_{\alpha}{}^{i} \frac{\partial \tilde{\mathcal{H}}_{\mathrm{D}}}{\partial e_{\mu}{}^{i}} \\
&= g^{\nu\alpha} e_{\alpha}{}^{i} \frac{\mathrm{i}M}{2} \left(\bar{\psi}\gamma_i \tilde{\kappa}^{\mu} - \tilde{\bar{\kappa}}^{\mu} \gamma_i \psi \right) + g^{\nu\alpha} e_{\alpha}{}^{i} (m - M) \bar{\psi}\psi e_i{}^{\mu} \varepsilon \\
&\quad - g^{\nu\alpha} e_{\alpha}{}^{i} \frac{3\mathrm{i}M}{\varepsilon} \left(\tilde{\bar{\kappa}}^{\mu} \tau_{ik} e_{\beta}{}^{k} \tilde{\kappa}^{\beta} + \tilde{\bar{\kappa}}^{\beta} e_{\beta}{}^{k} \tau_{ki} \tilde{\kappa}^{\mu} - \tilde{\bar{\kappa}}^{\gamma} e_{\gamma}{}^{k} \tau_{kn} e_{\beta}{}^{n} \tilde{\kappa}^{\beta} e_i{}^{\mu} \right).
\end{aligned}
$$

For a better readability we introduce the abbreviations

$$
\gamma_i e^{I}{}_{\alpha} \equiv \gamma_{\alpha},
$$

with analogous definitions for the σ and τ matrices. Then the energy-momentum tensor reads

$$
\begin{aligned}
\tilde{T}_{\mathrm{D}}^{\nu\mu} &= g^{\nu\alpha} \frac{iM}{2\varepsilon} \left[(\varepsilon\bar{\psi}\gamma_{\alpha} - 6\tilde{\bar{\kappa}}^{\beta}\tau_{\beta\alpha})\tilde{\kappa}^{\mu} - \tilde{\bar{\kappa}}^{\mu}(\varepsilon\gamma_{\alpha}\psi + 6\tau_{\alpha\beta}\tilde{\kappa}^{\beta}) \right] \\
&\quad - g^{\nu\mu} \left[(m - M)\bar{\psi}\psi\epsilon - \frac{3iM}{\varepsilon} \tilde{\bar{\kappa}}^{\gamma} \tau_{\gamma\beta} \tilde{\kappa}^{\beta} \right].
\end{aligned}
\tag{4.124}
$$

In order to express then energy-momentum tensor in terms of the spinor fields and its partial derivatives, we need to insert the explicit form of the canonical momenta which can be obtained from Eq. (4.79e) as explicitly shown in Sect. 4.5.3:

$$
\tilde{\kappa}^{\alpha} = \left[-\frac{\mathrm{i}}{2}\gamma^{j}\,\psi + \frac{\mathrm{i}}{3M}\,\sigma^{ji}\,e_i{}^{\beta} \left(\frac{\partial\psi}{\partial x^{\beta}} - \frac{\mathrm{i}}{4}\omega_{nm\beta}\,\sigma^{nm}\,\psi \right) \right] e_j{}^{\alpha} \varepsilon \tag{4.125a}
$$

$$
\tilde{\bar{\kappa}}^{\alpha} = \left[\ \frac{\mathrm{i}}{2}\bar{\psi}\,\gamma^{j} - \frac{\mathrm{i}}{3M} \left(\frac{\partial\bar{\psi}}{\partial x^{\beta}} + \frac{\mathrm{i}}{4}\omega_{nm\beta}\,\bar{\psi}\,\sigma^{nm} \right) \sigma^{ji}\,e_i{}^{\beta} \right] e_j{}^{\alpha} \varepsilon, \tag{4.125b}
$$

To keep the formulas as brief and readable as possible, we introduce the spinor covariant derivative defined in Eqs. (4.63a). Inserting the conjugate momenta gives

$$
\begin{aligned}
\tilde{T}_{\mathrm{D}}^{\nu\mu} &= \varepsilon g^{\nu\alpha} \Bigg[\bar{\psi}\overleftarrow{\mathcal{D}}_{\alpha} \left(-\frac{\mathrm{i}}{2}\gamma^{\mu}\,\psi + \frac{\mathrm{i}}{3M}\,\sigma^{\mu\beta}\overrightarrow{\mathcal{D}}_{\beta}\psi \right) \\
&\quad + \left(\frac{\mathrm{i}}{2}\bar{\psi}\,\gamma^{\mu} + \frac{\mathrm{i}}{3M}\bar{\psi}\overleftarrow{\mathcal{D}}_{\beta}\sigma^{\beta\mu} \right) \overrightarrow{\mathcal{D}}_{\alpha}\psi \Bigg] - g^{\nu\mu}\tilde{\mathcal{L}}_{\mathrm{D}}.
\end{aligned}
\tag{4.126}
$$

In the process we identified the Lagrangian of the Dirac system minimally coupled to gravitation

$$
\tilde{\mathcal{L}}_D = \left(\bar{\psi}\overleftarrow{\mathcal{D}}_{\beta} - \frac{\mathrm{i}M}{2}\bar{\psi}\,\gamma_{\beta} \right) \frac{\mathrm{i}\sigma^{\beta\alpha}}{3M} \left(\overrightarrow{\mathcal{D}}_{\alpha}\psi + \frac{\mathrm{i}M}{2}\gamma_{\alpha}\psi \right) - (m - M)\,\bar{\psi}\psi. \tag{4.127}
$$

One interesting feature is, that the skew symmetric part does not vanish in this case:

$$\tilde{T}_{\mathrm{D}[\mu\nu]} = \frac{\mathrm{i}\varepsilon}{4}\left(\bar{\psi}\gamma_\nu \overrightarrow{\mathcal{D}}_\mu\psi - \bar{\psi}\gamma_\mu \overrightarrow{\mathcal{D}}_\nu\psi - \bar{\psi}\overleftarrow{\mathcal{D}}_\mu\gamma_\nu\psi + \bar{\psi}\overleftarrow{\mathcal{D}}_\nu\gamma_\mu\psi\right) \\ + \frac{\mathrm{i}\varepsilon}{6M}\bar{\psi}\overleftarrow{\mathcal{D}}_\alpha\left(\sigma_\nu{}^\beta\delta^\alpha_\mu - \sigma_\mu{}^\beta\delta^\alpha_\nu - \sigma_\nu{}^\alpha\delta^\beta_\mu + \sigma_\mu{}^\alpha\delta^\beta_\nu\right)\overrightarrow{\mathcal{D}}_\beta\psi \neq 0. \tag{4.128}$$

For this reason, the Dirac systems does act as a source of torsion of spacetime, as will be laid out in the following.

4.4.4 Energy-Momentum Tensor for the Gravitational Field

In analogy to the accepted definition of the metric energy-momentum tensor of matter fields, the derivative of the Hamiltonian density (4.107) with respect to the vierbein is defined as the metric energy-momentum tensor density of spacetime:

$$\begin{aligned}\tilde{T}^{\mu\nu}_{\mathrm{Gr}} &= g^{\mu\alpha}\frac{\partial\tilde{\mathcal{H}}_{\mathrm{Gr}}}{\partial e_\nu{}^i}e_\alpha{}^i \\ &= -\frac{1}{g_1\varepsilon}\left(\tilde{q}_{lm\tau\alpha}\,\tilde{q}^{lm\tau\nu}g^{\mu\alpha} - \tfrac{1}{4}g^{\mu\nu}\,\tilde{q}_{lm\tau\beta}\,\tilde{q}^{lm\tau\beta}\right) \\ &\quad + \frac{2}{g_3\varepsilon}\left(\tilde{k}_{l\tau\alpha}\,\tilde{k}^{l\tau\nu}\,g^{\mu\alpha} - \tfrac{1}{4}g^{\mu\nu}\,\tilde{k}_{l\tau\beta}\,\tilde{k}^{l\tau\beta}\right) - 2g_2\,\tilde{q}_{li}{}^{\alpha\nu}e_\alpha{}^l\,\eta^{ij}e_j{}^\mu.\end{aligned} \tag{4.129}$$

Inserting Eqs. (4.109) and (4.110), the energy-momentum tensor $\tilde{T}^{\mu\nu}_{\mathrm{Gr}}$ is equivalently expressed in terms of the curvature and torsion tensors, R and S, as:

$$\begin{aligned}\tilde{T}^{\mu\nu}_{\mathrm{Gr}} &= -g_1\,\varepsilon\left(R_{lm\tau}{}^\nu\,R^{lm\tau}{}_\alpha g^{\mu\alpha} - \tfrac{1}{4}g^{\mu\nu}R_{lm\tau\beta}\,R^{lm\tau\beta}\right) \\ &\quad + 2g_1g_2\,\varepsilon\left(R_l{}^{j\beta}{}_\alpha\,e_\beta{}^l\,e_j{}^\nu g^{\alpha\mu} - \tfrac{1}{2}g^{\mu\nu}\,R_l{}^{n\alpha}{}_\beta\,e_\alpha{}^l\,e_n{}^\beta + 3g_2\,g^{\mu\nu}\right) \\ &\quad + 2g_3\,\varepsilon\left(S_{l\tau\alpha}\,S^{l\tau\nu}\,g^{\alpha\mu} - \tfrac{1}{4}g^{\mu\nu}\,S_{l\tau\beta}\,S^{l\tau\beta}\right).\end{aligned} \tag{4.130}$$

It is *not* symmetric. Its skew-symmetric portion is

$$\tilde{T}^{[\mu\nu]}_{\mathrm{Gr}} = \frac{1}{2}\frac{\partial\tilde{\mathcal{H}}_{\mathrm{Gr}}}{\partial e_\beta{}^n}\left(\delta^\nu_\beta e_\alpha{}^n g^{\alpha\mu} - \delta^\mu_\beta e_\alpha{}^n g^{\alpha\nu}\right) = g_1g_2\,\varepsilon\left(R_\alpha{}^{j\alpha i} - R_\alpha{}^{i\alpha j}\right)e_i{}^\mu e_j{}^\nu.$$

Re-writing the right-hand side as the skew-symmetric portion of the Ricci tensor $R^{\mu\nu} := R_\alpha{}^{i\alpha j}e_i{}^\mu e_j{}^\nu$ gives:

$$T^{[\mu\nu]}_{\mathrm{Gr}} = -2g_1g_2\,R^{[\mu\nu]}. \tag{4.131}$$

All other terms of (4.130) establish the symmetric portion of the energy-momentum tensor of the specific gravity ansatz $\tilde{\mathcal{H}}_{\mathrm{Gr}}$:

$$
\begin{aligned}
T_{\mathrm{Gr}}^{(\mu\nu)} = & -g_1 \left(R_{lm\tau}{}^{\nu}\, R^{lm\tau\mu} - \tfrac{1}{4} g^{\mu\nu} R_{lm\tau\beta}\, R^{lm\tau\beta} \right) \\
& + 2 g_1 g_2 \left(R^{(\mu\nu)} - \tfrac{1}{2} g^{\mu\nu}\, R_{\alpha}{}^{\alpha} + 3 g_2\, g^{\mu\nu} \right) \\
& + 2 g_3 \left(S_{l\tau}{}^{\mu}\, S^{l\tau\nu} - \tfrac{1}{4} g^{\mu\nu}\, S_{l\tau\beta}\, S^{l\tau\beta} \right). \qquad (4.132)
\end{aligned}
$$

4.5 Coupled Field Equations of Matter and Dynamic Spacetime

The diffeomorphism invariant field equations emerging from the covariant canonical transformation theory are now discussed for the particular Hamiltonians of the free scalar, vector, and spinor matter fields in conjunction with the particular model Hamiltonian for the free gravitational field.

The canonical equations as derived in the Hamiltonian formalism are first-order differential equations that determine the dynamics of the coupled conjugate pairs of matter and spacetime fields. In contrast, the field equation that emerge in the Lagrangian picture are second order differential equations where the momentum fields are substituted by the derivatives of the fields, in analogy to momenta in point mechanics being replaced by particle velocities. Hence, in order to get an equation that is comparable to Einstein's field equation, we need to carry out similar substitutions.

4.5.1 Klein-Gordon Equation in Curved Spacetime

In order to obtain the modified Klein-Gordon equation, we simply have to compute the canonical equations (4.79a) and (4.79b) for the Klein-Gordon Hamiltonian (4.90). This gives

$$
\frac{\partial \varphi}{\partial x^{\nu}} = \frac{\partial \tilde{\mathcal{H}}_{\mathrm{KG}}}{\partial \tilde{\pi}^{\nu}} = \frac{1}{\varepsilon} \tilde{\pi}^{\alpha} e_{\alpha}{}^{n} \eta_{nm} e_{\eta}{}^{m} = \pi_{\nu} \qquad (4.133a)
$$

$$
\frac{\partial \tilde{\pi}^{\alpha}}{\partial x^{\alpha}} = -\frac{\partial \tilde{\mathcal{H}}_{\mathrm{KG}}}{\partial \varphi} = -\varepsilon m^{2} \varphi \qquad (4.133b)
$$

Solving Eq. (4.133a) for $\tilde{\pi}^{\alpha}$ is now possible:

$$
\tilde{\pi}^{\alpha} = e_{i}{}^{\alpha}\, \eta^{ij}\, e_{j}{}^{\beta} \frac{\partial \varphi}{\partial x^{\beta}}\, \varepsilon = g^{\alpha\beta} \frac{\partial \varphi}{\partial x^{\beta}}\, \varepsilon,
$$

and the momentum field can be eliminated from the equation of motion to yield:

$$\frac{\partial \tilde{\pi}^{\alpha}}{\partial x^{\alpha}} = \frac{\partial}{\partial x^{\alpha}} \left(g^{\alpha\beta} \frac{\partial \varphi}{\partial x^{\beta}} \varepsilon \right) - \varepsilon\, m^2\, \varphi.$$

By executing the partial derivative of the metric and its determinant, we can cast the Klein-Gordon equation in dynamic spacetime into the form of a tensor equation:

$$g^{\alpha\beta} \varphi_{;\beta;\alpha} - 2 g^{\alpha\beta} S^{\xi}{}_{\alpha\xi} \varphi_{;\beta} + m^2 \varphi = 0, \tag{4.134}$$

where we have applied the covariant derivative of a general tensor in analogy to Eq. (2.18), meaning that every index is once contracted with either the affine or the spin connection.

4.5.2 *Maxwell-Proca Equation in Curved Spacetime*

The field equations for the Maxwell-Proca derive from the canonical equations (4.79c) and (4.79d). The first one, Eq. (4.94), gives $p^{\mu\nu} = f^{\mu\nu}$. The second one yields

$$\frac{\partial \tilde{p}^{\mu\alpha}}{\partial x^{\alpha}} = -\frac{\partial \tilde{\mathcal{H}}_{\mathrm{P}}}{\partial a_{\mu}} = m^2 \eta^{ij} e_i{}^{\alpha} e_j{}^{\mu} a_{\alpha} \varepsilon = m^2 g^{\mu\alpha} a_{\alpha} \varepsilon. \tag{4.135}$$

Since we demand metric compatibility, we can express the partial derivative of the metric with respect to the connection as

$$\frac{\partial g^{\mu\nu}}{\partial x^{\alpha}} = -\gamma^{\mu}{}_{\alpha\gamma} g^{\gamma\nu} - \gamma^{\nu}{}_{\alpha\gamma} g^{\mu\gamma},$$

where the affine connection is the function (4.54) of the spin connection. The derivative of the determinant of the metric is given by

$$\frac{\partial \sqrt{-g}}{\partial x^{\alpha}} = \sqrt{-g}\, \gamma^{\gamma}{}_{\alpha\gamma},$$

such that we get for the vector field equation

$$f^{\mu\alpha}{}_{;\alpha} - S^{\mu}{}_{\gamma\alpha} f^{\gamma\alpha} + 2 S^{\gamma}{}_{\alpha\gamma} f^{\mu\alpha} - m^2 a^{\mu} = 0, \tag{4.136}$$

where $S^{\mu}{}_{\gamma\alpha}$ is the torsion tensor according to Eqs. (4.61).

4.5.3 Dirac Equation in Curved Spacetime

In order to obtain the field equations for the spinor field, we just need to compute the canonical equations (4.79e) to (4.79h). They result in

$$\overrightarrow{\mathcal{D}}_{\mu}\psi = \frac{\partial\tilde{\mathcal{H}}_{\mathrm{D}}}{\partial\bar{\tilde{\kappa}}^{\mu}} = -\frac{\mathrm{i}M}{2}\left(\gamma_k\psi + \frac{6\tau_{kj}}{\varepsilon}e_{\alpha}{}^{j}\tilde{\kappa}^{\alpha}\right)e_{\mu}{}^{k} \tag{4.137a}$$

$$\overrightarrow{\mathcal{D}}_{\alpha}\tilde{\kappa}^{\alpha} = -\frac{\partial\tilde{\mathcal{H}}_{\mathrm{D}}}{\partial\bar{\psi}} = -\frac{\mathrm{i}M}{2}\gamma_k e_{\alpha}{}^{k}\tilde{\kappa}^{\alpha} - (m-M)\psi\varepsilon \tag{4.137b}$$

$$\bar{\psi}\overleftarrow{\mathcal{D}}_{\mu} = \frac{\partial\tilde{\mathcal{H}}_{D}}{\partial\tilde{\kappa}^{\mu}} = \frac{\mathrm{i}M}{2}\left(\bar{\psi}\gamma_j - \bar{\tilde{\kappa}}^{\alpha}\frac{6\tau_{kj}}{\varepsilon}e_{\alpha}{}^{k}\right)e_{\mu}{}^{j} \tag{4.137c}$$

$$\bar{\tilde{\kappa}}^{\alpha}\overleftarrow{\mathcal{D}}_{\alpha} = -\frac{\partial\tilde{\mathcal{H}}_{\mathrm{D}}}{\partial\psi} = \frac{\mathrm{i}M}{2}e_{\alpha}{}^{k}\bar{\tilde{\kappa}}^{\alpha}\gamma_k - (m-M)\bar{\psi}\varepsilon \tag{4.137d}$$

As we can see, due to the appearance of the gauge field $\omega_{ij\nu}$, these field equations are now properly covariant.

To express the coupled set of canonical equations as a field equation for the spinor ψ, Eqs. (4.137a) and (4.137c) must first be resolved for the spinor momentum fields $\tilde{\kappa}^{\alpha}$ and $\bar{\tilde{\kappa}}^{\alpha}$, respectively, giving Eqs. (4.125). Now we insert Eq. (4.125a) into Eq. (4.137b) to obtain:

$$\begin{aligned}
&\frac{\partial}{\partial x^{\alpha}}\left\{\left[-\tfrac{\mathrm{i}}{2}e_{j}{}^{\alpha}\gamma^{j}\psi + \frac{\mathrm{i}}{3M}e_{j}{}^{\alpha}\,\sigma^{ji}\,e_{i}{}^{\beta}\left(\frac{\partial\psi}{\partial x^{\beta}} - \tfrac{\mathrm{i}}{4}\omega_{nm\beta}\,\sigma^{nm}\,\psi\right)\right]\varepsilon\right\}\\
&= \left(\tfrac{\mathrm{i}}{2}M\gamma_k e_{\alpha}{}^{k} - \tfrac{\mathrm{i}}{4}\omega_{kl\alpha}\sigma^{kl}\right)\left[\tfrac{\mathrm{i}}{2}e_{j}{}^{\alpha}\gamma^{j}\psi - \frac{\mathrm{i}}{3M}e_{j}{}^{\alpha}\sigma^{ji}\,e_{i}{}^{\beta}\left(\frac{\partial\psi}{\partial x^{\beta}} - \tfrac{\mathrm{i}}{4}\omega_{nm\beta}\sigma^{nm}\psi\right)\right]\varepsilon\\
&\quad - (m-M)\,\psi\,\varepsilon.
\end{aligned} \tag{4.138}$$

The final version of the generalized Dirac equation (4.138) is worked out in the explicit calculation below. The result is:

$$\begin{aligned}
&\mathrm{i}\gamma^{\beta}\left(\overrightarrow{\mathcal{D}}_{\beta} - S^{\alpha}{}_{\beta\alpha}\right)\psi - m\psi = \\
&\quad -\frac{\mathrm{i}}{3M}\left[\left(2\sigma^{\xi\beta}\,S^{\alpha}{}_{\xi\alpha} + \sigma^{\alpha\xi}\,S^{\beta}{}_{\xi\alpha}\right)\overrightarrow{\mathcal{D}}_{\beta}\psi + \tfrac{\mathrm{i}}{4}\sigma^{\alpha\beta}\,\sigma^{\mu\nu}R_{\mu\nu\alpha\beta}\psi\right],
\end{aligned} \tag{4.139}$$

which shows that the spin coupling of the Dirac particle causes an effective mass correction term proportional to the Riemann tensor $R_{\mu\nu\alpha\beta}$. Neglecting torsion, this equation further simplifies to:

$$\mathrm{i}\,\gamma^{\beta}\overrightarrow{\mathcal{D}}_{\beta}\psi - \left(m + \frac{1}{24M}\sigma^{\alpha\beta}\,\sigma^{\mu\nu}\,\bar{R}_{\mu\nu\alpha\beta}\right)\psi = 0.$$

The contraction of the Riemann-Cartan tensor with the σ-matrices is shown below, in Eq. (5.70), to reduce to twice the Ricci scalar with respect to the Levi-Civita connection, $\bar{R}$, times the unit matrix in the spinor indices:

$$\mathrm{i}\,\gamma^{\beta}\overrightarrow{\mathcal{D}}_{\beta}\psi - \left(m + \frac{\bar{R}}{12M}\right)\psi = 0. \tag{4.140}$$

One thus encounters in the Dirac equation an additional mass term due to a direct interaction of ψ with the gravitational field.

✓ Explict Calculation

Equation (4.138) writes in expanded form:

$$
\begin{aligned}
& e_{\xi}^{\;k}\frac{\partial e_{k}^{\;\xi}}{\partial x^{\alpha}}\left[\tfrac{\mathrm{i}}{2}e_{j}^{\;\alpha}\gamma^{j}\psi - \frac{\mathrm{i}}{3M}e_{j}^{\;\alpha}\sigma^{ji}e_{i}^{\;\beta}\left(\frac{\partial\psi}{\partial x^{\beta}} - \tfrac{\mathrm{i}}{4}\omega_{nm\beta}\sigma^{nm}\psi\right)\right]\varepsilon \\
&- \tfrac{\mathrm{i}}{2}\frac{\partial e_{j}^{\;\alpha}}{\partial x^{\alpha}}\gamma^{j}\psi\,\varepsilon - \tfrac{\mathrm{i}}{2}e_{j}^{\;\alpha}\gamma^{j}\frac{\partial\psi}{\partial x^{\alpha}}\varepsilon \\
&+ \frac{\mathrm{i}}{3M}e_{j}^{\;\alpha}\sigma^{ji}e_{i}^{\;\beta}\left(\cancel{\frac{\partial^{2}\psi}{\partial x^{\beta}\partial x^{\alpha}}} - \tfrac{\mathrm{i}}{4}\frac{\partial\omega_{nm\beta}}{\partial x^{\alpha}}\sigma^{nm}\psi - \tfrac{\mathrm{i}}{4}\omega_{nm\beta}\sigma^{nm}\frac{\partial\psi}{\partial x^{\alpha}}\right)\varepsilon \\
&+ \frac{\mathrm{i}}{3M}\left(\frac{\partial e_{j}^{\;\alpha}}{\partial x^{\alpha}}\sigma^{ji}e_{i}^{\;\beta} + e_{j}^{\;\alpha}\sigma^{ji}\frac{\partial e_{i}^{\;\beta}}{\partial x^{\alpha}}\right)\left(\frac{\partial\psi}{\partial x^{\beta}} - \tfrac{\mathrm{i}}{4}\omega_{nm\beta}\sigma^{nm}\psi\right)\varepsilon \\
&= \tfrac{\mathrm{i}}{2}M\gamma_{k}e_{\alpha}^{\;k}\left[\cancel{\tfrac{\mathrm{i}}{2}e_{j}^{\;\alpha}\gamma^{j}\psi} - \frac{\mathrm{i}}{3M}e_{j}^{\;\alpha}\sigma^{ji}e_{i}^{\;\beta}\left(\frac{\partial\psi}{\partial x^{\beta}} - \tfrac{\mathrm{i}}{4}\omega_{nm\beta}\sigma^{nm}\psi\right)\right]\varepsilon \\
&- \tfrac{\mathrm{i}}{4}\,\omega_{kl\alpha}\sigma^{kl}\left[\tfrac{\mathrm{i}}{2}e_{j}^{\;\alpha}\gamma^{j}\psi - \frac{\mathrm{i}}{3M}e_{j}^{\;\alpha}\sigma^{ji}e_{i}^{\;\beta}\left(\frac{\partial\psi}{\partial x^{\beta}} - \tfrac{\mathrm{i}}{4}\omega_{nm\beta}\sigma^{nm}\psi\right)\right]\varepsilon \\
&- (m - \cancel{M})\,\psi\,\varepsilon,
\end{aligned}
$$

hence

$$
\begin{aligned}
\frac{\mathrm{i}}{3M}\Bigg[& - e_{j}^{\;\alpha}\sigma^{ji}e_{i}^{\;\beta}e_{\xi}^{\;k}\frac{\partial e_{k}^{\;\xi}}{\partial x^{\alpha}}\left(\frac{\partial\psi}{\partial x^{\beta}} - \tfrac{\mathrm{i}}{4}\omega_{nm\beta}\sigma^{nm}\psi\right) \\
& - \tfrac{\mathrm{i}}{4}e_{j}^{\;\alpha}\sigma^{ji}e_{i}^{\;\beta}\left(\frac{\partial\omega_{nm\beta}}{\partial x^{\alpha}}\sigma^{nm}\psi + \omega_{nm\beta}\sigma^{nm}\frac{\partial\psi}{\partial x^{\alpha}}\right) \\
& + \left(\frac{\partial e_{j}^{\;\alpha}}{\partial x^{\alpha}}\sigma^{ji}e_{i}^{\;\beta} + e_{j}^{\;\alpha}\sigma^{ji}\frac{\partial e_{i}^{\;\beta}}{\partial x^{\alpha}}\right)\left(\frac{\partial\psi}{\partial x^{\beta}} - \tfrac{\mathrm{i}}{4}\omega_{nm\beta}\sigma^{nm}\psi\right) \\
& + \tfrac{\mathrm{i}}{4}\,\omega_{kl\alpha}\sigma^{kl}e_{j}^{\;\alpha}\sigma^{ji}e_{i}^{\;\beta}\left(\frac{\partial\psi}{\partial x^{\beta}} - \tfrac{\mathrm{i}}{4}\omega_{nm\beta}\sigma^{nm}\psi\right)\Bigg]
\end{aligned}
$$

$$= \mathrm{i}\, e_j{}^{\alpha} \gamma^j \left(\frac{\partial \psi}{\partial x^{\alpha}} - \tfrac{\mathrm{i}}{4} \omega_{nm\alpha}\, \sigma^{nm}\, \psi \right) - m\, \psi + \tfrac{\mathrm{i}}{2} \gamma^j \left(\frac{\partial e_j{}^{\alpha}}{\partial x^{\alpha}} - e_j{}^{\alpha}\, e_\xi{}^k \frac{\partial e_k{}^{\xi}}{\partial x^{\alpha}} \right) \psi$$
$$- \tfrac{1}{8} \omega_{nm\alpha}\, e_j{}^{\alpha} \left(\gamma^j\, \sigma^{nm} - \sigma^{nm}\, \gamma^j \right) \psi.$$

The last line is converted according to:

$$\tfrac{1}{8}\, \omega_{nm\alpha}\, e_j{}^{\alpha} \left(\gamma^j \sigma^{nm} - \sigma^{nm} \gamma^j \right) = \tfrac{\mathrm{i}}{4}\, \omega_{nm\alpha}\, e_j{}^{\alpha} \left(\eta^{jn} \gamma^m - \eta^{mj} \gamma^n \right) = \tfrac{\mathrm{i}}{2} \gamma^j\, e_k{}^{\alpha}\, \omega^k{}_{j\alpha},$$

which yields

$$\frac{\mathrm{i}}{3M} \Bigg[- \tfrac{\mathrm{i}}{4} e_j{}^{\alpha}\, e_i{}^{\beta} \left(\sigma^{ji} \frac{\partial \omega_{nm\beta}}{\partial x^{\alpha}} + \tfrac{\mathrm{i}}{4}\, \omega_{kl\alpha}\, \sigma^{kl}\, \sigma^{ji}\, \omega_{nm\beta} \right) \sigma^{nm}\, \psi$$
$$+ \tfrac{\mathrm{i}}{4} e_j{}^{\alpha}\, e_i{}^{\beta}\, \omega_{nm\alpha} \left(\sigma^{nm} \sigma^{ji} - \sigma^{ji}\, \sigma^{nm} \right) \frac{\partial \psi}{\partial x^{\beta}}$$
$$+ \left(\frac{\partial e_j{}^{\alpha}}{\partial x^{\alpha}} \sigma^{ji}\, e_i{}^{\beta} + e_j{}^{\alpha}\, \sigma^{ji} \frac{\partial e_i{}^{\beta}}{\partial x^{\alpha}} - e_j{}^{\alpha}\, \sigma^{ji}\, e_i{}^{\beta}\, e_\xi{}^k \frac{\partial e_k{}^{\xi}}{\partial x^{\alpha}} \right) \left(\frac{\partial \psi}{\partial x^{\beta}} - \tfrac{\mathrm{i}}{4} \omega_{nm\beta}\, \sigma^{nm}\, \psi \right) \Bigg]$$
$$= \mathrm{i} \gamma^j\, e_j{}^{\alpha} \left(\frac{\partial \psi}{\partial x^{\alpha}} - \tfrac{\mathrm{i}}{4} \omega_{nm\alpha}\, \sigma^{nm}\, \psi \right) - m\, \psi + \tfrac{\mathrm{i}}{2} \gamma^j \left(\frac{\partial e_j{}^{\alpha}}{\partial x^{\alpha}} - e_k{}^{\alpha}\, \omega^k{}_{j\alpha} - e_j{}^{\alpha} e_\xi{}^k \frac{\partial e_k{}^{\xi}}{\partial x^{\alpha}} \right) \psi.$$

The first two lines are equivalently expressed as:

$$\frac{\mathrm{i}}{3M} \Bigg[- \tfrac{\mathrm{i}}{4} e_j{}^{\alpha}\, e_i{}^{\beta} \left(\sigma^{ji} \frac{\partial \omega_{nm\beta}}{\partial x^{\alpha}} + \tfrac{\mathrm{i}}{4}\, \omega_{kl\alpha}\, \sigma^{ji}\, \sigma^{kl}\, \omega_{nm\beta} \right) \sigma^{nm}\, \psi$$
$$+ \tfrac{\mathrm{i}}{4} e_j{}^{\alpha}\, e_i{}^{\beta}\, \omega_{kl\alpha} \left(\sigma^{kl} \sigma^{ji} - \sigma^{ji}\, \sigma^{kl} \right) \left(\frac{\partial \psi}{\partial x^{\beta}} - \tfrac{\mathrm{i}}{4} \omega_{nm\beta}\, \sigma^{nm}\, \psi \right)$$
$$+ \left(\frac{\partial e_j{}^{\alpha}}{\partial x^{\alpha}} \sigma^{ji}\, e_i{}^{\beta} + e_j{}^{\alpha}\, \sigma^{ji} \frac{\partial e_i{}^{\beta}}{\partial x^{\alpha}} - e_j{}^{\alpha}\, \sigma^{ji}\, e_i{}^{\beta}\, e_\xi{}^k \frac{\partial e_k{}^{\xi}}{\partial x^{\alpha}} \right) \left(\frac{\partial \psi}{\partial x^{\beta}} - \tfrac{\mathrm{i}}{4} \omega_{nm\beta}\, \sigma^{nm}\, \psi \right) \Bigg]$$
$$= \mathrm{i} \gamma^j\, e_j{}^{\alpha} \left(\frac{\partial \psi}{\partial x^{\alpha}} - \tfrac{\mathrm{i}}{4} \omega_{nm\alpha}\, \sigma^{nm}\, \psi \right) - m\, \psi + \tfrac{\mathrm{i}}{2} \gamma^j \left(\frac{\partial e_j{}^{\alpha}}{\partial x^{\alpha}} - e_k{}^{\alpha}\, \omega^k{}_{j\alpha} - e_j{}^{\alpha} e_\xi{}^k \frac{\partial e_k{}^{\xi}}{\partial x^{\alpha}} \right) \psi.$$

The product of two σ matrices in the first and second line is converted according to:

$$\tfrac{\mathrm{i}}{4} \omega_{kl\alpha} \left(\sigma^{kl} \sigma^{ji} - \sigma^{ji}\, \sigma^{kl} \right) = \omega^i{}_{k\alpha}\, \sigma^{kj} - \omega^j{}_{k\alpha}\, \sigma^{ki}$$

which yields

$$\frac{\mathrm{i}}{3M} \Bigg[- \tfrac{\mathrm{i}}{4} e_j{}^{\alpha}\, e_i{}^{\beta} \sigma^{ji} \left(\frac{\partial \omega_{nm\beta}}{\partial x^{\alpha}} + \tfrac{\mathrm{i}}{4}\, \sigma^{kl}\, \omega_{kl\alpha}\, \omega_{nm\beta} \right) \sigma^{nm}\, \psi$$
$$+ e_j{}^{\alpha}\, e_i{}^{\beta} \left(\omega^i{}_{k\alpha}\, \sigma^{kj} - \omega^j{}_{k\alpha}\, \sigma^{ki} \right) \left(\frac{\partial \psi}{\partial x^{\beta}} - \tfrac{\mathrm{i}}{4} \omega_{nm\beta}\, \sigma^{nm}\, \psi \right)$$

$$+\left(\frac{\partial e_j{}^{\alpha}}{\partial x^{\alpha}}\,\sigma^{ji}\,e_i{}^{\beta}+e_j{}^{\alpha}\,\sigma^{ji}\,\frac{\partial e_i{}^{\beta}}{\partial x^{\alpha}}-e_j{}^{\alpha}\,\sigma^{ji}\,e_i{}^{\beta}\,e_{\xi}{}^{k}\,\frac{\partial e_k{}^{\xi}}{\partial x^{\alpha}}\right)\left(\frac{\partial\psi}{\partial x^{\beta}}-\tfrac{\mathrm{i}}{4}\omega_{nm\beta}\,\sigma^{nm}\,\psi\right)\Bigg]$$

$$=\mathrm{i}\gamma^j\,e_j{}^{\alpha}\left(\frac{\partial\psi}{\partial x^{\alpha}}-\tfrac{\mathrm{i}}{4}\omega_{nm\alpha}\,\sigma^{nm}\,\psi\right)-m\,\psi+\tfrac{\mathrm{i}}{2}\gamma^j\left(\frac{\partial e_j{}^{\alpha}}{\partial x^{\alpha}}-e_k{}^{\alpha}\,\omega^k{}_{j\alpha}-e_j{}^{\alpha}e_{\xi}{}^{k}\,\frac{\partial e_k{}^{\xi}}{\partial x^{\alpha}}\right)\psi$$

By virtue of the identity

$$e_j{}^{\alpha}\,e_i{}^{\beta}\,\sigma^{ji}\,\tfrac{\mathrm{i}}{4}\,\omega_{kl\alpha}\,\sigma^{kl}\,\sigma^{nm}\,\omega_{nm\beta}=e_j{}^{\alpha}\,e_i{}^{\beta}\,\sigma^{ji}\,\sigma^{nm}\,\omega_{nk\alpha}\,\omega^k{}_{m\beta}$$

the generalized Dirac equation writes

$$\frac{\mathrm{i}}{3M}\sigma^{ji}\Bigg[\left(\frac{\partial e_j{}^{\alpha}}{\partial x^{\alpha}}\,e_i{}^{\beta}-e_k{}^{\alpha}\,\omega^k{}_{j\alpha}\,e_i{}^{\beta}+e_j{}^{\alpha}\,\frac{\partial e_i{}^{\beta}}{\partial x^{\alpha}}-e_j{}^{\alpha}\,e_k{}^{\beta}\,\omega^k{}_{i\alpha}-e_j{}^{\alpha}\,e_i{}^{\beta}\,e_{\xi}{}^{k}\,\frac{\partial e_k{}^{\xi}}{\partial x^{\alpha}}\right)$$
$$\times\left(\frac{\partial\psi}{\partial x^{\beta}}-\tfrac{\mathrm{i}}{4}\omega_{nm\beta}\,\sigma^{nm}\psi\right)-\tfrac{\mathrm{i}}{4}e_j{}^{\alpha}e_i{}^{\beta}\,\sigma^{nm}\left(\frac{\partial\omega_{nm\beta}}{\partial x^{\alpha}}+\omega_{nk\alpha}\,\omega^k{}_{m\beta}\right)\psi\Bigg]$$
$$=\mathrm{i}\gamma^j\,e_j{}^{\beta}\left(\frac{\partial\psi}{\partial x^{\beta}}-\frac{\mathrm{i}}{4}\omega_{nm\beta}\,\sigma^{nm}\,\psi\right)-m\,\psi$$
$$+\frac{\mathrm{i}}{2}\gamma^j\left(\frac{\partial e_j{}^{\alpha}}{\partial x^{\alpha}}-e_k{}^{\alpha}\,\omega^k{}_{j\alpha}-e_j{}^{\alpha}\,e_{\xi}{}^{k}\,\frac{\partial e_k{}^{\xi}}{\partial x^{\alpha}}\right)\psi.$$

The last term can be re-written using the identity (4.56) as

$$\frac{\partial e_j{}^{\alpha}}{\partial x^{\alpha}}-e_k{}^{\alpha}\omega^k{}_{j\alpha}-e_j{}^{\alpha}e_{\xi}{}^{k}\frac{\partial e_k{}^{\xi}}{\partial x^{\alpha}}=-e_j{}^{\xi}\gamma^{\alpha}{}_{\xi\alpha}-e_j{}^{\alpha}e_{\xi}{}^{k}\left(e_m{}^{\xi}\omega^m{}_{k\alpha}-\gamma^{\xi}{}_{\beta\alpha}e_k{}^{\beta}\right)$$
$$=-e_j{}^{\xi}\,\gamma^{\alpha}{}_{\xi\alpha}-\cancel{e_j{}^{\alpha}\omega^m{}_{m\alpha}}+e_j{}^{\xi}\,\gamma^{\alpha}{}_{\alpha\xi}$$
$$=2e_j{}^{\xi}\,S^{\alpha}{}_{\alpha\xi}.$$

The spin connection term vanishes due to the skew-symmetry in its first index pair. Similarly, by virtue of the skew-symmetry of σ^{ji}, we get the tensor expression

$$\sigma^{ji}\left(\frac{\partial e_j{}^{\alpha}}{\partial x^{\alpha}}\,e_i{}^{\beta}-e_k{}^{\alpha}\,\omega^k{}_{j\alpha}\,e_i{}^{\beta}+e_j{}^{\alpha}\,\frac{\partial e_i{}^{\beta}}{\partial x^{\alpha}}-e_j{}^{\alpha}\,e_k{}^{\beta}\,\omega^k{}_{i\alpha}-e_j{}^{\alpha}\,e_i{}^{\beta}\,e_{\xi}{}^{n}\,\frac{\partial e_n{}^{\xi}}{\partial x^{\alpha}}\right)$$
$$=\sigma^{ji}\left(-\gamma^{\alpha}{}_{\xi\alpha}\,e_j{}^{\xi}\,e_i{}^{\beta}-\gamma^{\beta}{}_{\xi\alpha}\,e_i{}^{\xi}\,e_j{}^{\alpha}+\gamma^{\alpha}{}_{\alpha\xi}\,e_i{}^{\beta}\,e_j{}^{\alpha}\right)$$
$$=\sigma^{ji}\left(2e_i{}^{\beta}\,e_j{}^{\xi}\,S^{\alpha}{}_{\alpha\xi}-e_i{}^{\xi}\,e_j{}^{\alpha}\,S^{\beta}{}_{\xi\alpha}\right).$$

It is now straightforward to simplify this equation to

$$
\begin{aligned}
&\mathrm{i}\gamma^{\beta}\overrightarrow{\mathcal{D}}_{\beta}\psi - m\psi = \\
&\quad -\frac{\mathrm{i}}{3M}\sigma^{ji}\left[\left(2e_i^{\ \beta}e_j^{\ \xi}S^{\alpha}{}_{\xi\alpha} + e_i^{\ \xi}e_j^{\ \alpha}S^{\beta}{}_{\xi\alpha}\right)\overrightarrow{\mathcal{D}}_{\beta}\psi\right. \\
&\qquad \left. + \tfrac{\mathrm{i}}{4}e_j^{\ \alpha}e_i^{\ \beta}\sigma^{nm}\left(\frac{\partial\omega_{nm\beta}}{\partial x^{\alpha}} + \omega_{nk\alpha}\,\omega^{k}{}_{m\beta}\right)\psi\right] + \mathrm{i}\gamma^{\xi}S^{\alpha}{}_{\xi\alpha}\psi
\end{aligned}
$$

This equation may be further simplified by noting that the term involving the derivative of the spin connection is just the Riemann tensor due to the antisymmetry of σ^{ij}. The result is (4.139).

4.5.4 Consistency Equation Revisited

We re-arrange first the terms of Eq. (4.81)

$$
\begin{aligned}
0 = &\ \tfrac{\mathrm{i}}{4}\frac{\partial}{\partial x^{\beta}}\left(\bar{\psi}\,\sigma_i^{\ j}\tilde{\kappa}^{\beta} - \bar{\tilde{\kappa}}^{\beta}\sigma_i^{\ j}\,\psi\right) + \tfrac{\mathrm{i}}{4}\left(\bar{\psi}\,\sigma_i^{\ n}\tilde{\kappa}^{\beta} - \bar{\tilde{\kappa}}^{\beta}\sigma_i^{\ n}\,\psi\right)\omega^{j}{}_{n\beta} \\
&- \tfrac{\mathrm{i}}{4}\left(\bar{\psi}\,\sigma_n^{\ j}\tilde{\kappa}^{\beta} - \bar{\tilde{\kappa}}^{\beta}\sigma_n^{\ j}\,\psi\right)\omega^{n}{}_{i\beta} + \left(\frac{\partial\tilde{k}_i^{\ [\alpha\beta]}}{\partial x^{\beta}} - \tilde{k}_n^{\ [\alpha\beta]}\omega^{n}{}_{i\beta}\right)e_{\alpha}^{\ j} \\
&+ \tilde{q}_i^{\ n[\alpha\beta]}\left(\frac{\partial\omega^{j}{}_{n\alpha}}{\partial x^{\beta}} + \omega^{j}{}_{m\beta}\omega^{m}{}_{n\alpha}\right) - \left(\frac{\partial\omega^{n}{}_{i\alpha}}{\partial x^{\beta}} - \omega^{n}{}_{m\alpha}\omega^{m}{}_{i\beta}\right)\tilde{q}_n^{\ j[\alpha\beta]} \\
&+ \tilde{k}_i^{\ [\alpha\beta]}\left(\frac{\partial e_{\alpha}^{\ j}}{\partial x^{\beta}} + \omega^{j}{}_{n\beta}\,e_{\alpha}^{\ n}\right).
\end{aligned}
$$

Inserting the abbreviations (4.61a) and (4.62b) for the last three terms gives then:

$$
\begin{aligned}
0 = &\left(\frac{\partial\tilde{k}_i^{\ [\mu\alpha]}}{\partial x^{\alpha}} - \tilde{k}_n^{\ [\mu\alpha]}\omega^{n}{}_{i\alpha}\right)e_{\mu}^{\ j} \\
&+ \tfrac{\mathrm{i}}{4}\frac{\partial}{\partial x^{\beta}}\left(\bar{\psi}\,\sigma_i^{\ j}\tilde{\kappa}^{\beta} - \bar{\tilde{\kappa}}^{\beta}\sigma_i^{\ j}\,\psi\right) + \tfrac{\mathrm{i}}{4}\left(\bar{\psi}\,\sigma_i^{\ n}\tilde{\kappa}^{\beta} - \bar{\tilde{\kappa}}^{\beta}\sigma_i^{\ n}\,\psi\right)\omega^{j}{}_{n\beta} \\
&- \tfrac{\mathrm{i}}{4}\left(\bar{\psi}\sigma_n^{\ j}\tilde{\kappa}^{\beta} - \bar{\tilde{\kappa}}^{\beta}\sigma_n^{\ j}\psi\right)\omega^{n}{}_{i\beta} - \tfrac{1}{2}\tilde{q}_i^{\ n\alpha\beta}R^{j}{}_{n\alpha\beta} + \tfrac{1}{2}R^{n}{}_{i\alpha\beta}\tilde{q}_n^{\ j\alpha\beta} + \tilde{k}_i^{\ [\alpha\beta]}S^{j}{}_{\alpha\beta}.
\end{aligned}
$$

The first term can now be replaced by the right-hand-side of the field Eq. (4.72), which, in turn, defines the metric energy-momentum tensors of the uncoupled systems described by $\tilde{\mathcal{H}}_0$ and $\tilde{\mathcal{H}}_{\mathrm{Gr}}$ according to Eqs. (4.120), (4.123), (4.126), and (4.129):

$$
\begin{aligned}
0 = &\tfrac{\mathrm{i}}{4} \frac{\partial}{\partial x^\beta} \left(\bar{\psi}\, \sigma_i{}^j \tilde{\kappa}^\beta - \bar{\tilde{\kappa}}^\beta \sigma_i{}^j\, \psi \right) + \tfrac{\mathrm{i}}{4} \left(\bar{\psi}\, \sigma_i{}^n \tilde{\kappa}^\beta - \bar{\tilde{\kappa}}^\beta \sigma_i{}^n\, \psi \right) \omega^j{}_{n\beta} \\
&- \tfrac{\mathrm{i}}{4} \left(\bar{\psi}\, \sigma_n{}^j \tilde{\kappa}^\beta - \bar{\tilde{\kappa}}^\beta \sigma_n{}^j\, \psi \right) \omega^n{}_{i\beta} - \tilde{T}_{0,i}{}^j - \tilde{T}_{\mathrm{Gr},i}{}^j \\
&+ \tfrac{1}{2} \left(R^n{}_{i\alpha\beta}\, \tilde{q}_n{}^{j\alpha\beta} - \tilde{q}_i{}^{n\alpha\beta}\, R^j{}_{n\alpha\beta} \right) + \tilde{k}_i{}^{\alpha\beta}\, S^j{}_{\alpha\beta}.
\end{aligned} \tag{4.141}
$$

The terms related to $\tilde{q}$ give due to the correlation to the Riemann tensor, Eq. (4.110):

$$
\begin{aligned}
&\tfrac{1}{2} \left(R^n{}_{i\alpha\beta}\, \tilde{q}_n{}^{j\alpha\beta} - \tilde{q}_i{}^{n\alpha\beta}\, R^j{}_{n\alpha\beta} \right) \\
= \;\; &\tfrac{1}{2} g_1 \varepsilon \left(\cancel{R^n{}_{i\alpha\beta}\, R_n{}^{j\alpha\beta}} - \cancel{R_i{}^{n\alpha\beta}\, R^j{}_{n\alpha\beta}} - R^n{}_{i\alpha\beta}\, \bar{R}_n{}^{j\alpha\beta} + R^j{}_{n\alpha\beta}\, \bar{R}_i{}^{n\alpha\beta} \right) \\
= &-\tfrac{1}{2} g_1 \varepsilon \left(R_{ni}{}^{\alpha\beta}\, \eta^{kj} - R^{jk\alpha\beta}\, \eta_{ni} \right) \bar{R}^n{}_{k\alpha\beta} \\
= &-\tfrac{1}{2} g_1 g_2\, \varepsilon \left(R_{ni}{}^{\alpha\beta}\, \eta^{kj} - R^{jk\alpha\beta}\, \eta_{ni} \right) \left(e^n{}_\alpha\, g_{\beta\lambda} - e^n{}_\beta\, g_{\alpha\lambda} \right) e_k{}^\lambda \\
= &-\tfrac{1}{2} g_1 g_2\, \varepsilon \left(R_{ni}{}^\alpha{}_\lambda \eta^{kj} - R_{ni\lambda}{}^\alpha \eta^{kj} - R^{jk\alpha}{}_\lambda \eta_{ni} + R^{jk}{}_\lambda{}^\alpha \eta_{ni} \right) e^n{}_\alpha\, e_k{}^\lambda \\
= &-g_1 g_2\, \varepsilon \left(R_{ni}{}^\alpha{}_\lambda \eta^{kj} - R^{jk\alpha}{}_\lambda \eta_{ni} \right) e^n{}_\alpha\, e_k{}^\lambda \\
= &-g_1 g_2\, \varepsilon \left(R_{ni}{}^\alpha{}_\lambda e^n{}_\alpha\, e^j{}_\beta\, g^{\beta\lambda} - R^{jn\alpha}{}_\lambda e_n{}^\lambda e_i{}^\beta\, g_{\beta\alpha} \right) \\
= &-g_1 g_2\, \varepsilon \left(R_{ni}{}^{\alpha\beta} e^n{}_\alpha\, e^j{}_\beta - R^{jn}{}_{\beta\alpha}\, e_n{}^\alpha e_i{}^\beta \right) \\
= &-g_1 g_2\, \varepsilon \left(R_i{}^j - R^j{}_i \right).
\end{aligned} \tag{4.142}
$$

In the last step, the Riemannn tensor was replaced by its contraction, namely the Ricci tensor.

In analogy to the energy-momentum tensor of matter, it is physically reasonable to define the *Spin-momentum tensor* as the derivative of the "gauged" matter Lagrangian with respect to the spin connection. From the coupling of the spinor fields in the gauge Hamiltonian (4.50) to the spin connection, we have

$$
\tilde{\Sigma}_i{}^{j\beta} := \frac{\partial \tilde{\mathcal{L}}_{\mathrm{matter}}}{\partial \omega^i{}_{j\beta}} = - \frac{\partial \tilde{\mathcal{H}}_{\mathrm{Gau}_2}}{\partial \omega^i{}_{j\beta}} \bigg|_{\psi,\bar{\psi}} = -\tfrac{\mathrm{i}}{4} \frac{\partial}{\partial \omega^i{}_{j\beta}} \left(\bar{\tilde{\kappa}}^\alpha\, \omega^n{}_{m\alpha}\, \sigma_n{}^m\, \psi - \bar{\psi}\, \omega^n{}_{m\alpha}\, \sigma_n{}^m\, \tilde{\kappa}^\alpha \right).
$$

This yields

$$
\tilde{\Sigma}^{ij\beta} = \tfrac{\mathrm{i}}{4} \left(\bar{\psi}\, \sigma^{ij} \tilde{\kappa}^\beta - \bar{\tilde{\kappa}}^\beta \sigma^{ij}\, \psi \right), \tag{4.143}
$$

which is skew-symmetric in i and j. Raising the index i in Eq. (4.141), and inserting (4.109) gives then:

$$\boxed{\frac{\partial \tilde{\Sigma}^{ij\beta}}{\partial x^{\beta}} + \tilde{\Sigma}^{in\beta} \omega^{j}{}_{n\beta} - \tilde{\Sigma}^{jn\beta} \omega^{i}{}_{n\beta} - 2g_1 g_2 \varepsilon R^{[ij]} + g_3 \varepsilon S^{i\alpha\beta} S^{j}{}_{\alpha\beta} = \tilde{T}^{ij}_{0} + \tilde{T}^{ij}_{\mathrm{Gr}}.} \tag{4.144}$$

This equation relates the spin-momentum tensor $\tilde{\Sigma}$ of the spinor fields to the metric energy-momentum tensor $\tilde{T}^{ij}_{0}$ of all source fields with the energy-momentum tensor of $\tilde{\mathcal{H}}_{\mathrm{Gr}}$. Equation (4.144) can now be split into skew-symmetric and symmetric portions in i and j. The skew-symmetric portion follows as:

$$\frac{\partial \tilde{\Sigma}^{ij\beta}}{\partial x^{\beta}} + \tilde{\Sigma}^{in\beta} \omega^{j}{}_{n\beta} - \tilde{\Sigma}^{jn\beta} \omega^{i}{}_{n\beta} - \tilde{T}^{[ij]}_{0} = \tilde{T}^{[ij]}_{\mathrm{Gr}} + 2g_1 g_2\, \varepsilon\, R^{[ij]}. \tag{4.145}$$

For the particular gravitational Hamiltonian, defined in Eq. (4.107), the Riemann tensor terms on the right-hand side of Eq. (4.145) cancel identically by virtue of the identity (4.131). As the energy-momentum tensors of the Klein-Gordon and Maxwell-Proca systems are symmetric, see Eqs. (4.120) and (4.123), only the spinor fields enter into the left-hand side. These terms cancel as derived in the following explicit calculation.

✓ Explict Calculation

In order to express the sum of the spin-momentum tensor terms in Eq. (4.145) by the spinor fields and their canonical momenta, we deduce from Eqs. (4.137b) and (4.137d):

$$\begin{aligned}\bar{\psi}\, \sigma_{i}{}^{j} \frac{\partial \tilde{\kappa}^{\beta}}{\partial x^{\beta}} - \frac{\partial \tilde{\bar{\kappa}}^{\beta}}{\partial x^{\beta}} \sigma_{i}{}^{j}\, \psi = &\left(-\frac{\mathrm{i}M}{2} \bar{\psi}\, \sigma_{i}{}^{j} \gamma_n\, e^{n}{}_{\beta} + \tfrac{\mathrm{i}}{4} \bar{\psi}\, \sigma_{i}{}^{j}\, \sigma_{n}{}^{m}\, \omega^{n}{}_{m\beta} \right) \tilde{\kappa}^{\beta} \\ &- \tilde{\bar{\kappa}}^{\beta} \left(\frac{\mathrm{i}M}{2} e^{n}{}_{\beta}\, \gamma_n\, \sigma_{i}{}^{j}\, \psi - \tfrac{\mathrm{i}}{4} \omega^{n}{}_{m\beta}\, \sigma_{n}{}^{m}\, \sigma_{i}{}^{j}\, \psi \right),\end{aligned}$$

and from Eqs. (4.137a) and (4.137c):

$$\begin{aligned}&\frac{\partial \bar{\psi}}{\partial x^{\beta}} \sigma_{i}{}^{j} \tilde{\kappa}^{\beta} - \tilde{\bar{\kappa}}^{\beta} \sigma_{i}{}^{j} \frac{\partial \psi}{\partial x^{\beta}} \\ &\quad = \left(\frac{\mathrm{i}M}{2} \bar{\psi} \gamma_n\, e^{n}{}_{\beta} - \tilde{\bar{\kappa}}^{\alpha}\, e^{n}{}_{\alpha} \frac{3\mathrm{i}M \tau_{nm}}{\varepsilon} e^{m}{}_{\beta} - \tfrac{\mathrm{i}}{4} \bar{\psi}\, \omega_{nm\beta}\, \sigma^{nm} \right) \sigma_{i}{}^{j} \tilde{\kappa}^{\beta} \\ &\quad\quad + \tilde{\bar{\kappa}}^{\beta} \sigma_{i}{}^{j} \left(\frac{\mathrm{i}M}{2} \gamma_n\, e^{n}{}_{\beta}\, \psi + e^{n}{}_{\beta} \frac{3\mathrm{i}M \tau_{nm}}{\varepsilon} e^{m}{}_{\alpha}\, \tilde{\kappa}^{\alpha} - \frac{\mathrm{i}}{4} \omega_{nm\beta}\, \sigma^{nm}\, \psi \right),\end{aligned}$$

thus

$$\frac{\partial \tilde{\Sigma}_{i}{}^{j\beta}}{\partial x^{\beta}} + \tilde{\Sigma}_{i}{}^{n\beta} \omega^{j}{}_{n\beta} - \tilde{\Sigma}_{n}{}^{j\beta} \omega^{n}{}_{i\beta}$$

$$
\begin{aligned}
&= \frac{\mathrm{i}}{4}\bar{\psi}\left(\sigma_i{}^{n}\,\omega^{j}{}_{n\beta} - \omega^{n}{}_{i\beta}\,\sigma_n{}^{j} - \frac{\mathrm{i}M}{2}\sigma_i{}^{j}\gamma_n\,e^{n}{}_{\beta} + \tfrac{\mathrm{i}}{4}\omega_{nm\beta}\,\sigma_i{}^{j}\,\sigma^{nm}\right)\tilde{\kappa}^{\beta} \\
&\quad - \frac{\mathrm{i}}{4}\bar{\tilde{\kappa}}^{\beta}\left(\frac{\mathrm{i}M}{2}e^{n}{}_{\beta}\,\gamma_n\,\sigma_i{}^{j} - \tfrac{\mathrm{i}}{4}\omega_{nm\beta}\,\sigma^{nm}\,\sigma_i{}^{j} + \sigma_i{}^{n}\,\omega^{j}{}_{n\beta} - \omega^{n}{}_{i\beta}\,\sigma_n{}^{j}\right)\psi \\
&\quad + \frac{\mathrm{i}}{4}\left(\frac{\mathrm{i}M}{2}\bar{\psi}\,\gamma_n\,e^{n}{}_{\beta} - \bar{\tilde{\kappa}}^{\alpha}\,e^{n}{}_{\alpha}\,\frac{3\mathrm{i}M\tau_{nm}}{\varepsilon}\,e^{m}{}_{\beta} - \tfrac{\mathrm{i}}{4}\,\bar{\psi}\,\omega_{nm\beta}\,\sigma^{nm}\right)\sigma_i{}^{j}\,\tilde{\kappa}^{\beta} \\
&\quad - \frac{\mathrm{i}}{4}\bar{\tilde{\kappa}}^{\beta}\sigma_i{}^{j}\left(-\frac{\mathrm{i}M}{2}\gamma_n\,e^{n}{}_{\beta}\,\psi - e^{n}{}_{\beta}\,\frac{3\mathrm{i}M\tau_{nm}}{\varepsilon}\,e^{m}{}_{\alpha}\,\tilde{\kappa}^{\alpha} + \tfrac{\mathrm{i}}{4}\,\omega_{nm\beta}\,\sigma^{nm}\,\psi\right),
\end{aligned}
$$

hence

$$
\begin{aligned}
&\frac{\partial\tilde{\Sigma}_i{}^{j\beta}}{\partial x^{\beta}} + \tilde{\Sigma}_i{}^{n\beta}\omega^{j}{}_{n\beta} - \tilde{\Sigma}_n{}^{j\beta}\omega^{n}{}_{i\beta} \\
&= \frac{\mathrm{i}}{4}\bar{\psi}\left[\sigma_i{}^{n}\omega^{j}{}_{n\beta} - \omega^{n}{}_{i\beta}\sigma_n{}^{j} + \tfrac{\mathrm{i}}{4}\omega^{n}{}_{m\beta}\left(\sigma_i{}^{j}\sigma_n{}^{m} - \sigma_n{}^{m}\sigma_i{}^{j}\right) - \frac{\mathrm{i}M}{2}e^{n}{}_{\beta}\left(\sigma_i{}^{j}\gamma_n - \gamma_n\sigma_i{}^{j}\right)\right]\tilde{\kappa}^{\beta} \\
&\quad - \frac{\mathrm{i}}{4}\bar{\tilde{\kappa}}^{\beta}\left[\sigma_i{}^{n}\omega^{j}{}_{n\beta} - \omega^{n}{}_{i\beta}\sigma_n{}^{j} + \tfrac{\mathrm{i}}{4}\omega^{n}{}_{m\beta}\left(\sigma_i{}^{j}\,\sigma_n{}^{m} - \sigma_n{}^{m}\,\sigma_i{}^{j}\right) - \frac{\mathrm{i}M}{2}e^{n}{}_{\beta}\left(\sigma_i{}^{j}\gamma_n - \gamma_n\sigma_i{}^{j}\right)\right]\psi \\
&\quad - \frac{3M}{4\varepsilon}\bar{\tilde{\kappa}}^{\alpha}\,e^{n}{}_{\alpha}\left(\sigma_i{}^{j}\,\tau_{nm} - \tau_{nm}\,\sigma_i{}^{j}\right)e^{m}{}_{\beta}\,\tilde{\kappa}^{\beta}.
\end{aligned}
$$

Contracting the commutator of the σ-matrices with $\frac{\mathrm{i}}{4}\omega^{n}{}_{m\beta}$, c.f. (4.85), gives:

$$
\begin{aligned}
\tfrac{\mathrm{i}}{4}\left(\sigma_i{}^{j}\sigma_n{}^{m} - \sigma_n{}^{m}\sigma_i{}^{j}\right)\omega^{n}{}_{m\beta} &= \tfrac{1}{2}\left(\sigma^{jm}\omega_{im\beta} - \sigma_i{}^{m}\omega^{j}{}_{m\beta} - \sigma_{ni}\,\omega^{nj}{}_{\beta} + \sigma_n{}^{j}\omega^{n}{}_{i\beta}\right) \\
&= \omega^{n}{}_{i\beta}\,\sigma_n{}^{j} - \sigma_i{}^{n}\,\omega^{j}{}_{n\beta}.
\end{aligned}
$$

The spin tensor terms thus simplify to:

$$
\begin{aligned}
&\frac{\partial\tilde{\Sigma}_i{}^{j\beta}}{\partial x^{\beta}} + \tilde{\Sigma}_i{}^{n\beta}\omega^{j}{}_{n\beta} - \tilde{\Sigma}_n{}^{j\beta}\omega^{n}{}_{i\beta} \\
&= \frac{M}{8}\left[\bar{\psi}\,e^{n}{}_{\beta}\left(\sigma_i{}^{j}\gamma_n - \gamma_n\sigma_i{}^{j}\right)\tilde{\kappa}^{\beta} - \bar{\tilde{\kappa}}^{\beta}\,e^{n}{}_{\beta}\left(\sigma_i{}^{j}\gamma_n - \gamma_n\sigma_i{}^{j}\right)\psi\right] \\
&\quad - \frac{3M}{4\varepsilon}\bar{\tilde{\kappa}}^{\alpha}\,e^{n}{}_{\alpha}\left(\sigma_i{}^{j}\,\tau_{nm} - \tau_{nm}\,\sigma_i{}^{j}\right)e^{m}{}_{\beta}\,\tilde{\kappa}^{\beta}.
\end{aligned}
$$

Remarkably, all couplings of the spinor fields to the gauge field, i.e. the spin connection $\omega^{j}{}_{n\beta}$, cancel. This equation can be further simplifies by means of the Dirac algebra identities,

$$
\begin{aligned}
\sigma_i{}^{j}\,\gamma_n - \gamma_n\,\sigma_i{}^{j} &\equiv 2\mathrm{i}\left(\delta^{j}_{n}\,\gamma_i - \gamma^{j}\,\eta_{ni}\right) \\
\sigma_i{}^{j}\,\tau_{nm} - \tau_{nm}\,\sigma_i{}^{j} &\equiv 2\mathrm{i}\left(\eta_{ni}\,\tau_m{}^{j} - \delta^{j}_{n}\,\tau_{mi} + \eta_{mi}\,\tau^{j}{}_{n} - \delta^{j}_{m}\,\tau_{in}\right) \\
&\equiv -2\mathrm{i}\left(\eta_{ni}\,\tau^{j}{}_{m} - \delta^{j}_{n}\,\tau_{im} + \eta_{mi}\,\tau_n{}^{j} - \delta^{j}_{m}\,\tau_{ni}\right)
\end{aligned}
$$

$$\tau_n{}^j + \tau^j{}_n \equiv \frac{4\mathrm{i}}{3}\delta^j_n \mathbb{1},$$

to yield

$$\begin{aligned}
&\frac{\partial \tilde{\Sigma}_i{}^{j\beta}}{\partial x^\beta} + \tilde{\Sigma}_i{}^{n\beta}\omega^j{}_{n\beta} - \tilde{\Sigma}_n{}^{j\beta}\omega^n{}_{i\beta} \\
&= \frac{\mathrm{i}M}{4}\left[\bar{\psi}\, e^n{}_\beta \left(\delta^j_n\,\gamma_i - \gamma^j\,\eta_{ni}\right)\tilde{\kappa}^\beta - \bar{\tilde{\kappa}}^\beta\, e^n{}_\beta \left(\delta^j_n\,\gamma_i - \gamma^j\,\eta_{ni}\right)\psi\right] \\
&+ \frac{3\mathrm{i}M}{2\varepsilon}\bar{\tilde{\kappa}}^\alpha\, e^n{}_\alpha \left(\eta_{ni}\,\tau^j{}_m - \delta^j_n\,\tau_{im} + \eta_{mi}\,\tau_n{}^j - \delta^j_m\,\tau_{ni}\right) e^m{}_\beta\, \tilde{\kappa}^\beta .
\end{aligned}$$

Hence

$$\begin{aligned}
&\frac{\partial \tilde{\Sigma}_i{}^{j\beta}}{\partial x^\beta} + \tilde{\Sigma}_i{}^{n\beta}\omega^j{}_{n\beta} - \tilde{\Sigma}_n{}^{j\beta}\omega^n{}_{i\beta} \\
&= \frac{\mathrm{i}M}{4}\left(\bar{\psi}\,\gamma_i\,\tilde{\kappa}^j - \bar{\tilde{\kappa}}^j\,\gamma_i\,\psi - \bar{\psi}\,\gamma^j\,\tilde{\kappa}_i + \bar{\tilde{\kappa}}_i\,\gamma^j\,\psi\right) \\
&- \frac{3\mathrm{i}M}{2\varepsilon}\left(\bar{\tilde{\kappa}}^j\,\tau_{in}\,\tilde{\kappa}^n + \bar{\tilde{\kappa}}^n\,\tau_{ni}\,\tilde{\kappa}^j - \bar{\tilde{\kappa}}^n\,\tau_n{}^j\,\tilde{\kappa}_i - \bar{\tilde{\kappa}}_i\,\tau^j{}_n\,\tilde{\kappa}^n\right) \\
&= \frac{1}{2}\left(\frac{\partial \tilde{\mathcal{H}}_\mathrm{D}}{\partial e^I{}_\alpha} e^J{}_\alpha - \frac{\partial \tilde{\mathcal{H}}_\mathrm{D}}{\partial e^n{}_\alpha} e^m{}_\alpha \eta^{nj}\eta_{mi}\right) = \tfrac{1}{2}\left(\tilde{T}_{\mathrm{D}i}{}^j - \tilde{T}_\mathrm{D}{}^j{}_i\right),
\end{aligned}$$

where in the last line the fields are expressed in terms of the Hamiltonian form of the energy-momentum tensor of the Dirac system from Eq. (4.126). Raising the index i finally yields:

$$\begin{aligned}
&\frac{\partial \tilde{\Sigma}^{ij\beta}}{\partial x^\beta} + \tilde{\Sigma}^{in\beta}\omega^j{}_{n\beta} - \tilde{\Sigma}^{jn\beta}\omega^i{}_{n\beta} \\
&= \frac{1}{2}\frac{\partial \tilde{\mathcal{H}}_\mathrm{D}}{\partial e^n{}_\alpha}\left(e^J{}_\alpha \eta^{ni} - e^I{}_\alpha \eta^{nj}\right) = \tfrac{1}{2}\left(\tilde{T}^{ij}_\mathrm{D} - \tilde{T}^{ji}_\mathrm{D}\right) \\
&= \tilde{T}^{[ij]}_\mathrm{D}.
\end{aligned}$$

This gives is an identity.

Hence, we are left with the identity

$$\frac{\partial \tilde{\Sigma}^{ij\beta}}{\partial x^\beta} + \tilde{\Sigma}^{in\beta}\omega^j{}_{n\beta} - \tilde{\Sigma}^{jn\beta}\omega^i{}_{n\beta} \equiv \varepsilon\left(\Sigma^{ij\beta}{}_{;\beta} - 2S^\mu{}_{\beta\mu}\,\Sigma^{ij\beta}\right) = \tilde{T}^{[ij]}_\mathrm{D}. \tag{4.146}$$

such that (4.145) is identically satisfied for the given Hamiltonians and does not provide any information on the field dynamics.

On the other hand, the symmetric portion of Eq. (4.144) written in metric coordinates gives:

$$T_0^{(\mu\nu)} = g_3\, S^{\mu\alpha\beta}\, S^{\nu}{}_{\alpha\beta} - T_{\mathrm{Gr}}^{(\mu\nu)}. \tag{4.147}$$

The torsion term $\propto g_3$ obviously contributes to the total energy-momentum density of spacetime. With (4.132) for $T_{\mathrm{Gr}}^{(\mu\nu)}$, and setting $R := R_\alpha{}^\alpha$ this gives in metric coordinates:

$$\begin{aligned} g_1 &\left(R_{\alpha\beta\gamma}{}^{\nu}\, R^{\alpha\beta\gamma\mu} - \tfrac{1}{4} g^{\mu\nu} R_{\alpha\beta\gamma\delta}\, R^{\alpha\beta\gamma\delta}\right) - 2g_1 g_2 \left(R^{(\mu\nu)} - \tfrac{1}{2} g^{\mu\nu}\, R + 3g_2\, g^{\mu\nu}\right) \\ &- g_3 \left(S_{\alpha\beta}{}^{\mu}\, S^{\alpha\beta\nu} - \tfrac{1}{2} g^{\mu\nu}\, S_{\alpha\beta\gamma}\, S^{\alpha\beta\gamma}\right) = T_0^{(\mu\nu)}. \end{aligned} \tag{4.148}$$

This is the so called *CCGG field equation* extending the Einstein field equation by the quadratic Riemann-Cartan tensor concomitant (quadratic trace-free Kretschmann) and a quadratic torsion term.

In the weak-field limit, the term quadratic in $R^{\alpha\beta\gamma\mu}$ is small as compared to the linear term and can thus be neglected. If also the torsion is discarded, Eq. (4.148) reduces to the classical Einstein equation with the cosmological constant term, $\lambda := 3g_2$, and with $2g_1g_2 = (8\pi G)^{-1} \equiv M_p^2$ where G is the Newton gravitational constant and M_p the reduced Planck mass:

$$R^{(\nu\mu)} - g^{\nu\mu}\left(\tfrac{1}{2}R - \lambda\right) = -8\pi G\, T_0^{(\mu\nu)}. \tag{4.149}$$

This identification gives then the final form of the CCGG equation:

$$\begin{aligned} g_1 &\left(R_{\alpha\beta\gamma}{}^{\nu}\, R^{\alpha\beta\gamma\mu} - \tfrac{1}{4} g^{\mu\nu} R_{\alpha\beta\gamma\delta}\, R^{\alpha\beta\gamma\delta}\right) - \frac{1}{8\pi G}\left(R^{(\mu\nu)} - \tfrac{1}{2} g^{\mu\nu}\, R + \lambda g^{\mu\nu}\right) \\ &- g_3 \left(S_{\alpha\beta}{}^{\mu}\, S^{\alpha\beta\nu} - \tfrac{1}{2} g^{\mu\nu}\, S_{\alpha\beta\gamma}\, S^{\alpha\beta\gamma}\right) = T_0^{(\mu\nu)}. \end{aligned} \tag{4.150}$$

The CCGG equation can be interpreted as a local energy cancellation if the right-hand side of (4.147) is interpreted as the total (symmetric) energy-momentum of spacetime, similar to the *Strain-energy tensor* in elastic media. Then we can write for the left-hand side of (4.150):

$$\begin{aligned} -\Theta_{\mathrm{Gr}}^{(\mu\nu)} :=& g_1 \left(R_{\alpha\beta\gamma}{}^{\nu}\, R^{\alpha\beta\gamma\mu} - \tfrac{1}{4} g^{\mu\nu} R_{\alpha\beta\gamma\delta}\, R^{\alpha\beta\gamma\delta}\right) - \frac{1}{8\pi G}\left(R^{(\mu\nu)} - \tfrac{1}{2} g^{\mu\nu}\, R + \lambda g^{\mu\nu}\right) \\ &- g_3 \left(S_{\alpha\beta}{}^{\mu}\, S^{\alpha\beta\nu} - \tfrac{1}{2} g^{\mu\nu}\, S_{\alpha\beta\gamma}\, S^{\alpha\beta\gamma}\right). \end{aligned} \tag{4.151}$$

Consequently, $T_0^{(\mu\nu)}$ corresponds to the *Stress-energy tensor* and with

$$\Theta_{\mathrm{Gr}}^{(\mu\nu)} + T_0^{(\mu\nu)} = 0, \tag{4.152}$$

we encounter a Universe with everywhere locally vanishing total energy. That recovers the conjecture of *Zero-Energy Universe* which has been discussed long ago by Lorentz [27], Levi-Civita [28], Jordan [29], and, independently in each case, by Sciama [30], Feynman [31], Tryon [32], Rosen [33], Cooperstock [34], Hamada [35], Melia [36]and Hawking [37], based on various physical reasonings.

It is important to stress here that the cosmological constant term, λ, is not introduced "by hand" but actually emerges here as a geometrical term from the gauge procedure [38], and is in this interpretation the semiclassical *Vacuum energy of spacetime*. It is independent of the vacuum energy of matter that will, though, counteract the observed cosmological constant [25]. This, and the torsional origin of dark energy [25, 39, 40], will be discussed in the context of cosmological implications in Chap. 8.

Now by switching on and off the different terms in (4.150), the Einstein-Cartan limit ($g_1 = 0$ while $g_1\, g_2 = \text{const}$) and the Einstein-Hilbert limit ($g_1 = g_3 = 0$) are obtained. As in the torsion-free limit ($g_3 = 0$) the is a solution of (4.150), as laid out in Chap. 7 following Kehm et al. [41], it is consistent with solar-scale observations.

4.5.5 Spin-Curvature Tensor Coupling Equation

We replace in the canonical field Eq. (4.791) the momentum field $\tilde{k}_i^{\ \alpha\mu}$ by the result of the canonical equation (4.109) for the "free gravity" Hamiltonian (4.107), and the $\tilde{q}_i^{\ j\mu\alpha}$-dependent terms according to Eq. (4.110). Equation (4.791) is then equivalently expressed as

$$g_1 \left[\left(R_i^{\ j\mu\alpha} - \bar{R}_i^{\ j\mu\alpha} \right)_{;\alpha} - \left(R_i^{\ j\xi\alpha} - \bar{R}_i^{\ j\xi\alpha} \right) S^{\mu}{}_{\xi\alpha} - 2 \left(R_i^{\ j\mu\xi} - \bar{R}_i^{\ j\mu\xi} \right) S^{\alpha}{}_{\xi\alpha} \right]$$
$$= g_3\ S_i^{\ \alpha\mu} e_\alpha^{\ j} + \Sigma_i^{\ j\mu}.$$

Inserting $\bar{R}_i^{\ j\mu\alpha}$ gives finally:

$$g_1 \left(R_i^{\ j\mu\alpha}{}_{;\alpha} - R_i^{\ j\xi\alpha} S^{\mu}{}_{\xi\alpha} - 2R_i^{\ j\mu\xi} S^{\alpha}{}_{\xi\alpha} \right) - g_3\, S_i^{\ \alpha\mu} e_\alpha^{\ j}$$
$$+ 2g_1 g_2 \left[e_k^{\ \alpha}\, e_i^{\ \xi}\, S^{\mu}{}_{\xi\alpha} + \left(e_k^{\ \xi}\, e_i^{\ \mu} - e_k^{\ \mu}\, e_i^{\ \xi} \right) S^{\alpha}{}_{\xi\alpha} \right] \eta^{kj} = \Sigma_i^{\ j\mu}. \tag{4.153}$$

This equation shows that the spinor fields act as the . It becomes more obvious when torsion is neglected:

$$g_1\, R_i^{\ j\mu\alpha}{}_{;\alpha} = \Sigma_i^{\ j\mu}. \tag{4.154}$$

To obtain the coordinate space representation of Eq. (4.153), we contract it with $e_\nu^{\ i}\, e_j^{\ \beta}$:

$$g_1\left(R_\nu{}^{\beta\mu\alpha}{}_{;\alpha} - R_\nu{}^{\beta\xi\alpha}S^\mu{}_{\xi\alpha} - 2R_\nu{}^{\beta\mu\xi}S^\alpha{}_{\xi\alpha}\right)$$
$$+ M_p^2\left(S^\mu{}_\nu{}^\beta + \delta^\mu_\nu S^{\alpha\beta}{}_\alpha - g^{\beta\mu}S^\alpha{}_{\nu\alpha}\right) - g_3\, S_\nu{}^{\beta\mu} = \Sigma_\nu{}^{\beta\mu}.$$

Raising finally the index ν, we observe that merely the skew-symmetric portion of the term proportional to g_3 contributes as all other terms are already skew-symmetric in ν and β:

$$g_1\left(R^{\nu\beta\mu\alpha}{}_{;\alpha} - R^{\nu\beta\xi\alpha}S^\mu{}_{\xi\alpha} - 2R^{\nu\beta\mu\xi}S^\alpha{}_{\xi\alpha}\right)$$
$$+ M_p^2\left(S^{\mu\nu\beta} + g^{\nu\mu}S^{\alpha\beta}{}_\alpha - g^{\beta\mu}S^{\alpha\nu}{}_\alpha\right) - g_3\, S^{[\nu\beta]\mu} = \Sigma^{\nu\beta\mu}. \tag{4.155}$$

On the other hand, this proves that with this ansatz for gauge gravity the symmetric term proportional to g_3 has to vanish identically, which implies that torsion must be restricted to a completely anti-symmetric form (aka pseudo-vector torsion)[2]:

$$\boxed{S^{(\alpha\beta)\mu} = 0.} \tag{4.156}$$

This dynamic constraint resolves many discussions on the structure of torsion, cf. e.g. [44], and also ensures that the are identical.[3] We shall in the following take this constraint into account and be led to considerably simplified formulae. For Eq. (4.155) this gives

$$\boxed{g_1\left(R^{\nu\beta\mu\alpha}{}_{;\alpha} - R^{\nu\beta\xi\alpha}S^\mu_{\xi\alpha}\right) + (M_p^2 - g_3)\, S^{\nu\beta\mu} = \Sigma^{\nu\beta\mu}.} \tag{4.157}$$

Now for metric-compatible geometries the affine connection satisfies the identity

$$\gamma^\lambda{}_{\mu\nu} = \left\{ {\lambda \atop \mu\nu} \right\} + K^\lambda{}_{\mu\nu} \tag{4.158}$$

as discussed in detail in Sects. 2.6. The contortion $K^\lambda{}_{\mu\nu}$ is identical to torsion,

$$K^\lambda{}_{\mu\nu} \equiv S^\lambda{}_{\mu\nu},$$

if the latter is totally anti-symmetric. Hence we have

$$\gamma^\lambda{}_{\mu\nu} = \left\{ {\lambda \atop \mu\nu} \right\} + S^\lambda{}_{\mu\nu}.$$

[2] For similar consideration on the symmetry of the torsion implied by the cosmological principle see [42, 43].

[3] Then also the null curve that photons follow according to the eikonal model coincides with both the geodetic and autoparallel trajectories [45].

Denoting the covariant derivative with respect to the Levi-Civita connection $\left\{ {\lambda \atop \mu\nu} \right\}$ by $\bar{\nabla}$, we finally arrive at

$$\boxed{g_1 \left(\bar{\nabla}_\alpha R^{\nu\beta\mu\alpha} + R^{\xi\beta\mu\alpha} S^{\nu}{}_{\xi\alpha} - R^{\xi\nu\mu\alpha} S^{\beta}{}_{\xi\alpha} \right) + (M_p^2 - g_3)\, S^{\nu\beta\mu} = \Sigma^{\nu\beta\mu}.} \tag{4.159}$$

Notice that with $g_1 = g_3 = 0$ this equation reduces to the so called Cartan equation, an algebraic equation between spin momentum and (non-propagating) torsion. We remark that the Riemann-Cartan curvature can be split into a Riemann and a torsion-dependent "Cartan" portions:

$$R^{\xi\beta\mu\alpha} = \bar{R}^{\xi\beta\mu\alpha} + P^{\xi\beta\mu\alpha}.$$

Then for the purpose of finding a solution the metric (vierbein) and torsion dependent terms in that equation can be isolated.

4.5.6 Spin-Torsion Tensor Coupling Equation

Inserting into Eq. (4.79k) the relation (4.109) gives

$$g_3 \left(S^{\nu\mu\alpha}{}_{;\alpha} - S^{\nu\beta\alpha}\, S^{\mu}{}_{\beta\alpha} \right) + T^{\nu\mu}_{\text{Gr}} = -T^{\nu\mu}_{0}. \tag{4.160}$$

The symmetric portion in ν, μ of this equation is then

$$T_0^{(\nu\mu)} = g_3 S^{\nu\beta\alpha} S^{\mu}{}_{\beta\alpha} - T_{\text{Gr}}^{(\nu\mu)}, \tag{4.161}$$

as the symmetric part of $S^{\nu\mu\alpha}$ vanishes. This coincides with the consistency Eq. (4.147), and gives in its explicit form (4.148).

The skew-symmetric portion of Eq. (4.160) is, on the other hand,

$$g_3\, \varepsilon S^{\nu\mu\alpha}{}_{;\alpha} + \tilde{T}^{[\nu\mu]}_{\text{Gr}} = -\tilde{T}^{[\nu\mu]}_{\text{D}},$$

and simplifies with Eq. (4.131) to:

$$M_p^2\, R_{[\nu\mu]} - g_3 S_{\nu\mu}{}^{\alpha}{}_{;\alpha} = T_{\text{D}\,[\nu\mu]}. \tag{4.162}$$

Using the identity

$$R_{[\nu\mu]} \equiv S^{\alpha}{}_{\nu\mu;\alpha} - 2S^{\beta}{}_{\nu\mu} S^{\alpha}{}_{\beta\alpha} + S^{\alpha}{}_{\nu\alpha;\mu} - S^{\alpha}{}_{\mu\alpha;\nu}, \tag{4.163}$$

simplifies finally to

$$M_p^2 S^\alpha{}_{\nu\mu;\alpha} - g_3 S_{\nu\mu}{}^\alpha{}_{;\alpha} = T_{\mathrm{D}\,[\nu\mu]}. \tag{4.164}$$

This is a proper source equation for the torsion tensor. With (4.146) the right-hand side can be alternatively substituted by the divergence of the spin-momentum density:

$$\left(M_p^2 - g_3\right) S_{\nu\mu}{}^\alpha{}_{;\alpha} = T_{\mathrm{D}\,[\nu\mu]} = \Sigma_{\nu\mu}{}^\alpha{}_{;\alpha}. \tag{4.165}$$

which turns out to be a conservation law for torsion-spin:

$$\boxed{\left[\left(M_p^2 - g_3\right) S^{\nu\mu\alpha} - \Sigma^{\nu\mu\alpha}\right]_{;\alpha} = 0.} \tag{4.166}$$

The expression in brackets, an algebraic relation between torsion and spin density, vanishes in the Einstein-Cartan theory [46]. Here, in contrast, it is a propagating field.

In terms of the Levi-Civita derivative, (4.166) becomes

$$\bar{\nabla}_\alpha \left[\left(M_p^2 - g_3\right) S^{\nu\mu\alpha} - \Sigma^{\nu\mu\alpha}\right] = S^\mu{}_{\beta\alpha}\, \Sigma^{\nu\beta\alpha} - S^\nu{}_{\beta\alpha}\, \Sigma^{\mu\beta\alpha}. \tag{4.167}$$

4.5.7 Field Equations of Matter with Anti-symmetric Torsion

4.5.7.1 Scalars

Replacing the covariant derivative in the Klein-Gordon Eq. (4.134) the Levi-Civita derivative $\bar{\nabla}$ and the torsion term leads to

$$\boxed{g^{\alpha\beta}\bar{\nabla}_\alpha \bar{\nabla}_\beta \varphi + m^2 \varphi = 0.} \tag{4.168}$$

We see, that torsion drops out of the field equation of the scalar field.

4.5.7.2 Massive Vectors

Similarly, the Maxwell-Proca equation (4.136) becomes

$$\boxed{\bar{\nabla}_\alpha f^{\mu\alpha} - m^2 a^\mu = 0,} \tag{4.169}$$

since all torsion dependent terms cancel out. The Proca-Maxwell equation is thus equal to that known from the standard Einsteinian gravity. This means that while gauge fields influence curvature via their energy-momentum content, they do not “feel” torsion in their own propagation. It is straightforward to extend this conclusion also to non-abelian gauge fields. However, for fields not subject to SU(N)

symmetry, choosing the Maxwell-type exterior derivative for the free Lagrangian is not mandatory.

4.5.7.3 Spinors

Also the Dirac equation simplifies for a totally anti-symmetric torsion:

$$\mathrm{i}\left(\gamma^{\beta} - \frac{1}{3M}\sigma^{\xi\alpha}S^{\beta}{}_{\xi\alpha}\right)\overrightarrow{\mathcal{D}}_{\beta}\psi - \left(m + \frac{1}{24M}\sigma^{\alpha\beta}\sigma^{nm}R_{nm\alpha\beta}\right)\psi = 0. \tag{4.170}$$

Compared to the standard Dirac equation in curved spacetime, two "anomalous" corrections related to the mass parameter M arise. Firstly, a term proportional to torsion modifies the dynamic affine connection $\gamma^{\beta} = \gamma^{i}\, e_{i}{}^{\beta}$ that is projected via the vierbein field to the base manifold. Secondly, there is a mass correction proportional to the Riemann tensor. A further, regular correction arising from torsion is encountered once we realize that the spin connection can be, according to the vierbein postulate (4.60), split up into a Levi-Civita portion $\bar{\omega}^{i}{}_{j\mu}$ (Ricci rotation coefficients) and a torsion term:

$$\omega^{i}{}_{j\mu} = \bar{\omega}^{i}{}_{j\mu} + S^{i}{}_{j\mu}\,.$$

Then the covariant derivative (4.63a) too has a (overbared) Levi-Civita and a torsional portion:

$$\overrightarrow{\mathcal{D}}_{\mu} = \frac{\overrightarrow{\partial}}{\partial x^{\mu}} - \tfrac{\mathrm{i}}{4}\sigma^{lj}\,\omega_{lj\mu} = \overrightarrow{\bar{\mathcal{D}}}_{\mu} - \tfrac{\mathrm{i}}{4}\sigma^{lj}\,S_{lj\mu}. \tag{4.171}$$

Plugging this into (4.5.7.3) gives finally in coordinate notation:

$$\boxed{\mathrm{i}\left(\gamma^{\beta} - \frac{1}{3M}\sigma^{\xi\alpha}S_{\xi\alpha}{}^{\beta}\right)\left(\overrightarrow{\bar{\mathcal{D}}}_{\beta} - \tfrac{\mathrm{i}}{4}\sigma^{\xi\alpha}\,S_{\xi\alpha\beta}\right)\psi - \left(m + \frac{1}{24M}\sigma^{\xi\alpha}\sigma^{\tau\beta}R_{\tau\beta\xi\alpha}\right)\psi = 0.} \tag{4.172}$$

4.6 Conclusion

4.6.1 Summary

The Covariant Canonical Gauge Gravity (CCGG) is a semi-classical relativistic field theory in the De Donder-Weyl formulation in the Hamiltonian picture. For implementing local symmetries, the framework of the canonical transformation theory is deployed, providing a rigorous mathematical blueprint for implementing gauge fields for any kind of symmetries, starting off from a limited set of fundamental assumptions.

In that framework, the famous boundary term in the field-theoretical action functional, that is needed for ensuring invariance of the system dynamics under any kind of local transformations of the involved fields, is counter-acted by a corrective term and absorbed into a new, form-invariant version of the action endowed with newly introduced gauge fields. The interaction between the original fields and the gauge fields is thereby unambiguously fixed.

After proving in the past that the Yang-Mills theories can be reproduced in this way, the approach is applied here to the external Lorentz and diffeomorphism symmetry on principal frame bundles. The key initial postulates are the Hamiltonian principle of least action, and the principles of General Relativity and Equivalence. The sole input is then the transformation behaviour of the involved matter fields, here the real Klein-Gordon, Maxwell-Proca, and the Dirac fields. In order to establish a 1 : 1 correspondence between the Lagranian and Hamiltonian pictures, we request the Hamiltonians to be non-degenerate, i.e. to contain quadratic terms in the momentum fields. The framework enables in the first place to *derive* the field-gravity interactions and generic field equations of motion for any matter fields in any kind of spacetime geometry.

For the $SO(1,3) \times \mathrm{Diff}(M)$ symmetry addressed here in the vierbein formalism, we identify the spin connection to be the gauge field for any theory of gravity. This generalizes past work going back to Sciama, Kibble et al. that was based on the linear Einstein-Cartan ansatz, and sheds a new light on the so called Poincare Gauge Theory in its quadratic realization pursued by Hehl et al.

The definition of a particular form of matter dynamics, and of a reasonable phenomenologically sound gravity ansatz, enables then the generic dynamic equations to be made specific for the selected particles and geometry. The results are a set of field equations for matter fields in curved geometry that, for scalar and vector fields, confirm the so called comma-to-semicolon rule with the Levi-Civita connection.

For the Dirac field, in contrast, we apply the Gasiorowicz quadratic formulation, that leads to an $\ell = {}^1\!/_M$ and to an with torsion and a correction.

The power of the covariant Hamiltonian approach is manifested in further findings as well. For the selected linear-quadratic gravity ansatz these are for example:

- Einstein's field equation is extended by a quadratic trace-free Kretschmann term and quadratic torsion.
- The source term on the right-hand side of that equation turns out to be the total symmetrized metric (Hilbert) energy-momentum tensor of matter.
- The total energy-momentum of matter and spacetime vanishes locally everywhere ("Zero-Energy Universe"), similar to the stress-strain relation in elastic media. The total energy in any finite section of spacetime is thus zero as well. Its portion that can be contributed to the mass of a 3-D object thus depends on the observer's frame, similar to point mechanics where we for example distinguish between center-of-mass and lab energies.
- The Einstein-Cartan or Einstein-Hilbert theories are obtained as limiting cases by setting respectively $g_1 = 0$ or $g_1 = g_3 = 0$

- The extended theory does not contradict solar-scale observations as the usual is its valid solution
- The ground state of geometry is the maximally symmetric, de Sitter or Anti-de Sitter, configuration
- The quadratic term in the gravity Hamiltonian lends inertia to spacetime w.r.t. deformations versus de Sitter or Anti-de Sitter geometry
- The dynamics of curvature is given by covariant source equations à la Gauss law of electrodynamics, with source terms based on fermion spin only
- Torsion of spacetime is totally anti-symmetric (pseudo-vector only) field
- A linear combination of torsion of spacetime and spin-momentum density is a covariantly conserved propagating field generalizing the algebraic relation established for Einstein-Cartan theories
- Torsional energy density acts as dark energy and contributes to the total energy density of spacetime
- The cosmological constant acquires a contribution of geometrical origin that is independent of the vacuum energy of matter, and can be interpreted as the
- After embedding in curved geometry, an arises in the Dirac theory
- Fermions interact with torsion and acquire a curvature-dependent mass correction.

Below, for the sake of convenience, the key dynamic equations for both, matter and spacetime, based on the CCGG ansatz, are collected:

- Klein-Gordon equation
  $g^{\alpha\beta}\bar{\nabla}_\alpha\bar{\nabla}_\beta\varphi + m^2\varphi = 0$
- Proca equation
  $\bar{\nabla}_\alpha f^{\mu\alpha} - m^2 a^\mu = 0$
- Dirac equation
  $\mathrm{i}\left(\gamma^\beta - \frac{1}{3M}\sigma^{\xi\alpha}S^\beta{}_{\xi\alpha}\right)\left(\overrightarrow{\mathcal{D}}_\mu - \frac{\mathrm{i}}{4}\sigma^{lj}\,S_{lj\mu}\right) - \left(m + \frac{1}{24M}\sigma^{\alpha\beta}\sigma^{nm}R_{nm\alpha\beta}\right)\psi = 0$
- Curvature equation
  $g_1\left(\bar{\nabla}_\alpha R^{\nu\beta\mu\alpha} + R^{\xi\beta\mu\alpha}S^\nu_{\xi\alpha} - R^{\xi\nu\mu\alpha}S^\beta_{\xi\alpha}\right) + (M_p^2 - g_3)\,S^{\nu\beta\mu} = \Sigma^{\nu\beta\mu}$
- Spin-torsion equation
  $S^{\alpha\beta\mu} \equiv S^{[\alpha\beta\mu]} \qquad \bar{\nabla}_\alpha\, S^\alpha{}_{\nu\mu} = \frac{1}{M_p^2 - g_3}T_{\mathrm{D}\,[\nu\mu]} \;\Leftrightarrow\; \bar{\nabla}_\alpha\left[\left(M_p^2 - g_3\right)S^{\nu\mu\alpha} - \Sigma^{\nu\mu\alpha}\right] = S^\mu{}_{\beta\alpha}\,\Sigma^{\nu\beta\alpha} - S^\nu{}_{\beta\alpha}\,\Sigma^{\mu\beta\alpha}$
- CCGG field equation
  $g_1\left(R_{\alpha\beta\gamma}{}^\mu\, R^{\alpha\beta\gamma\nu} - \frac{1}{4}g^{\mu\nu}R_{\alpha\beta\gamma\delta}\,R^{\alpha\beta\gamma\delta}\right) - M_p^2\left(R^{(\mu\nu)} - \frac{1}{2}g^{\mu\nu}\,R + \lambda g^{\mu\nu}\right) \quad - g_3 \left(S_{\alpha\beta}{}^\mu\, S^{\alpha\beta\nu} - \frac{1}{2}g^{\mu\nu}\,S_{\alpha\beta\gamma}\,S^{\alpha\beta\gamma}\right) = T_{\mathrm{KG}}^{(\mu\nu)} + T_{\mathrm{P}}^{(\mu\nu)} + T_{\mathrm{D}}^{(\mu\nu)}$

The left-hand side of this equation is interpreted as the strain-energy tensor of spacetime, (4.132). The formulas for the energy-momentum tensors are discussed in Sect. 4.4, the source terms, spin-momentum $\Sigma^{\nu\beta\mu}$ and skew-symmetric energy-momentum $T_{\mathrm{D}}^{[\nu\mu]}$, are defined in Eqs. (4.143) and (4.128), respectively.

Applying common terminology we might call the CCGG and spin-torsion equations *Constraint equations* expressing respectively the and the conservation of the total torsion-spin, and the remaining field equations *Evolution equations*.

4.6.2 Discussion

The formulation presented here has a generic and a specific aspect. Its specific features are incorporated in the choice of the matter and gravity Hamiltonians, and for both some ambiguities exist. For the Klein-Gordon field we might for example include a Higgs-type potential term replacing the mass term.

The question whether a massive vector field a_μ acts as a source for torsion of spacetime was found to depend on the model for the free vector-field dynamics, as was discussed in Sect. 4.1.7.1. Here we defined the field tensor in the Lagrangian as the *exterior derivative* of the vector field a_μ, aka Maxwell-Proca field, and hence the momentum field tensor in the corresponding Hamiltonian was defined to be *skew-symmetric*. Then the respective term in the action (4.53) did *not* couple directly to the gauge field $\omega^i{}_{j\nu}$. Consequently, vector field terms do not turn up in the gauge Hamiltonian (4.50) and do not contribute to the skew-symmetric portion (4.145) of the consistency equation, nor to the spin-curvature tensor Eq. (4.155) and in the spin-torsion tensor Eq. (4.162).

The other option for the Proca field was discussed earlier in Refs. [10, 47], where the diffeomorphism-invariance of the final action was achieved by converting the partial derivative of the vector field in the initial action (4.32) into a *covariant derivative* in the course of the gauge procedure. In that case, the spin density of the vector field does not vanish and acts as a source of torsion of spacetime.

Novel features are found when applying the quadratic Gasiorowicz-Dirac Lagrangian. While it reproduces the standard Dirac equation in the free case, anomalous terms arise once interactions are turned on. A Fermi-like interaction term arises in the electromagnetic case, and for gravity novel interactions with torsion and curvature are found. Both are connected by the emerging length parameter, and so could, in principle, be subject of independent measurements. The spin-curvature interaction should have immediate consequences in scenarios with $R \gg 0$, e.g. inflation [12] and neutron star mergers. On the other hand, depending on the size of the parameter M, even the ambient curvature might affect massless ferminons, i.e. influence the neutrino mass. An independent estimate of the parameter M is needed, e.g. from the electron and the muon $g-2$ experiments [48]. However, there is a priori no reason to expect the value of M to be independent of the particle flavour, and since neutrino experiments are notoriously difficult, it does not seem likely that there will be estimates any time soon.

Last not least, the so far omitted interactions between matter fields has to be taken into account in a combined way following the process sketched above. The presented gauge procedure leading to the final action (4.53) can in particular be extended to include also internal symmetries.

Our anticipation based on previous work [49] is that the action of a $\mathrm{SO}(1,3) \times \mathrm{SU}(N) \times \mathrm{Diff}(M)$ gauge theory becomes by a straightforward analogy:

$$
\begin{aligned}
S = \int_V \mathrm{d}^4 x \Bigg\{ & \tilde{\pi}^{\nu} \frac{\partial \varphi}{\partial x^{\nu}} + \bar{\tilde{\kappa}}_{cI}{}^{\nu} \left(\frac{\partial \psi^{cI}}{\partial x^{\nu}} - \mathrm{i} g \, a^I{}_{J\nu} \, \psi^{cJ} - \tfrac{\mathrm{i}}{4} \omega^i{}_{j\nu} \, \sigma^c{}_{bi}{}^j \, \psi^{bI} \right) \\
& + \left(\frac{\partial \bar{\psi}_{cI}}{\partial x^{\nu}} + \mathrm{i} g \, \bar{\psi}_{cJ} \, a^J{}_{I\nu} + \tfrac{\mathrm{i}}{4} \bar{\psi}_{bI} \, \omega^i{}_{j\nu} \, \sigma^b{}_{ci}{}^j \right) \tilde{\kappa}^{cI\nu} \\
& + \tfrac{1}{2} \tilde{p}_I{}^{J\mu\nu} \left(\frac{\partial a^I{}_{J\mu}}{\partial x^{\nu}} - \frac{\partial a^I{}_{J\nu}}{\partial x^{\mu}} - \mathrm{i} g \left(a^I{}_{N\nu} \, a^N{}_{J\mu} - a^I{}_{N\mu} \, a^N{}_{J\nu} \right) \right) \\
& + \tfrac{1}{2} \tilde{k}_i{}^{\mu\nu} \left(\frac{\partial e_{\mu}{}^i}{\partial x^{\nu}} - \frac{\partial e_{\nu}{}^i}{\partial x^{\mu}} + \omega^i{}_{j\nu} \, e_{\mu}{}^j - \omega^i{}_{j\mu} \, e_{\nu}{}^j \right) \\
& + \tfrac{1}{2} \tilde{q}_i{}^{j\mu\nu} \left(\frac{\partial \omega^i{}_{j\mu}}{\partial x^{\nu}} - \frac{\partial \omega^i{}_{j\nu}}{\partial x^{\mu}} + \omega^i{}_{n\nu} \, \omega^n{}_{j\mu} - \omega^i{}_{n\mu} \, \omega^n{}_{j\nu} \right) - \tilde{\mathcal{H}}_0 - \tilde{\mathcal{H}}_{\mathrm{Gr}} \Bigg\} .
\end{aligned}
\tag{4.173}
$$

Herein, the spinor indices are written explicitly using lower case Latin letters b and c, whereas the $\mathrm{SU}(N)$ indices are denoted by upper case Latin letters. As before, the lower case Latin letters i, j, and n stand for the Lorentz indices and Greek letters for the coordinate space indices. The Hamiltonian $\tilde{\mathcal{H}}_0$ is supposed here to also describe the dynamics of the Hermitian matrix $a^I{}_{J\mu}$ of *massless* vector gauge fields. An example is the strict (unbroken) color symmetry $\mathrm{SU}(3)_c$ of chromodynamics. By analogy with Yang-Mills theories, the theory (4.173) being a quadratic Palatini theory might be void of ghosts [50] and also renormalizable [51].

References

1. T. De Donder, *Théorie Invariantive Du Calcul des Variations* (Gaulthier-Villars & Cie, Paris, 1930)
2. H. Weyl, Geodesic fields in the calculus of variation for multiple integrals. Ann. Math. **36.3**, 607–629 (1935). https://doi.org/10.2307/1968645. https://www.jstor.org/stable/1968645
3. H. Weyl, Eine neue Erweiterung der Relativitästheorie, in *Annalen der Physik IV Folge*, vol 59, p. 101 (1919). https://doi.org/10.1002/andp.19193641002
4. Yang, Mills, Conservation of isotopic spin and isotopic gauge invariance. Phys. Rev. **96**, 191 (1954)
5. R. Utiyama, Invariant theoretical interpretation of interaction. Phys. Rev. **101.5**, 1597 (1956)
6. D.W. Sciama, The analogy between charge and spin in general relativity, in *Recent Developments in Festschrift for Infeld* (Pergamon Press, Oxford; PWN, Warsaw, 1962), pp. 415–439
7. T.W.B. Kibble, Lorentz invariance and the gravitational field. J. Math. Phys. **2**, 212–221 (1961)
8. K. Hayashi, T. Shirafuji, Gravity from the Poincaré gauge theory of fundamental interactions. Prog. Theor. Phys. **64.3**, 866, 883, 1435, 2222 (1980). https://doi.org/10.1143/PTP.64.866
9. F.W. Hehl et al., General relativity with spin and torsion: foundations and prospects. Rev. Mod. Phys. **48.3**, 393 (1976). https://doi.org/10.1103/revmodphys.48.393
10. J. Struckmeier et al., Canonical transformation path to gauge theories of gravity. Phys. Rev. D **95**, 124048 (2017). https://doi.org/10.1103/PhysRevD.95.124048. arXiv: 1704.07246
11. J. Struckmeier et al., Pauli-type coupling between spinors and curved spacetime (2020). arXiv:1812.09669

12. D. Benisty et al., Inflation from fermions with curvature-dependent mass. Phys. Rev. D **D100.4**, 043523 (2019). https://doi.org/10.1103/PhysRevD.100.043523. arXiv: 1905.03731 [gr-qc]
13. D. Vasak et al., Covariant canonical gauge gravitation and cosmology. J. Phys. Conf. Ser. **1194**, 012108 (2019). arXiv:1812.00578
14. D. Vasak, J. Kirsch, J. Struckmeier, Locally contorted space-time invokes inflation, dark energy, and a non-singular Big Bang (2019). https://arxiv.org/abs/1910.01088
15. Ch.W. Misner, K.S. Thorne, J.A. Wheeler, *Gravitation* (W. H. Freeman and Company, New York, 1973)
16. F.W. Hehl, Four Lectures on Poincaré Gauge Field Theory, in *Cosmology and Gravitation: Spin*, ed. by Peter G. Bergmann, V. De Sabbata (Springer, Boston, MA, USA, 1980), pp. 5–61. ISBN: 978-1-4613-3123-0. https://doi.org/10.1007/978-1-4613-3123-0_2. arXiv: 2303.05366 [gr-qc]. https://doi.org/10.1007/978-1-4613-3123-0_2
17. F.W. Hehl, *Gauge theory of gravity and spacetime* (2014). https://doi.org/10.1007/978-1-4939-3210-8_5. arXiv:1204.3672
18. A.A. Tseytlin, On the Poincare and De Sitter Gauge theories of gravity with propagating torsion. Phys. Rev. D **26**, 3327 (1982). https://doi.org/10.1103/PhysRevD.26.3327
19. M.E. Peskin, D.V. Schroeder, *Quantum Field Theory* (Westview Press, Boulder, Colorado, 1995). 978-0-201-50397-5
20. James York, Boundary terms in the action principles of general relativity. Found. Phys. **16**, 249–257 (1986). https://doi.org/10.1007/BF01889475
21. S. Gasiorowicz, *Elementary Particle Physics* (Wiley, New York, 1966)
22. J. Struckmeier, A. Redelbach, Covariant hamiltonian field theory. Int. J. Mod. Phys. E **17**, 435–491 (2008). https://doi.org/10.1142/s0218301308009458. arXiv: 0811.0508
23. J. Struckmeier, H. Reichau, General U(N) gauge transformations in the realm of covariant Hamiltonian field theory, in *Exciting Interdisciplinary Physics*, ed. by W. Greiner, FIAS Interdisciplinary Science Series (Springer International Publishing, Switzerland, May 29, 2013), p. 367. ISBN: 978-3-319-00046-6. https://doi.org/10.1007/978-3-319-00047-3_31. arXiv: 1205.5754. https://arxiv.org/abs/1205.5754
24. D. Benisty et al., Quadratic curvature theories formulated as Covariant Canonical Gauge theories of Gravity. Phys. Rev. D **98**, 106021 (2018)
25. D. Vasak et al., On the cosmological constant in the deformed Einstein-Cartan gauge gravity in De Donder-Weyl Hamiltonian formulation. Astron. Nachr. (2022). https://doi.org/10.1002/asna.20220069,2022. arXiv: 2209.00501 [gr-qc]
26. S. Carroll, *Spacetime and Geometry* (Prentice Hall, 11 Aug. 2013). ISBN: 1292026634
27. H. Lorentz, Over Einstein's theorie der zwaartekracht (iii). Koninklijke Akademie van Wetenschappen **25**, 468–486 (1916)
28. T. Levi-Civita, On the analytic expression that must be given to the gravitational tensor in Einstein's theory. Trans. by S. Antoci, A. Loinger. In: Atti della Accademia Nazionale dei Lincei, vol. 26.4 (1917)
29. P. Jordan, Bemerkungen zur Kosmologie. Annalen der Physik **428.1**, 64–70 (1939)
30. D.W. Sciama, On the origin of inertia, in *Monthly Notices of the Royal Astronomical Society*, vol. 113, p. 34 (1953)
31. R. Feynman, W. Morinigo, W. Wagner, *Feynman Lectures On Gravitation (Frontiers in Physics)* (Westview Press, Boulder, Colorado, 2002). ISBN: 978-0813340388
32. E.P. Tryon, Is the universe a vacuum fluctuation. *Nature* **246**, 396 (1973). https://doi.org/10.1038/246396a0
33. N. Rosen, The energy of the universe. Gen. Rel. Grav. **26**, 319–321 (1994)
34. F.I. Cooperstock, M. Israelit, The energy of the universe. Found. Phys. **25**(4) (1995)
35. K.-J. Hamada, *Revealing a Trans-Planckian World Solves the Cosmological Constant Problem*. arXiv:2204.03914
36. F. Melia, Initial energy of a spatially flat universe: a hint of its possible origin. Astron. Nachr. **343.3**, e224010 (2022)
37. S. Hawking, *The Theory of Everything* (New Millenium Press, 2003). ISBN: 978-1-59777-611-0

38. J. Struckmeier, D. Vasak, Covariant canonical gauge theory of gravitation for fermions. Astron. Nachr. (2021). https://doi.org/10.1002/asna.202113991. arXiv:2101.04467
39. A. van de Venn et al., Torsional dark energy in quadratic gauge gravity. European Phys. J. (2023). https://doi.org/10.1140/epjc/s10052-023-11397-y. arXiv: 2211.11868 [gr-qc]
40. J. Kirsch et al., Torsion driving cosmic expansion. European Phys. J. C (2023). https://doi.org/10.1140/epjc/s10052-023-11571-2. https://link.springer.com/content/pdf/10.1140/epjc/s10052-023-11571-2.pdf
41. D. Kehm et al., Violation of Birkhoff's theorem for pure quadratic gravity action. Astron. Nachr./AN 338.9-10, pp. 1015–1018 (2017). https://doi.org/10.1002/asna.201713421. https://doi.org/10.1002/asna.201713421
42. M. Tsamparlis, Methods for deriving solutions in generalized theories of gravitation: the Einstein-Cartan theory. Phys. Rev. D **24**, 1451 (1981). https://doi.org/10.1103/PhysRevD.24.1451. https://link.aps.org/doi/10.1103/PhysRevD.24.1451
43. Y. Yoon, Conformally coupled induced gravity with gradient torsion. Phys. Rev. D **59**, 127501 (1999). https://doi.org/10.1103/PhysRevD.59.127501. https://link.aps.org/doi/10.1103/PhysRevD.59.127501
44. S. Capozziello, G. Lambiase, C. Stornaiolo, Geometric classification of the torsion tensor in space-time. Ann. Phys. **10**, 713–727 (2001). https://doi.org/10.1002/1521-3889(200108)10:8<713::AID-ANDP713>3.0.CO;2-2. (arXiv: gr-qc/0101038)
45. L.T. Santana et al., How does light move in a generic metric-affine background? Phys. Rev. D **95.6**, 061501 (2017). https://doi.org/10.1103/PhysRevD.95.061501. arXiv: 1703.10871 [gr-qc]
46. F.W. Hehl, B.K. Datta, Nonlinear spinor equation and asymmetric connection in general relativity. J. Math. Phys. **12**, 1334–1339 (1971). https://doi.org/10.1063/1.1665738
47. J. Struckmeier, D. Vasak, J. Kirsch, Generic theory of geometrodynamics from noether's theorem for the diff(m) symmetry group, in *Discoveries at the Frontiers* ed. by J. Kirsch et al. (Springer Nature Switzerland AG, 2020), pp. 143–181. https://doi.org/10.1007/978-3-030-34234-0_12. arXiv: 1807.03000 [gr-qc]. https://arxiv.org/abs/1807.03000
48. R.R. Sastry, Quantum electrodynamics with the Pauli term (1999). arXiv: hep-th/9903179
49. J. Struckmeier, D. Vasak, H. Stoecker, Covariant hamiltonian representation of Noether's therorem and its application to SU(N) Gauge theories, in *New Horizons in Fundamental*, ed. by S. Schramm, M. Schaefer, FIAS Interdisciplinary Science Series (Springer International Publishing Switzerland, 2017). ISBN: 978-3-319-44165-8. https://doi.org/10.1007/978-3-319-44165-8. URL: https://arxiv.org/abs/1608.01151
50. T. Koivisto, Covariant conservation of energy-momentum in modified gravities. Class. Quant. Grav. **23**, 4289 (2006). https://arxiv.org/abs/grqc/0505128
51. K.S. Stelle, Renormalization of higher-derivative quantum gravity. Phys. Rev. D **16.4**, 953 (1977). https://doi.org/10.1103/physrevd.16.953

Chapter 5
Spinor Representation of the Gauge Theory of Gravity for Fermions

The purpose of this chapter is to present a self-contained equivalent formulation of the approach derived in Chap. 4 using the spinor representation. Only the spin-1/2 fields are considered first for simplicity. The spinorial representation of the Dirac theory is developed in Sect. 5.1. The idea is to convert the static Dirac matrices $\boldsymbol{\gamma}^k$ occurring in the Dirac equation in flat spacetime into general spacetime-dependent matrices $\boldsymbol{\gamma}^\mu(x)$, which can be thought of as the contraction of $\boldsymbol{\gamma}^k$ with the vierbein ${e_k}^\mu$. This way, the Lorentz index—denoted by the Latin index k—no longer emerges explicitly. Rather the spinor indices a, b in $\boldsymbol{\gamma}^\mu \equiv {\gamma^a}_{b^\mu}$ are directly associated with the field transformation in the inertial frame. The gist of gauge theory of gravity, i.e. the recipe to convert the action of a Lorentz-invariant system $\mathcal{H}_\mathrm{D}$ into the corresponding generally covariant system, is analogously pursued in this formalism in Sect. 5.2. Once the formalism has been developed for the spin-1/2 field, it is straightforward to apply it to spin-3/2 fields as well as shown in Sect. 5.3.

This chapter can be skipped by readers who have neither a special interest in the spinorial formulation of gauge theory, nor in spin-3/2 fields in curved spacetimes.

5.1 Dirac Lagrangian for a Spin-1/2 Particle Field

5.1.1 Dirac Lagrangian in Flat Spacetime

The Dirac Lagrangian (4.98)

$$\mathcal{L}_\mathrm{D} = \frac{\mathrm{i}}{2}\left(\bar{\psi}\,\boldsymbol{\gamma}^\alpha \frac{\partial\psi}{\partial x^\alpha} - \frac{\partial\bar{\psi}}{\partial x^\alpha}\boldsymbol{\gamma}^\alpha\psi\right) - m\,\bar{\psi}\psi \tag{5.1}$$

involves the mixed (1, 1)-spinor-(0, 1)-tensors $\boldsymbol{\gamma}_\mu \equiv {\gamma^a}_{b_\mu}$ —commonly referred to as Dirac matrices—defined by the property

D. Vasak et al., *Covariant Canonical Gauge Gravity*, FIAS Interdisciplinary Science Series, https://doi.org/10.1007/978-3-031-43717-5_5

$$\tfrac{1}{2}\left(\boldsymbol{\gamma}_\mu\boldsymbol{\gamma}_\nu + \boldsymbol{\gamma}_\nu\boldsymbol{\gamma}_\mu\right) = g_{\mu\nu}\mathbb{1}, \qquad \tfrac{1}{8}\,\mathrm{Tr}\left\{\boldsymbol{\gamma}_\mu\boldsymbol{\gamma}_\nu + \boldsymbol{\gamma}_\nu\boldsymbol{\gamma}_\mu\right\} = \tfrac{1}{4}\,\mathrm{Tr}\left\{\boldsymbol{\gamma}_\mu\boldsymbol{\gamma}_\nu\right\} = g_{\mu\nu}. \tag{5.2}$$

While $g_{\mu\nu}$ denotes the metric tensor, $\mathbb{1} \equiv \delta^a_b$ is the $(1, 1)$-spinor unit matrix. For clarity we rewrite Eq. (5.2) in full index notation with Latin letters $a, b, c, \ldots$ denoting the spinor indices:

$$\tfrac{1}{2}\left(\gamma^a{}_{c\mu}\,\gamma^c{}_{b\nu} + \gamma^a{}_{c\nu}\,\gamma^c{}_{b\mu}\right) = g_{\mu\nu}\,\delta^a_b, \qquad \tfrac{1}{2}\left(\gamma^a{}_{c}{}^{\mu}\,\gamma^c{}_{b\nu} + \gamma^a{}_{c\nu}\,\gamma^c{}_{b}{}^{\mu}\right) = \delta^\mu_\nu\,\delta^a_b.$$

For the contractions we get

$$\tfrac{1}{2}\left(\gamma^a{}_{c}{}^{\mu}\,\gamma^c{}_{a\nu} + \gamma^a{}_{c\nu}\,\gamma^c{}_{a}{}^{\mu}\right) = \gamma^a{}_{c}{}^{\mu}\,\gamma^c{}_{a\nu} = \delta^a_a\,\delta^\mu_\nu = 4\delta^\mu_\nu \quad\Leftrightarrow\quad \mathrm{Tr}\left\{\boldsymbol{\gamma}^\mu\boldsymbol{\gamma}_\nu\right\} = 4\delta^\mu_\nu,$$

and

$$\tfrac{1}{2}\left(\gamma^a{}_{c}{}^{\mu}\,\gamma^c{}_{b\mu} + \gamma^a{}_{c\mu}\,\gamma^c{}_{b}{}^{\mu}\right) = \gamma^a{}_{c}{}^{\mu}\,\gamma^c{}_{b\mu} = \delta^a_b\,\delta^\mu_\mu = 4\delta^a_b \quad\Leftrightarrow\quad \boldsymbol{\gamma}^\mu\boldsymbol{\gamma}_\mu = 4\mathbb{1}.$$

The Euler-Lagrange equations

$$\frac{\partial}{\partial x^\alpha}\frac{\partial\mathcal{L}_\mathrm{D}}{\partial\left(\frac{\partial\bar{\psi}}{\partial x^\alpha}\right)} - \frac{\partial\mathcal{L}_\mathrm{D}}{\partial\bar{\psi}} = 0, \qquad \frac{\partial}{\partial x^\alpha}\frac{\partial\mathcal{L}_\mathrm{D}}{\partial\left(\frac{\partial\psi}{\partial x^\alpha}\right)} - \frac{\partial\mathcal{L}_\mathrm{D}}{\partial\psi} = 0$$

then provide the Dirac equations for the spinor ψ^b and its adjoint, $\bar{\psi}_b = \psi^\dagger_a\gamma^a{}_{b0}$:

$$\mathrm{i}\gamma^a{}_{b}{}^{\alpha}\,\frac{\partial\psi^b}{\partial x^\alpha} - m\psi^a = 0, \qquad \frac{\partial\bar{\psi}_a}{\partial x^\alpha}\mathrm{i}\gamma^a{}_{b}{}^{\alpha} + m\bar{\psi}_b = 0. \tag{5.3}$$

5.1.2 Condition for the Invariance of the Dirac Equation in Flat Spacetime

We first ask the question how a spinor $\psi(x)$ transforms under a global transformation $x \mapsto X$, which is determined by the set of 16 coefficients $\partial x^\alpha/\partial X^\beta = \text{const}$ and the four constant $(1, 1)$-spinor-$(1, 0)$-tensors $\boldsymbol{\gamma}^\beta = \text{const}$. This means to determine the $(1, 1)$-spinor-$(0, 0)$-tensor $\boldsymbol{S}(x) = (S^a{}_b(x))$ that provides

$$\Psi(X) = \boldsymbol{S}\psi(x), \qquad \bar{\Psi}(X) = \bar{\psi}(x)\,\boldsymbol{S}^{-1} \tag{5.4}$$

by requiring the Dirac equation (5.3) to be form-invariant:

$$
\begin{aligned}
0 &= \mathrm{i}\boldsymbol{\Gamma}^{\alpha}\frac{\partial \Psi}{\partial X^{\alpha}} - m\Psi \\
&= \mathrm{i}\boldsymbol{\Gamma}^{\alpha}\frac{\partial x^{\beta}}{\partial X^{\alpha}}\left(\boldsymbol{S}\frac{\partial \psi}{\partial x^{\beta}} + \frac{\partial \boldsymbol{S}}{\partial x^{\beta}}\psi\right) - m\,\boldsymbol{S}\psi.
\end{aligned}
$$

Multiplication with $\boldsymbol{S}^{-1}$ from the left now yields:

$$
\mathrm{i}\left(\boldsymbol{S}^{-1}\boldsymbol{\Gamma}^{\alpha}\boldsymbol{S}\frac{\partial x^{\beta}}{\partial X^{\alpha}}\right)\frac{\partial \psi}{\partial x^{\beta}} + \left(\mathrm{i}\boldsymbol{S}^{-1}\boldsymbol{\Gamma}^{\alpha}\frac{\partial \boldsymbol{S}}{\partial x^{\beta}}\frac{\partial x^{\beta}}{\partial X^{\alpha}} - m\right)\psi = 0,
$$

which warrants the form-invariance of the Dirac equation in a flat spacetime provided that:

$$
\boldsymbol{S}^{-1}\boldsymbol{\Gamma}^{\alpha}\boldsymbol{S}\frac{\partial x^{\beta}}{\partial X^{\alpha}} = \boldsymbol{\gamma}^{\beta}, \qquad \frac{\partial \boldsymbol{S}}{\partial x^{\beta}} \equiv 0, \qquad \frac{\partial^2 x^{\beta}}{\partial X^{\alpha}\partial X^{\xi}} \equiv 0. \tag{5.5}
$$

5.1.3 Dirac Hamiltonian in a Flat Spacetime

The regularized Dirac Lagrangian is defined by amending the conventional Dirac Lagrangian (5.1) by the Gasiorowicz term ([1]), as discussed in Sect. 4.3.3, Eq. (4.99):

$$
\mathcal{L}_{\mathrm{D}} = \frac{\mathrm{i}}{2}\left(\bar{\psi}\,\boldsymbol{\gamma}^{\alpha}\frac{\partial \psi}{\partial x^{\alpha}} - \frac{\partial \bar{\psi}}{\partial x^{\alpha}}\boldsymbol{\gamma}^{\alpha}\psi\right) - m\,\bar{\psi}\psi + \frac{\mathrm{i}}{3M}\frac{\partial \bar{\psi}}{\partial x^{\alpha}}\boldsymbol{\sigma}^{\alpha\beta}\frac{\partial \psi}{\partial x^{\beta}}, \tag{5.6}
$$

with M a coupling constant of mass dimension, and $\boldsymbol{\sigma}^{\alpha\beta}$ defined as the commutator of the matrix product $\boldsymbol{\gamma}^{\alpha}\boldsymbol{\gamma}^{\beta}$

$$
\boldsymbol{\sigma}^{\alpha\beta} = \frac{\mathrm{i}}{2}\left(\boldsymbol{\gamma}^{\alpha}\boldsymbol{\gamma}^{\beta} - \boldsymbol{\gamma}^{\beta}\boldsymbol{\gamma}^{\alpha}\right) \quad \Leftrightarrow \quad \sigma^{a}{}_{b\alpha\beta} = \frac{\mathrm{i}}{2}\left(\gamma^{a}{}_{c\alpha}\,\gamma^{c}{}_{b\beta} - \gamma^{a}{}_{c\beta}\,\gamma^{c}{}_{b\alpha}\right). \tag{5.7}
$$

Due to the skew-symmetry of $\boldsymbol{\sigma}^{\alpha\beta}$, the “Gasiorowicz term” does not contribute to the Euler-Lagrange field equations and hence reproduces the Dirac equations (5.3). Yet, the Lagrangian (5.6) allows a Legendre transformation to set up the Hamiltonian formulation of the action principle for a complex spinor field ψ and its adjoint $\bar{\psi}$, in conjunction with their respective conjugate fields $\bar{\kappa}^{\alpha}$ and κ^{α}

$$
S_0 = \int_V \mathcal{L}_{\mathrm{D}}\,\mathrm{d}^4x = \int_V \left(\bar{\kappa}^{\alpha}\frac{\partial \psi}{\partial x^{\alpha}} + \frac{\partial \bar{\psi}}{\partial x^{\alpha}}\kappa^{\alpha} - \mathcal{H}_{\mathrm{D}}\right)\mathrm{d}^4x, \qquad \delta S_0 \stackrel{!}{=} 0. \tag{5.8}
$$

All spinor indices are fully contracted in (5.8). With the canonical momentum fields, defined by

$$
\bar{\kappa}^{\mu} = \frac{\partial \mathcal{L}_{\mathrm{D}}}{\partial\left(\frac{\partial \psi}{\partial x^{\mu}}\right)}, \qquad \kappa^{\mu} = \frac{\partial \mathcal{L}_{\mathrm{D}}}{\partial\left(\frac{\partial \bar{\psi}}{\partial x^{\mu}}\right)}, \tag{5.9}
$$

a regular Dirac Hamiltonian $\mathcal{H}_{\mathrm{D}}$ is obtained as the Legendre transform of $\mathcal{L}_{\mathrm{D}}$ [1–3]:

$$\mathcal{H}_{\mathrm{D}}(\psi, \bar{\psi}, \kappa^{\mu}, \bar{\kappa}^{\mu}) = \frac{\mathrm{i}M}{2}\left(\bar{\psi}\,\boldsymbol{\gamma}_{\beta}\,\kappa^{\beta} - \bar{\kappa}^{\alpha}\,6\boldsymbol{\tau}_{\alpha\beta}\,\kappa^{\beta} - \bar{\kappa}^{\alpha}\,\boldsymbol{\gamma}_{\alpha}\,\psi\right) + (m-M)\,\bar{\psi}\psi. \tag{5.10}$$

The (1, 1)-spinor-(0, 2)-tensor $\boldsymbol{\tau}_{\alpha\beta}$ in Eq. (5.10), defined by

$$\boldsymbol{\tau}_{\mu\beta} = \frac{\mathrm{i}}{6}\left(\boldsymbol{\gamma}_{\mu}\boldsymbol{\gamma}_{\beta} + 3\boldsymbol{\gamma}_{\beta}\boldsymbol{\gamma}_{\mu}\right), \qquad \boldsymbol{\tau}_{\mu\beta}\boldsymbol{\sigma}^{\beta\nu} = \boldsymbol{\sigma}^{\nu\beta}\boldsymbol{\tau}_{\beta\mu} = \delta^{\nu}_{\mu}\,\mathbb{1}, \tag{5.11}$$

represents the inverse of the (1, 1)-spinor-(2, 0)-tensor $\boldsymbol{\sigma}^{\alpha\beta}$. Regular here means that the Hesse matrix $\partial^2\mathcal{H}_{\mathrm{D}}/\partial\bar{\kappa}^{\mu}\partial\kappa^{\nu}$ is non-degenerate and hence that the Legendre transformation exists and thus allows a swapping between the covariant Hamiltonian and the Lagrangian pictures. Only then the canonical equations provide a unique correlation of momenta and their duals, namely the derivatives of the fields. Under this precondition, the canonical field equations as well as the canonical transformation formalism can be worked out.

Explicitly, the set of canonical field equations emerging from the action principle (5.8) is obtained from the Dirac Hamiltonian $\mathcal{H}_{\mathrm{D}}$ in a modified notation as compared to Eqs. (4.133) from Sect. 4.5.3

$$\frac{\partial\bar{\psi}}{\partial x^{\nu}} = \frac{\partial\mathcal{H}_{\mathrm{D}}}{\partial\kappa^{\nu}} = \frac{\mathrm{i}M}{2}\bar{\psi}\boldsymbol{\gamma}_{\nu} - 3\mathrm{i}M\bar{\kappa}^{\beta}\boldsymbol{\tau}_{\beta\nu} \tag{5.12a}$$

$$\frac{\partial\psi}{\partial x^{\nu}} = \frac{\partial\mathcal{H}_{\mathrm{D}}}{\partial\bar{\kappa}^{\nu}} = -\frac{\mathrm{i}M}{2}\boldsymbol{\gamma}_{\nu}\psi - 3\mathrm{i}M\boldsymbol{\tau}_{\nu\beta}\kappa^{\beta} \tag{5.12b}$$

$$\frac{\partial\bar{\kappa}^{\beta}}{\partial x^{\beta}} = -\frac{\partial\mathcal{H}_{\mathrm{D}}}{\partial\psi} = \frac{\mathrm{i}M}{2}\bar{\kappa}^{\beta}\boldsymbol{\gamma}_{\beta} - (m-M)\,\bar{\psi} \tag{5.12c}$$

$$\frac{\partial\kappa^{\beta}}{\partial x^{\beta}} = -\frac{\partial\mathcal{H}_{\mathrm{D}}}{\partial\bar{\psi}} = -\frac{\mathrm{i}M}{2}\boldsymbol{\gamma}_{\beta}\kappa^{\beta} - (m-M)\,\psi. \tag{5.12d}$$

Making use of the identities

$$\boldsymbol{\gamma}_{\alpha}\boldsymbol{\sigma}^{\alpha\mu} \equiv \boldsymbol{\sigma}^{\mu\alpha}\boldsymbol{\gamma}_{\alpha} \equiv 3\mathrm{i}\,\boldsymbol{\gamma}^{\mu},$$

Equations (5.12a) and (5.12b) are solved for the momentum spinors by contracting with $\boldsymbol{\sigma}^{\nu\mu}$:

$$\bar{\kappa}^{\beta} = \frac{\mathrm{i}}{2}\bar{\psi}\boldsymbol{\gamma}^{\beta} + \frac{\mathrm{i}}{3M}\frac{\partial\bar{\psi}}{\partial x^{\alpha}}\boldsymbol{\sigma}^{\alpha\beta} \tag{5.13a}$$

$$\kappa^{\beta} = -\frac{\mathrm{i}}{2}\boldsymbol{\gamma}^{\beta}\psi + \frac{\mathrm{i}}{3M}\boldsymbol{\sigma}^{\beta\alpha}\frac{\partial\psi}{\partial x^{\alpha}}. \tag{5.13b}$$

The momentum tensors can now be eliminated by inserting Eqs. (5.13a) and (5.13b) into Eqs. (5.12c) and (5.12d), respectively,

$$
\begin{aligned}
\frac{\partial \bar{\kappa}^{\beta}}{\partial x^{\beta}} &= \frac{\mathrm{i}}{2}\frac{\partial \bar{\psi}}{\partial x^{\beta}}\boldsymbol{\gamma}^{\beta} + \frac{\mathrm{i}}{3M}\frac{\partial^{2}\bar{\psi}}{\partial x^{\alpha}\partial x^{\beta}}\boldsymbol{\sigma}^{\alpha\beta} = \frac{\mathrm{i}}{2}\frac{\partial \bar{\psi}}{\partial x^{\beta}}\boldsymbol{\gamma}^{\beta} \\
\frac{\partial \kappa^{\beta}}{\partial x^{\beta}} &= -\frac{\mathrm{i}}{2}\boldsymbol{\gamma}^{\beta}\frac{\partial \psi}{\partial x^{\beta}} + \frac{\mathrm{i}}{3M}\boldsymbol{\sigma}^{\beta\alpha}\frac{\partial^{2}\psi}{\partial x^{\alpha}\partial x^{\beta}} = -\frac{\mathrm{i}}{2}\boldsymbol{\gamma}^{\beta}\frac{\partial \psi}{\partial x^{\beta}}.
\end{aligned}
$$

The second partial derivative terms of the spinors cancel due to the skew-symmetry of $\boldsymbol{\sigma}^{\alpha\beta}$. Thus

$$
\begin{aligned}
\frac{\mathrm{i}}{2}\frac{\partial \bar{\psi}}{\partial x^{\beta}}\boldsymbol{\gamma}^{\beta} &= \frac{\mathrm{i}M}{2}\left(\frac{\mathrm{i}}{2}\bar{\psi}\boldsymbol{\gamma}^{\beta} + \frac{\mathrm{i}}{3M}\frac{\partial \bar{\psi}}{\partial x^{\alpha}}\boldsymbol{\sigma}^{\alpha\beta}\right)\boldsymbol{\gamma}_{\beta} - (m - M)\,\bar{\psi} \\
&= \frac{\mathrm{i}M}{2}\left(2\mathrm{i}\bar{\psi} + \frac{\mathrm{i}}{3M}\frac{\partial \bar{\psi}}{\partial x^{\alpha}}3\mathrm{i}\boldsymbol{\gamma}^{\alpha}\right) - (m - M)\,\bar{\psi}
\end{aligned}
$$

and

$$
\begin{aligned}
-\frac{\mathrm{i}}{2}\boldsymbol{\gamma}^{\beta}\frac{\partial \psi}{\partial x^{\beta}} &= -\frac{\mathrm{i}M}{2}\boldsymbol{\gamma}_{\beta}\left(-\frac{\mathrm{i}}{2}\boldsymbol{\gamma}^{\beta}\psi + \frac{\mathrm{i}}{3M}\boldsymbol{\sigma}^{\beta\alpha}\frac{\partial \psi}{\partial x^{\alpha}}\right) - (m - M)\,\psi \\
&= -\frac{\mathrm{i}M}{2}\left(-2\mathrm{i}\psi + \frac{\mathrm{i}}{3M}3\mathrm{i}\boldsymbol{\gamma}^{\alpha}\frac{\partial \psi}{\partial x^{\alpha}}\right) - (m - M)\,\psi,
\end{aligned}
$$

finally reproduce the Dirac equations (5.3) for ψ and the adjoint spinor, $\bar{\psi}$:

$$
\mathrm{i}\boldsymbol{\gamma}^{\beta}\frac{\partial \psi}{\partial x^{\beta}} - m\psi = 0, \qquad \mathrm{i}\frac{\partial \bar{\psi}}{\partial x^{\beta}}\boldsymbol{\gamma}^{\beta} + m\bar{\psi} = 0.
$$

5.2 Canonical Gauge Theory for Spin-1/2 Particle Fields

5.2.1 Transformations Rules for Spinors in Curved spacetime

In a curved spacetime, all quantities in (5.5) become arbitrary functions of the spacetime coordinates:

$$
\boldsymbol{S} = \boldsymbol{S}(x), \quad \boldsymbol{\gamma} = \boldsymbol{\gamma}(x), \quad \boldsymbol{\Gamma} = \boldsymbol{\Gamma}(X), \quad x^{\mu} = x^{\mu}(X) \qquad \Rightarrow \qquad \frac{\partial^{2} x^{\beta}}{\partial X^{\alpha}\partial X^{\xi}} \neq 0.
$$

Contracting then Eq. (5.5) with $\bar{\psi}$ from the left and with ψ from the right yields

$$
\bar{\psi}\,\boldsymbol{\gamma}^{\beta}\psi = \bar{\psi}\,\boldsymbol{S}^{-1}\boldsymbol{\Gamma}^{\alpha}\,\boldsymbol{S}\psi\frac{\partial x^{\beta}}{\partial X^{\alpha}} = \bar{\Psi}\,\boldsymbol{\Gamma}^{\alpha}\,\Psi\frac{\partial x^{\beta}}{\partial X^{\alpha}},
$$

which shows that $\bar{\psi}\boldsymbol{\gamma}^{\beta}\psi \equiv \bar{\psi}_{a}\gamma^{a}{}_{b\beta}\,\psi^{b}$ transforms like an ordinary $(1, 0)$-tensor as all spinor indices are contracted.

The general form of the transformation rule (5.5) for the $\boldsymbol{\gamma}_\mu$ matrices under a diffeomorphism,

$$\boldsymbol{\Gamma}_\mu(X) = \boldsymbol{S}(x)\,\boldsymbol{\gamma}_\alpha(x)\,\boldsymbol{S}^{-1}(x)\frac{\partial x^\alpha}{\partial X^\mu}, \tag{5.14}$$

induces the following transformation rule $g_{\mu\nu}(x) \mapsto G_{\mu\nu}(X)$ for the metric by virtue of the definition of the Dirac algebra (5.2)

$$\begin{aligned}
G_{\mu\nu}(X)\,\mathbb{1} &= \tfrac{1}{2}\left(\boldsymbol{\Gamma}_\mu\boldsymbol{\Gamma}_\nu + \boldsymbol{\Gamma}_\nu\boldsymbol{\Gamma}_\mu\right)\\
&= \tfrac{1}{2}\boldsymbol{S}\left(\boldsymbol{\gamma}_\alpha\boldsymbol{S}\boldsymbol{S}^{-1}\boldsymbol{\gamma}_\beta + \boldsymbol{\gamma}_\beta\boldsymbol{S}\boldsymbol{S}^{-1}\boldsymbol{\gamma}_\alpha\right)\boldsymbol{S}^{-1}\frac{\partial x^\alpha}{\partial X^\mu}\frac{\partial x^\beta}{\partial X^\nu} = \boldsymbol{S}g_{\alpha\beta}\mathbb{1}\,\boldsymbol{S}^{-1}\frac{\partial x^\alpha}{\partial X^\mu}\frac{\partial x^\beta}{\partial X^\nu}\\
&= g_{\alpha\beta}(x)\frac{\partial x^\alpha}{\partial X^\mu}\frac{\partial x^\beta}{\partial X^\nu}\,\mathbb{1}.
\end{aligned}$$

This is, of course, a relation equivalent to Eq. (2.6) for the vierbeins. The information on the transformation property of the metric tensor $g_{\mu\nu}(x)$ is thus encoded in the transformation rule (5.14) for $\boldsymbol{\gamma}_\mu(x)$, which allows us to replace the metric tensor by the dynamic Dirac matrices in the gauge formalism to be presented in the following sections.

The requirement of form-invariance of the variation of the for a system of a complex $(1,0)$-spinor field $\psi(x)$ in a dynamic spacetime, whose dynamics is described by the Hamiltonian $\tilde{\mathcal{H}}_\mathrm{D}(\psi, \bar{\psi}, \tilde{\kappa}^\mu, \tilde{\bar{\kappa}}^\mu, \boldsymbol{\gamma}_\mu)$ is given by

$$\begin{aligned}
&\delta\int\limits_{V_x}\left(\tilde{\bar{\kappa}}^\alpha\frac{\partial\psi}{\partial x^\alpha} + \frac{\partial\bar{\psi}}{\partial x^\alpha}\tilde{\kappa}^\alpha + \mathrm{Tr}\left\{\tilde{\boldsymbol{k}}^{\beta\alpha}\frac{\partial\boldsymbol{\gamma}_\beta}{\partial x^\alpha}\right\} - \tilde{\mathcal{H}}_\mathrm{D}\right)\mathrm{d}^4x\\
&\stackrel{!}{=}\delta\int\limits_{V_X}\left(\tilde{\bar{\mathcal{K}}}^\alpha\frac{\partial\Psi}{\partial X^\alpha} + \frac{\partial\bar{\Psi}}{\partial X^\alpha}\tilde{\mathcal{K}}^\alpha + \mathrm{Tr}\left\{\tilde{\boldsymbol{K}}^{\beta\alpha}\frac{\partial\boldsymbol{\Gamma}_\beta}{\partial x^\alpha}\right\} - \tilde{\mathcal{H}}_\mathrm{D}'\right)\mathrm{d}^4X,
\end{aligned} \tag{5.15}$$

The $(1,0)$-spinor-$(1,0)$-tensor density $\tilde{\kappa}^\alpha = \kappa^\alpha\sqrt{-g}$ stands for the conjugate momentum of $\bar{\psi}$, and, correspondingly, the $(0,1)$-spinor-$(1,0)$-tensor density $\tilde{\bar{\kappa}}^\alpha = (\tilde{\kappa}^\alpha)^\dagger\boldsymbol{\gamma}^0$ the conjugate momentum field of the spinor ψ. The coordinate dependent Dirac matrix $\boldsymbol{\gamma}_\mu(x)$ describes the spacetime coupling of the spinor fields. Finally, the $(1,1)$-spinor-$(2,0)$-tensor density is the conjugate field of the $(1,1)$-spinor-$(0,1)$-tensor $\boldsymbol{\gamma}_\beta \equiv \gamma^a{}_{b\beta}$, which replaces the metric or the vierbeins in the actual spinor description.

The system Hamiltonian $\tilde{\mathcal{H}}_\mathrm{D}$ is supposed to be form-invariant under global rotations in the spin space, i.e. w.r.t. the spinor representation SL(2, $\mathbb{C}$) of the Lorentz group, mediated by constant transformations $\boldsymbol{S}$. The objective is now to modify that system such that the action functional in Eq. (5.15) becomes form-invariant under local rotations in the spin space and diffeomorphisms, i.e. w.r.t. the transformation group SL(2, $\mathbb{C}$)×Diff(M), mediated by the function $\boldsymbol{S}(x)$. (Notice that here and in the following we deploy utilize the equivalence of requiring the form-invariance of the equations of motion with the form-invariance of the action functional itself as shown in Chap. 4.)

5.2.2 Step 1: Generating Function for the Canonical Transformation of ψ, $\bar{\psi}$, and $\boldsymbol{\gamma}_\nu$

In contrast to Chap. 4 we deploy here the type-2 generating function of a canonical transformation in the realm of field theories [3]. That generating vector function, $\tilde{\mathcal{F}}_2^\mu\big(\psi, \tilde{\bar{\mathcal{K}}}, \bar{\psi}, \tilde{\mathcal{K}}, \boldsymbol{\gamma}, \tilde{\boldsymbol{K}}, x\big)$, which defines the mapping (5.18) of a Hamiltonian system of a complex spinor field $\psi(x)$, and the Dirac (1, 1)-spinor-(0, 1)-tensor $\boldsymbol{\gamma}_\alpha(x)$ from a space-time event x to X, is easily constructed:

$$\tilde{\mathcal{F}}_2^\mu\Big|_x = \left[\tilde{\bar{\mathcal{K}}}^\beta(X)\,\boldsymbol{S}\psi(x) + \bar{\psi}(x)\,\boldsymbol{S}^{-1}\tilde{\mathcal{K}}^\beta(X) + \mathrm{Tr}\left\{\tilde{\boldsymbol{K}}^{\alpha\beta}(X)\,\boldsymbol{S}\boldsymbol{\gamma}_\xi(x)\,\boldsymbol{S}^{-1}\frac{\partial x^\xi}{\partial X^\alpha}\right\}\right]\frac{\partial x^\mu}{\partial X^\beta}\left|\frac{\partial X}{\partial x}\right| \tag{5.16}$$

The general canonical transformation rules applied to the particular generating function (5.16) are:

$$\tilde{\bar{\kappa}}^\mu(x) = \frac{\partial \tilde{\mathcal{F}}_2^\mu(x)}{\partial \psi(x)} = \tilde{\bar{\mathcal{K}}}^\beta(X)\,\boldsymbol{S}\frac{\partial x^\mu}{\partial X^\beta}\left|\frac{\partial X}{\partial x}\right| \tag{5.17a}$$

$$\tilde{\kappa}^\mu(x) = \frac{\partial \tilde{\mathcal{F}}_2^\mu(x)}{\partial \bar{\psi}(x)} = \boldsymbol{S}^{-1}\tilde{\mathcal{K}}^\beta(X)\frac{\partial x^\mu}{\partial X^\beta}\left|\frac{\partial X}{\partial x}\right| \tag{5.17b}$$

$$\delta_\nu^\mu\bar{\Psi}(X) = \frac{\partial \tilde{\mathcal{F}}_2^\kappa(x)}{\partial \tilde{\mathcal{K}}^\nu(X)}\frac{\partial X^\mu}{\partial x^\kappa}\left|\frac{\partial x}{\partial X}\right| = \bar{\psi}(x)\,\boldsymbol{S}^{-1}\delta_\nu^\beta\frac{\partial x^\kappa}{\partial X^\beta}\frac{\partial X^\mu}{\partial x^\kappa} = \delta_\nu^\mu\bar{\psi}(x)\,\boldsymbol{S}^{-1} \tag{5.17c}$$

$$\delta_\nu^\mu\Psi(X) = \frac{\partial \tilde{\mathcal{F}}_2^\kappa(x)}{\partial \tilde{\bar{\mathcal{K}}}^\nu(X)}\frac{\partial X^\mu}{\partial x^\kappa}\left|\frac{\partial x}{\partial X}\right| = \boldsymbol{S}\psi(x)\delta_\nu^\beta\frac{\partial x^\kappa}{\partial X^\beta}\frac{\partial X^\mu}{\partial x^\kappa} = \delta_\nu^\mu\boldsymbol{S}\psi(x) \tag{5.17d}$$

$$\tilde{\boldsymbol{k}}^{\xi\mu}(x) = \frac{\partial \tilde{\mathcal{F}}_2^\mu(x)}{\partial \boldsymbol{\gamma}_\xi(x)} = \boldsymbol{S}^{-1}\tilde{\boldsymbol{K}}^{\alpha\beta}(X)\,\boldsymbol{S}\frac{\partial x^\xi}{\partial X^\alpha}\frac{\partial x^\mu}{\partial X^\beta}\left|\frac{\partial X}{\partial x}\right| \tag{5.17e}$$

$$\delta_\nu^\mu\boldsymbol{\Gamma}_\alpha(X) = \frac{\partial \tilde{\mathcal{F}}_2^\kappa(x)}{\partial \tilde{\boldsymbol{K}}^{\alpha\nu}(X)}\frac{\partial X^\mu}{\partial x^\kappa}\left|\frac{\partial x}{\partial X}\right| = \delta_\nu^\mu\boldsymbol{S}\boldsymbol{\gamma}_\xi(x)\,\boldsymbol{S}^{-1}\frac{\partial x^\xi}{\partial X^\alpha}. \tag{5.17f}$$

By construction of the generating function (5.16), the canonical transformation rules are:

$$\Psi(X) = \boldsymbol{S}\psi(x) \qquad \tilde{\mathcal{K}}^\mu(X) = \boldsymbol{S}\tilde{\kappa}^\beta(x)\frac{\partial X^\mu}{\partial x^\beta}\left|\frac{\partial x}{\partial X}\right| \tag{5.18a}$$

$$\bar{\Psi}(X) = \bar{\psi}(x)\,\boldsymbol{S}^{-1} \qquad \tilde{\bar{\mathcal{K}}}^{\mu}(X) = \tilde{\bar{\kappa}}^{\beta}(x)\,\boldsymbol{S}^{-1}\frac{\partial X^{\mu}}{\partial x^{\beta}}\left|\frac{\partial x}{\partial X}\right| \tag{5.18b}$$

$$\boldsymbol{\Gamma}_{\nu}(X) = \boldsymbol{S}\boldsymbol{\gamma}_{\alpha}(x)\,\boldsymbol{S}^{-1}\frac{\partial x^{\alpha}}{\partial X^{\nu}} \qquad \tilde{\boldsymbol{K}}^{\nu\mu}(X) = \boldsymbol{S}\,\tilde{\boldsymbol{k}}^{\alpha\beta}(x)\,\boldsymbol{S}^{-1}\frac{\partial X^{\nu}}{\partial x^{\alpha}}\frac{\partial X^{\mu}}{\partial x^{\beta}}\left|\frac{\partial x}{\partial X}\right|. \tag{5.18c}$$

This reproduces the required mappings (5.18a) and (5.18b) of the spinors and their conjugates, as well as that of $\boldsymbol{\gamma}_{\xi}$ from Eq. (5.18c).

The transformation rule for the Hamiltonian is again obtained from the divergence of the explicit x-dependence of the generating function $\tilde{\mathcal{F}}_2^{\mu}$:

$$\tilde{\mathcal{H}}_{\mathrm{D}}' = \left(\tilde{\mathcal{H}}_{\mathrm{D}} + \left.\frac{\partial \tilde{\mathcal{F}}_2^{\alpha}}{\partial x^{\alpha}}\right|_{\mathrm{expl}}\right)\left|\frac{\partial x}{\partial X}\right|. \tag{5.19}$$

✓ Explicit Calculation

According to the chain rule, we have with $\left|\frac{\partial X}{\partial x}\right| = \left|\frac{\partial x}{\partial X}\right|^{-1}$

$$\frac{\partial}{\partial x^{\alpha}}\left(\frac{\partial x^{\alpha}}{\partial X^{\beta}}\left|\frac{\partial X}{\partial x}\right|\right) = \left(\frac{\partial^2 x^{\alpha}}{\partial X^{\beta}\partial X^{\xi}}\frac{\partial X^{\xi}}{\partial x^{\alpha}} - \left|\frac{\partial X}{\partial x}\right|\frac{\partial\left|\frac{\partial x}{\partial X}\right|}{\partial X^{\beta}}\right)\left|\frac{\partial X}{\partial x}\right|. \tag{5.20}$$

The derivative of the determinant is obtained from the general identity for the derivative of the determinant of a matrix $A = (a_{\mu}{}^{\nu})$ with respect to a matrix element $a_{\mu}{}^{\nu}$:

$$\frac{\partial \det A}{\partial a_{\mu}{}^{\nu}} = \left(A^{-1}\right)_{\nu}{}^{\mu}\det A \quad\Rightarrow\quad \frac{\partial\left|\frac{\partial x}{\partial X}\right|}{\partial\left(\frac{\partial x^{\alpha}}{\partial X^{\xi}}\right)} = \frac{\partial X^{\xi}}{\partial x^{\alpha}}\left|\frac{\partial x}{\partial X}\right|, \tag{5.21}$$

the X^{β}-derivative of $\ln|\partial x/\partial X|$ in Eq. (5.20) is converted into

$$\left|\frac{\partial X}{\partial x}\right|\frac{\partial\left|\frac{\partial x}{\partial X}\right|}{\partial X^{\beta}} = \left|\frac{\partial X}{\partial x}\right|\frac{\partial\left|\frac{\partial x}{\partial X}\right|}{\partial\left(\frac{\partial x^{\alpha}}{\partial X^{\xi}}\right)}\frac{\partial\left(\frac{\partial x^{\alpha}}{\partial X^{\xi}}\right)}{\partial X^{\beta}} = \frac{\partial X^{\xi}}{\partial x^{\alpha}}\frac{\partial^2 x^{\alpha}}{\partial X^{\xi}\partial X^{\beta}}\left|\frac{\partial X}{\partial x}\right|. \tag{5.22}$$

Inserting Eq. (5.22) into (5.20) then yields

$$\frac{\partial}{\partial x^{\alpha}}\left(\frac{\partial x^{\alpha}}{\partial X^{\beta}}\left|\frac{\partial X}{\partial x}\right|\right) \equiv 0. \tag{5.23}$$

Making use of the identity (5.23), the derivatives of the explicitly spacetime-dependent coefficients of the generating function (5.16) sum up to

$$\left.\frac{\partial\tilde{\mathcal{F}}_2^\alpha}{\partial x^\alpha}\right|_{\text{expl}} = \frac{\partial x^\alpha}{\partial X^\beta}\left|\frac{\partial X}{\partial x}\right|\Bigg[\bar{\tilde{\mathcal{K}}}^\beta(X)\frac{\partial \boldsymbol{S}(x)}{\partial x^\alpha}\psi(x)+\bar{\psi}(x)\frac{\partial \boldsymbol{S}^{-1}(x)}{\partial x^\alpha}\tilde{\mathcal{K}}^\beta(X)$$
$$+\operatorname{Tr}\left\{\tilde{\boldsymbol{K}}^{(\eta\beta)}(X)\left(\frac{\partial \boldsymbol{S}}{\partial x^\alpha}\boldsymbol{\gamma}_\xi\,\boldsymbol{S}^{-1}\frac{\partial x^\xi}{\partial X^\eta}+\boldsymbol{S}\boldsymbol{\gamma}_\xi\frac{\partial \boldsymbol{S}^{-1}}{\partial x^\alpha}\frac{\partial x^\xi}{\partial X^\eta}+\boldsymbol{S}\boldsymbol{\gamma}_\xi\,\boldsymbol{S}^{-1}\frac{\partial^2 x^\xi}{\partial X^\eta\partial X^\lambda}\frac{\partial X^\lambda}{\partial x^\alpha}\right)\right\}$$
$$+\operatorname{Tr}\left\{\tilde{\boldsymbol{K}}^{[\eta\beta]}(X)\left(\frac{\partial \boldsymbol{S}}{\partial x^\alpha}\boldsymbol{\gamma}_\xi\,\boldsymbol{S}^{-1}\frac{\partial x^\xi}{\partial X^\eta}+\boldsymbol{S}\boldsymbol{\gamma}_\xi\frac{\partial \boldsymbol{S}^{-1}}{\partial x^\alpha}\frac{\partial x^\xi}{\partial X^\eta}\right)\right\}\Bigg]. \tag{5.24}$$

The Hamiltonian $\tilde{\mathcal{H}}_{\text{D}}$ is no longer invariant under the transformation generated by (5.16). The system's form-invariance under local Lorentz and diffeomorphism transformations can be restored only at the expense of adding new degrees of freedom to the system by amending the original Hamiltonian $\tilde{\mathcal{H}}_{\text{D}}$ with a particular "gauge Hamiltonian" $\tilde{\mathcal{H}}_{\text{Gauge}}$. The final "gauged" system $\tilde{\mathcal{H}}_{\text{D}}+\tilde{\mathcal{H}}_{\text{Gauge}}$ must not depend on the parameters of a particular symmetry transformation and be form-invariant under the local symmetry group.

After having set up the x^α-derivative of the explicit spacetime-dependent terms in the generating function $\tilde{\mathcal{F}}_2^\alpha$, the transformed momentum fields can then be replaced by the original momentum fields according to the canonical transformation rules (5.17a), (5.17b), and (5.17e):

$$\left.\frac{\partial\tilde{\mathcal{F}}_2^\alpha}{\partial x^\alpha}\right|_{\text{expl}} = \operatorname{Tr}\left\{\tilde{\boldsymbol{K}}^{(\beta\alpha)}\frac{\partial \boldsymbol{\Gamma}_\beta}{\partial X^\alpha}\right\}\left|\frac{\partial X}{\partial x}\right|-\operatorname{Tr}\left\{\tilde{\boldsymbol{k}}^{(\beta\alpha)}\frac{\partial \boldsymbol{\gamma}_\beta}{\partial x^\alpha}\right\} \tag{5.25}$$
$$+\bar{\tilde{\kappa}}^\alpha\boldsymbol{S}^{-1}\frac{\partial \boldsymbol{S}}{\partial x^\alpha}\psi-\bar{\psi}\boldsymbol{S}^{-1}\frac{\partial \boldsymbol{S}}{\partial x^\alpha}\tilde{\kappa}^\alpha+\operatorname{Tr}\left\{\left(\boldsymbol{\gamma}_\beta\,\tilde{\boldsymbol{k}}^{[\beta\alpha]}-\tilde{\boldsymbol{k}}^{[\beta\alpha]}\boldsymbol{\gamma}_\beta\right)\boldsymbol{S}^{-1}\frac{\partial \boldsymbol{S}}{\partial x^\alpha}\right\}.$$

The parameters of the particular diffeomorphism $x\mapsto X$ are now represented by the spinor transformation matrix $\boldsymbol{S}$ and its derivative only. At this point, a gauge field—also referred to as Spinor connection and denoted by $\boldsymbol{\omega}_\alpha(x)\equiv\omega^b{}_{c\alpha}(x)$—must be introduced. The gauge field is supposed to be a physical field with components exactly matching the index structure of the terms $\boldsymbol{S}^{-1}\partial\boldsymbol{S}/\partial x^\alpha$ in (5.25). Moreover, the transformation property $\boldsymbol{\omega}_\alpha(x)\mapsto\boldsymbol{\Omega}_\lambda(X)$ of the gauge field must render the transformation rule of the Hamiltonian (5.25) symmetric regarding its dependence on the original and the transformed fields in order for the gauged system to be form-invariant under the canonical transformation generated by (5.16). This is achieved if the gauge field $\boldsymbol{\omega}_\alpha(x)$ transforms inhomogeneously as

$$\boldsymbol{S}^{-1}\frac{\partial \boldsymbol{S}}{\partial x^\alpha}=\boldsymbol{\omega}_\alpha(x)-\boldsymbol{S}^{-1}\boldsymbol{\Omega}_\lambda(X)\,\boldsymbol{S}\frac{\partial X^\lambda}{\partial x^\alpha}, \tag{5.26}$$

hence

$$\boldsymbol{\Omega}_\mu(X)=\left(\boldsymbol{S}\boldsymbol{\omega}_\alpha(x)\,\boldsymbol{S}^{-1}-\frac{\partial \boldsymbol{S}}{\partial x^\alpha}\boldsymbol{S}^{-1}\right)\frac{\partial x^\alpha}{\partial X^\mu}. \tag{5.27}$$

Inserting (5.26) into (5.25) yields

$$\left.\frac{\partial \tilde{\mathcal{F}}_2^\alpha}{\partial x^\alpha}\right|_{\text{expl}} = \operatorname{Tr}\left\{\tilde{\boldsymbol{K}}^{(\beta\alpha)}\frac{\partial \boldsymbol{\Gamma}_\beta}{\partial X^\alpha}\right\}\left|\frac{\partial X}{\partial x}\right| - \operatorname{Tr}\left\{\tilde{\boldsymbol{k}}^{(\beta\alpha)}\frac{\partial \boldsymbol{\gamma}_\beta}{\partial x^\alpha}\right\}$$
$$+\tilde{\bar{\kappa}}^\alpha\left(\boldsymbol{\omega}_\alpha - \boldsymbol{S}^{-1}\boldsymbol{\Omega}_\lambda\,\boldsymbol{S}\frac{\partial X^\lambda}{\partial x^\alpha}\right)\psi - \bar{\psi}\left(\boldsymbol{\omega}_\alpha - \boldsymbol{S}^{-1}\boldsymbol{\Omega}_\lambda\,\boldsymbol{S}\frac{\partial X^\lambda}{\partial x^\alpha}\right)\tilde{\kappa}^\alpha$$
$$+\operatorname{Tr}\left\{\left(\boldsymbol{\gamma}_\beta\tilde{\boldsymbol{k}}^{[\beta\alpha]} - \tilde{\boldsymbol{k}}^{[\beta\alpha]}\boldsymbol{\gamma}_\beta\right)\left(\boldsymbol{\omega}_\alpha - \boldsymbol{S}^{-1}\boldsymbol{\Omega}_\lambda\,\boldsymbol{S}\frac{\partial X^\lambda}{\partial x^\alpha}\right)\right\}.$$

This has the following equivalent form if all terms depending on the transformed gauge field $\boldsymbol{\Omega}_\lambda$ are expressed in terms of the transformed momenta according to the canonical transformation rules (5.17):

$$\left.\frac{\partial \tilde{\mathcal{F}}_2^\alpha}{\partial x^\alpha}\right|_{\text{expl}} = \operatorname{Tr}\left\{\tilde{\boldsymbol{K}}^{(\beta\alpha)}\frac{\partial \boldsymbol{\Gamma}_\beta}{\partial X^\alpha}\right\}\left|\frac{\partial X}{\partial x}\right| - \operatorname{Tr}\left\{\tilde{\boldsymbol{k}}^{(\beta\alpha)}\frac{\partial \boldsymbol{\gamma}_\beta}{\partial x^\alpha}\right\}$$
$$+\tilde{\bar{\kappa}}^\alpha\boldsymbol{\omega}_\alpha\psi - \bar{\psi}\,\boldsymbol{\omega}_\alpha\tilde{\kappa}^\alpha - \tilde{\bar{\mathcal{K}}}^\alpha\,\boldsymbol{\Omega}_\alpha\,\Psi\left|\frac{\partial X}{\partial x}\right| + \bar{\Psi}\,\boldsymbol{\Omega}_\alpha\,\tilde{\mathcal{K}}^\alpha\left|\frac{\partial X}{\partial x}\right|$$
$$+\operatorname{Tr}\left\{\left(\boldsymbol{\gamma}_\beta\tilde{\boldsymbol{k}}^{[\beta\alpha]} - \tilde{\boldsymbol{k}}^{[\beta\alpha]}\boldsymbol{\gamma}_\alpha\right)\boldsymbol{\omega}_\alpha\right\} - \operatorname{Tr}\left\{\left(\boldsymbol{\Gamma}_\beta\tilde{\boldsymbol{K}}^{[\beta\alpha]} - \tilde{\boldsymbol{K}}^{[\beta\alpha]}\boldsymbol{\Gamma}_\beta\right)\boldsymbol{\Omega}_\alpha\right\}\left|\frac{\partial X}{\partial x}\right|$$
$$= \tilde{\mathcal{H}}'_{\text{Gauge}_1} + \tfrac{1}{2}\operatorname{Tr}\left\{\tilde{\boldsymbol{K}}^{\beta\alpha}\left(\frac{\partial \boldsymbol{\Gamma}_\beta}{\partial X^\alpha} + \frac{\partial \boldsymbol{\Gamma}_\alpha}{\partial X^\beta}\right)\right\}\left|\frac{\partial X}{\partial x}\right|$$
$$- \tilde{\mathcal{H}}_{\text{Gauge}_1} - \tfrac{1}{2}\operatorname{Tr}\left\{\tilde{\boldsymbol{k}}^{\beta\alpha}\left(\frac{\partial \boldsymbol{\gamma}_\beta}{\partial x^\alpha} + \frac{\partial \boldsymbol{\gamma}_\alpha}{\partial x^\beta}\right)\right\}. \tag{5.28}$$

A form-invariant gauge Hamiltonian emerges here as

$$\tilde{\mathcal{H}}_{\text{Gauge}_1} := -\tilde{\bar{\kappa}}^\alpha\boldsymbol{\omega}_\alpha\,\psi + \bar{\psi}\,\boldsymbol{\omega}_\alpha\,\tilde{\kappa}^\alpha - \tfrac{1}{2}\operatorname{Tr}\left\{\tilde{\boldsymbol{k}}^{\beta\alpha}\left(\boldsymbol{\omega}_\alpha\boldsymbol{\gamma}_\beta - \boldsymbol{\gamma}_\beta\,\boldsymbol{\omega}_\alpha - \boldsymbol{\omega}_\beta\,\boldsymbol{\gamma}_\alpha + \boldsymbol{\gamma}_\alpha\,\boldsymbol{\omega}_\beta\right)\right\}, \tag{5.29}$$

with $\tilde{\mathcal{H}}'_{\text{Gauge}_1}$ having the same form in the transformed fields. One observes that the transformation rule for the Hamiltonian is now symmetrically expressed in terms of the original and the transformed fields. Its previous dependence on the arbitrary parameters of the symmetry transformation—$\partial X^\mu/\partial x^\nu$ and the coefficients of $\boldsymbol{S}$—has now been replaced by the dependence on the gauge fields $\boldsymbol{\omega}_\alpha$ and $\boldsymbol{\Omega}_\alpha$.

The amended Hamiltonian system $\tilde{\mathcal{H}}_1$ at x is then described by the sum

$$\tilde{\mathcal{H}}_1 = \tilde{\mathcal{H}}_{\text{D}} + \tilde{\mathcal{H}}_{\text{Gauge}_1}. \tag{5.30}$$

The form-invariant gauged action is now given by:

$$
\begin{aligned}
S &= \int\limits_{V_x} \left[\bar{\tilde{\kappa}}^{\alpha} \frac{\partial \psi}{\partial x^{\alpha}} + \frac{\partial \bar{\psi}}{\partial x^{\alpha}} \tilde{\kappa}^{\alpha} + \tfrac{1}{2} \operatorname{Tr} \left\{ \tilde{\boldsymbol{k}}^{\beta\alpha} \left(\frac{\partial \boldsymbol{\gamma}_{\beta}}{\partial x^{\alpha}} - \frac{\partial \boldsymbol{\gamma}_{\alpha}}{\partial x^{\beta}} \right) \right\} - \tilde{\mathcal{H}}_1 \right] \mathrm{d}^4 x \\
&= \int\limits_{V_x} \left[\bar{\tilde{\kappa}}^{\alpha} \left(\frac{\partial \psi}{\partial x^{\alpha}} + \boldsymbol{\omega}_{\alpha} \psi \right) + \left(\frac{\partial \bar{\psi}}{\partial x^{\alpha}} - \bar{\psi}\, \boldsymbol{\omega}_{\alpha} \right) \tilde{\kappa}^{\alpha} \right. \\
&\quad \left. + \tfrac{1}{2} \operatorname{Tr} \left\{ \tilde{\boldsymbol{k}}^{\beta\alpha} \left(\frac{\partial \boldsymbol{\gamma}_{\beta}}{\partial x^{\alpha}} - \frac{\partial \boldsymbol{\gamma}_{\alpha}}{\partial x^{\beta}} + \boldsymbol{\omega}_{\alpha} \boldsymbol{\gamma}_{\beta} - \boldsymbol{\omega}_{\beta} \boldsymbol{\gamma}_{\alpha} - \boldsymbol{\gamma}_{\beta} \boldsymbol{\omega}_{\alpha} + \boldsymbol{\gamma}_{\alpha} \boldsymbol{\omega}_{\beta} \right) \right\} - \tilde{\mathcal{H}}_{\mathrm{D}} \right] \mathrm{d}^4 x .
\end{aligned}
\tag{5.31}
$$

5.2.3 Step 2: Generating Function Including the Gauge Field Transformation

With Eq. (5.27), the inhomogeneous transformation rule for the gauge field $\boldsymbol{\omega}_{\mu}(x)$ renders the action (5.31) form-invariant under the symmetry transformation group $\mathrm{SO}(1,3)\times\mathrm{Diff}(M)$. Yet, the gauge field $\boldsymbol{\omega}_{\mu}$ represents an external field, whose physical properties remain undetermined as its dynamics is not described by the Hamiltonian $\tilde{\mathcal{H}}_{\mathrm{D}}$. We thus need to make the gauge field $\boldsymbol{\omega}_{\mu}$ an internal dynamical quantity in order to finally end up with the gauge-invariant Hamiltonian of a closed system of fields. The way to include the description of the dynamics of the gauge field and its conjugate is to incorporate the transformation rules (5.27) for the gauge field into the canonical transformation formalism. This is achieved by implementing in the generating function (5.16) as the transformation property (5.27):

$$
\begin{aligned}
\tilde{\mathcal{F}}_2^{\mu}\Big|_x &= \left[\bar{\tilde{\mathcal{K}}}^{\beta}(X)\, \boldsymbol{S} \psi(x) + \bar{\psi}(x)\, \boldsymbol{S}^{-1} \tilde{\mathcal{K}}^{\beta}(X) + \operatorname{Tr} \left\{ \tilde{\boldsymbol{K}}^{\alpha\beta}(X)\, \boldsymbol{S} \boldsymbol{\gamma}_{\xi}(x)\, \boldsymbol{S}^{-1} \frac{\partial x^{\xi}}{\partial X^{\alpha}} \right\} \right. \\
&\quad \left. + \operatorname{Tr} \left\{ \tilde{\boldsymbol{Q}}^{\alpha\beta}(X) \left(\boldsymbol{S} \boldsymbol{\omega}_{\xi}(x)\, \boldsymbol{S}^{-1} - \frac{\partial \boldsymbol{S}}{\partial x^{\xi}} \boldsymbol{S}^{-1} \right) \frac{\partial x^{\xi}}{\partial X^{\alpha}} \right\} \right] \frac{\partial x^{\mu}}{\partial X^{\beta}} \left| \frac{\partial X}{\partial x} \right| .
\end{aligned}
\tag{5.32}
$$

With $\tilde{\boldsymbol{Q}}^{\alpha\beta}$ the conjugate momentum field of the transformed gauge field $\boldsymbol{\Omega}_{\alpha}$ is formally introduced. The pair of additional transformation rules follows as:

$$
\tilde{\boldsymbol{q}}^{\nu\mu}(x) = \frac{\partial \tilde{\mathcal{F}}_2^{\mu}(x)}{\partial \boldsymbol{\omega}_{\nu}(x)} = \boldsymbol{S}^{-1} \tilde{\boldsymbol{Q}}^{\alpha\beta}(X)\, \boldsymbol{S} \frac{\partial x^{\nu}}{\partial X^{\alpha}} \frac{\partial x^{\mu}}{\partial X^{\beta}} \left| \frac{\partial X}{\partial x} \right| \tag{5.33a}
$$

$$
\delta_{\nu}^{\mu}\, \boldsymbol{\Omega}_{\alpha}(X) = \frac{\partial \tilde{\mathcal{F}}_2^{\kappa}(x)}{\partial \tilde{\boldsymbol{Q}}^{\alpha\nu}(X)} \frac{\partial X^{\mu}}{\partial x^{\kappa}} \left| \frac{\partial x}{\partial X} \right| = \delta_{\nu}^{\beta} \left(\boldsymbol{S} \boldsymbol{\omega}_{\xi}(x)\, \boldsymbol{S}^{-1} - \frac{\partial \boldsymbol{S}}{\partial x^{\xi}} \boldsymbol{S}^{-1} \right) \frac{\partial x^{\xi}}{\partial X^{\alpha}} \frac{\partial x^{\kappa}}{\partial X^{\beta}} \frac{\partial X^{\mu}}{\partial x^{\kappa}} . \tag{5.33b}
$$

The additional transformation rule (5.33b) for the spinor connection $\boldsymbol{\omega}_{\mu}(x)$ reproduces the inhomogeneous rule from Eq. (5.27), as required. Its conjugate, $\tilde{\boldsymbol{q}}^{\xi\mu}(x)$, transforms homogeneously:

$$\boldsymbol{S}\tilde{\boldsymbol{q}}^{\nu\mu}(x)\,\boldsymbol{S}^{-1} = \tilde{\boldsymbol{Q}}^{\alpha\beta}(X)\frac{\partial x^{\nu}}{\partial X^{\alpha}}\frac{\partial x^{\mu}}{\partial X^{\beta}}\left|\frac{\partial X}{\partial x}\right|. \tag{5.34}$$

Making again use of the identity (5.23), the divergence of the generating function (5.32) with respect to the coefficients of the canonical transformation, i.e., with respect to its explicitly x-dependent terms, follows as

$$\begin{aligned}
\left.\frac{\partial\tilde{\mathcal{F}}_2^{\mu}}{\partial x^{\mu}}\right|_{\mathrm{expl}} &= \Bigg[\tilde{\bar{\mathcal{K}}}^{\beta}\frac{\partial\boldsymbol{S}}{\partial x^{\mu}}\psi + \bar{\psi}\frac{\partial\boldsymbol{S}^{-1}}{\partial x^{\mu}}\tilde{\mathcal{K}}^{\beta} \\
&+\mathrm{Tr}\left\{\tilde{\boldsymbol{K}}^{(\alpha\beta)}\left(\frac{\partial\boldsymbol{S}}{\partial x^{\mu}}\boldsymbol{\gamma}_{\xi}\boldsymbol{S}^{-1}\frac{\partial x^{\xi}}{\partial X^{\alpha}} + \boldsymbol{S}\boldsymbol{\gamma}_{\xi}\frac{\partial\boldsymbol{S}^{-1}}{\partial x^{\mu}}\frac{\partial x^{\xi}}{\partial X^{\alpha}} + \boldsymbol{S}\boldsymbol{\gamma}_{\xi}\boldsymbol{S}^{-1}\frac{\partial^2 x^{\xi}}{\partial X^{\alpha}\partial X^{\tau}}\frac{\partial X^{\tau}}{\partial x^{\mu}}\right)\right\} \\
&+\mathrm{Tr}\left\{\tilde{\boldsymbol{K}}^{[\alpha\beta]}\left(\frac{\partial\boldsymbol{S}}{\partial x^{\mu}}\boldsymbol{\gamma}_{\xi}\boldsymbol{S}^{-1}\frac{\partial x^{\xi}}{\partial X^{\alpha}} + \boldsymbol{S}\boldsymbol{\gamma}_{\xi}\frac{\partial\boldsymbol{S}^{-1}}{\partial x^{\mu}}\frac{\partial x^{\xi}}{\partial X^{\alpha}}\right)\right\} \\
&+\mathrm{Tr}\left\{\tilde{\boldsymbol{Q}}^{(\alpha\beta)}\left[\left(\frac{\partial\boldsymbol{S}}{\partial x^{\mu}}\boldsymbol{\omega}_{\xi}\boldsymbol{S}^{-1} + \boldsymbol{S}\boldsymbol{\omega}_{\xi}\frac{\partial\boldsymbol{S}^{-1}}{\partial x^{\mu}} - \frac{\partial\boldsymbol{S}}{\partial x^{\xi}}\frac{\partial\boldsymbol{S}^{-1}}{\partial x^{\mu}} - \frac{\partial^2\boldsymbol{S}}{\partial x^{\xi}\partial x^{\mu}}\boldsymbol{S}^{-1}\right)\frac{\partial x^{\xi}}{\partial X^{\alpha}}\right.\right. \\
&\qquad\left.\left.+\left(\boldsymbol{S}\boldsymbol{\omega}_{\xi}\boldsymbol{S}^{-1} - \frac{\partial\boldsymbol{S}}{\partial x^{\xi}}\boldsymbol{S}^{-1}\right)\frac{\partial^2 x^{\xi}}{\partial X^{\alpha}\partial X^{\lambda}}\frac{\partial X^{\lambda}}{\partial x^{\mu}}\right]\right\} \\
&+\mathrm{Tr}\left\{\tilde{\boldsymbol{Q}}^{[\alpha\beta]}\left(\frac{\partial\boldsymbol{S}}{\partial x^{\mu}}\boldsymbol{\omega}_{\xi}\boldsymbol{S}^{-1} + \boldsymbol{S}\boldsymbol{\omega}_{\xi}\frac{\partial\boldsymbol{S}^{-1}}{\partial x^{\mu}} - \frac{\partial\boldsymbol{S}}{\partial x^{\xi}}\frac{\partial\boldsymbol{S}^{-1}}{\partial x^{\mu}}\right)\frac{\partial x^{\xi}}{\partial X^{\alpha}}\right\}\Bigg]\frac{\partial x^{\mu}}{\partial X^{\beta}}\left|\frac{\partial X}{\partial x}\right|.
\end{aligned}$$

The terms proportional to $\tilde{\boldsymbol{Q}}^{(\alpha\beta)}$ can be replaced by the X^{β}-derivative of the identity embodied by the inhomogeneous transformation rule (5.27) for the spinor connection:

$$\begin{aligned}
\frac{\partial\boldsymbol{\Omega}_{\alpha}}{\partial X^{\beta}} - \boldsymbol{S}\frac{\partial\boldsymbol{\omega}_{\xi}}{\partial x^{\mu}}\boldsymbol{S}^{-1}\frac{\partial x^{\xi}}{\partial X^{\alpha}}\frac{\partial x^{\mu}}{\partial X^{\beta}} &= \left(\boldsymbol{S}\boldsymbol{\omega}_{\xi}(x)\,\boldsymbol{S}^{-1} - \frac{\partial\boldsymbol{S}}{\partial x^{\xi}}\boldsymbol{S}^{-1}\right)\frac{\partial^2 x^{\xi}}{\partial X^{\alpha}\partial X^{\beta}} \\
+\left(\frac{\partial\boldsymbol{S}}{\partial x^{\mu}}\boldsymbol{\omega}_{\xi}\boldsymbol{S}^{-1} + \boldsymbol{S}\boldsymbol{\omega}_{\xi}\frac{\partial\boldsymbol{S}^{-1}}{\partial x^{\mu}} \right.&\left.- \frac{\partial\boldsymbol{S}}{\partial x^{\xi}}\frac{\partial\boldsymbol{S}^{-1}}{\partial x^{\mu}} - \frac{\partial^2\boldsymbol{S}}{\partial x^{\xi}\partial x^{\mu}}\boldsymbol{S}^{-1}\right)\frac{\partial x^{\xi}}{\partial X^{\alpha}}\frac{\partial x^{\mu}}{\partial X^{\beta}}.
\end{aligned} \tag{5.35}$$

Inserting Eq. (5.35), the transformation rule for the Hamiltonian then simplifies to:

$$
\begin{aligned}
\left.\frac{\partial \tilde{\mathcal{F}}_2^{\mu}}{\partial x^{\mu}}\right|_{\text{expl}} = \Bigg[& \left(\bar{\tilde{\mathcal{K}}}^{\beta} \frac{\partial \boldsymbol{S}}{\partial x^{\mu}} \psi + \bar{\psi} \frac{\partial \boldsymbol{S}}{\partial x^{\mu}} \tilde{\mathcal{K}}^{\beta} \right) \frac{\partial x^{\mu}}{\partial X^{\beta}} \\
& + \operatorname{Tr}\left\{ \tilde{\boldsymbol{K}}^{(\alpha\beta)} \left(\frac{\partial \boldsymbol{\Gamma}_{\alpha}}{\partial X^{\beta}} - \boldsymbol{S} \frac{\partial \boldsymbol{\gamma}_{\xi}}{\partial x^{\tau}} \boldsymbol{S}^{-1} \frac{\partial x^{\xi}}{\partial X^{\alpha}} \frac{\partial x^{\tau}}{\partial X^{\beta}} \right) \right\} \\
& + \operatorname{Tr}\left\{ \tilde{\boldsymbol{K}}^{[\alpha\beta]} \left(\frac{\partial \boldsymbol{S}}{\partial x^{\mu}} \boldsymbol{\gamma}_{\xi} \boldsymbol{S}^{-1} \frac{\partial x^{\xi}}{\partial X^{\alpha}} \frac{\partial x^{\mu}}{\partial X^{\beta}} + \boldsymbol{S} \boldsymbol{\gamma}_{\xi} \frac{\partial \boldsymbol{S}^{-1}}{\partial x^{\mu}} \frac{\partial x^{\xi}}{\partial X^{\alpha}} \frac{\partial x^{\mu}}{\partial X^{\beta}} \right) \right\} \\
& + \operatorname{Tr}\left\{ \tilde{\boldsymbol{Q}}^{(\alpha\beta)} \left[\frac{\partial \boldsymbol{\Omega}_{\alpha}}{\partial X^{\beta}} - \boldsymbol{S} \frac{\partial \boldsymbol{\omega}_{\xi}}{\partial x^{\mu}} \boldsymbol{S}^{-1} \frac{\partial x^{\xi}}{\partial X^{\alpha}} \frac{\partial x^{\mu}}{\partial X^{\beta}} \right] \right\} \\
& + \operatorname{Tr}\left\{ \tilde{\boldsymbol{Q}}^{[\alpha\beta]} \left(\frac{\partial \boldsymbol{S}}{\partial x^{\mu}} \boldsymbol{\omega}_{\xi} \boldsymbol{S}^{-1} + \boldsymbol{S} \boldsymbol{\omega}_{\xi} \frac{\partial \boldsymbol{S}^{-1}}{\partial x^{\mu}} - \frac{\partial \boldsymbol{S}}{\partial x^{\xi}} \frac{\partial \boldsymbol{S}^{-1}}{\partial x^{\mu}} \right) \frac{\partial x^{\xi}}{\partial X^{\alpha}} \frac{\partial x^{\mu}}{\partial X^{\beta}} \right\} \Bigg] \left| \frac{\partial X}{\partial x} \right| .
\end{aligned}
$$

Making use of the canonical transformation rules (5.17) for the momentum fields this gives:

$$
\begin{aligned}
\left.\frac{\partial \tilde{\mathcal{F}}_2^{\alpha}}{\partial x^{\alpha}}\right|_{\text{expl}} = & \operatorname{Tr}\left\{ \tilde{\boldsymbol{K}}^{(\beta\alpha)} \frac{\partial \boldsymbol{\Gamma}_{\beta}}{\partial X^{\alpha}} \right\} \left| \frac{\partial X}{\partial x} \right| - \operatorname{Tr}\left\{ \tilde{\boldsymbol{k}}^{(\beta\alpha)} \frac{\partial \boldsymbol{\gamma}_{\beta}}{\partial x^{\alpha}} \right\} \\
& + \bar{\tilde{\kappa}}^{\alpha} \boldsymbol{S}^{-1} \frac{\partial \boldsymbol{S}}{\partial x^{\alpha}} \psi - \bar{\psi} \boldsymbol{S}^{-1} \frac{\partial \boldsymbol{S}}{\partial x^{\alpha}} \tilde{\kappa}^{\alpha} + \operatorname{Tr}\left\{ \left(\boldsymbol{\gamma}_{\beta} \tilde{\boldsymbol{k}}^{[\beta\alpha]} - \tilde{\boldsymbol{k}}^{[\beta\alpha]} \boldsymbol{\gamma}_{\beta} \right) \boldsymbol{S}^{-1} \frac{\partial \boldsymbol{S}}{\partial x^{\alpha}} \right\} \\
& + \operatorname{Tr}\Bigg\{ \tilde{\boldsymbol{q}}^{[\beta\alpha]} \left(\boldsymbol{S}^{-1} \frac{\partial \boldsymbol{S}}{\partial x^{\alpha}} \boldsymbol{\omega}_{\beta} - \boldsymbol{\omega}_{\beta} \boldsymbol{S}^{-1} \frac{\partial \boldsymbol{S}}{\partial x^{\alpha}} - \frac{\partial \boldsymbol{S}^{-1}}{\partial x^{\beta}} \frac{\partial \boldsymbol{S}}{\partial x^{\alpha}} \right) \\
& \quad + \tilde{\boldsymbol{Q}}^{(\beta\alpha)} \frac{\partial \boldsymbol{\Omega}_{\beta}}{\partial X^{\alpha}} \left| \frac{\partial X}{\partial x} \right| - \tilde{\boldsymbol{q}}^{(\beta\alpha)} \frac{\partial \boldsymbol{\omega}_{\beta}}{\partial x^{\alpha}} \Bigg\} .
\end{aligned}
\tag{5.36}
$$

The derivative of $\boldsymbol{S}$ is now again replaced according to Eq. (5.26). So, only the additional terms proportional to the skew-symmetric part $\tilde{\boldsymbol{q}}^{[\beta\alpha]}$ of the conjugate momentum tensor of the gauge field $\boldsymbol{\omega}_{\beta}$ remain to be discussed here:

$$
\begin{aligned}
& \operatorname{Tr}\left\{ \tilde{\boldsymbol{q}}^{[\beta\alpha]} \left(\boldsymbol{S}^{-1} \frac{\partial \boldsymbol{S}}{\partial x^{\alpha}} \boldsymbol{\omega}_{\beta} - \boldsymbol{\omega}_{\beta} \boldsymbol{S}^{-1} \frac{\partial \boldsymbol{S}}{\partial x^{\alpha}} - \frac{\partial \boldsymbol{S}^{-1}}{\partial x^{\beta}} \frac{\partial \boldsymbol{S}}{\partial x^{\alpha}} \right) \right\} \\
& = \operatorname{Tr}\Bigg\{ \tilde{\boldsymbol{q}}^{[\beta\alpha]} \Bigg[\boldsymbol{\omega}_{\alpha} \boldsymbol{\omega}_{\beta} - \boldsymbol{\omega}_{\beta} \boldsymbol{\omega}_{\alpha} - \boldsymbol{S}^{-1} \boldsymbol{\Omega}_{\eta} \boldsymbol{S} \boldsymbol{\omega}_{\beta} \frac{\partial X^{\eta}}{\partial x^{\alpha}} + \boldsymbol{\omega}_{\beta} \boldsymbol{S}^{-1} \boldsymbol{\Omega}_{\eta} \boldsymbol{S} \frac{\partial X^{\eta}}{\partial x^{\alpha}} \\
& \qquad + \left(\boldsymbol{\omega}_{\beta} \boldsymbol{S}^{-1} - \boldsymbol{S}^{-1} \boldsymbol{\Omega}_{\xi} \frac{\partial X^{\xi}}{\partial x^{\beta}} \right) \left(\boldsymbol{S} \boldsymbol{\omega}_{\alpha} - \boldsymbol{\Omega}_{\eta} \boldsymbol{S} \frac{\partial X^{\eta}}{\partial x^{\alpha}} \right) \Bigg] \Bigg\} \\
& = \operatorname{Tr}\Bigg\{ \tilde{\boldsymbol{q}}^{[\beta\alpha]} \Bigg[\boldsymbol{\omega}_{\alpha} \boldsymbol{\omega}_{\beta} - \cancel{\boldsymbol{\omega}_{\beta} \boldsymbol{\omega}_{\alpha}} - \cancel{\boldsymbol{S}^{-1} \boldsymbol{\Omega}_{\eta} \boldsymbol{S} \boldsymbol{\omega}_{\beta} \frac{\partial X^{\eta}}{\partial x^{\alpha}}} + \cancel{\boldsymbol{\omega}_{\beta} \boldsymbol{S}^{-1} \boldsymbol{\Omega}_{\eta} \boldsymbol{S} \frac{\partial X^{\eta}}{\partial x^{\alpha}}} \\
& \qquad + \cancel{\boldsymbol{\omega}_{\beta} \boldsymbol{\omega}_{\alpha}} - \cancel{\boldsymbol{\omega}_{\beta} \boldsymbol{S}^{-1} \boldsymbol{\Omega}_{\eta} \boldsymbol{S} \frac{\partial X^{\eta}}{\partial x^{\alpha}}} + \boldsymbol{S}^{-1} \boldsymbol{\Omega}_{\xi} \boldsymbol{\Omega}_{\eta} \boldsymbol{S} \frac{\partial X^{\xi}}{\partial x^{\beta}} \frac{\partial X^{\eta}}{\partial x^{\alpha}} - \cancel{\boldsymbol{S}^{-1} \boldsymbol{\Omega}_{\eta} \boldsymbol{S} \boldsymbol{\omega}_{\alpha} \frac{\partial X^{\eta}}{\partial x^{\beta}}} \Bigg] \Bigg\} \\
& = \operatorname{Tr}\left\{ \tilde{\boldsymbol{q}}^{[\beta\alpha]} \boldsymbol{\omega}_{\alpha} \boldsymbol{\omega}_{\beta} - \tilde{\boldsymbol{Q}}^{[\beta\alpha]} \boldsymbol{\Omega}_{\alpha} \boldsymbol{\Omega}_{\beta} \left| \frac{\partial X}{\partial x} \right| \right\} .
\end{aligned}
\tag{5.37}
$$

The terms in the last line of (5.37) again no longer depend on $\boldsymbol{S}$ but has been transformed into expressions of the gauge field and its conjugate symmetric in the original and transformed systems. In summary, the divergence of the explicitly spacetime-dependent terms of the generating function (5.32) can be completely expressed in terms of the original and the transformed fields:

$$\begin{aligned}
\left.\frac{\partial \tilde{\mathcal{F}}_2^{\alpha}}{\partial x^{\alpha}}\right|_{\text{expl}} &= \left(\bar{\Psi}\,\boldsymbol{\Omega}_{\alpha}\tilde{\mathcal{K}}^{\alpha} - \tilde{\bar{\mathcal{K}}}^{\alpha}\boldsymbol{\Omega}_{\alpha}\Psi\right)\left|\frac{\partial X}{\partial x}\right| - \left(\bar{\psi}\,\boldsymbol{\omega}_{\alpha}\tilde{\kappa}^{\alpha} - \tilde{\bar{\kappa}}^{\alpha}\boldsymbol{\omega}_{\alpha}\psi\right)\\
&+ \mathrm{Tr}\left\{\tilde{\boldsymbol{K}}^{(\beta\alpha)}\frac{\partial \boldsymbol{\Gamma}_{\beta}}{\partial X^{\alpha}} + \tilde{\boldsymbol{K}}^{[\beta\alpha]}\left(\boldsymbol{\Gamma}_{\beta}\boldsymbol{\Omega}_{\alpha} - \boldsymbol{\Omega}_{\alpha}\boldsymbol{\Gamma}_{\beta}\right)\right\}\left|\frac{\partial X}{\partial x}\right|\\
&- \mathrm{Tr}\left\{\tilde{\boldsymbol{k}}^{(\beta\alpha)}\frac{\partial \boldsymbol{\gamma}_{\beta}}{\partial x^{\alpha}} + \tilde{\boldsymbol{k}}^{[\beta\alpha]}\left(\boldsymbol{\gamma}_{\beta}\boldsymbol{\omega}_{\alpha} - \boldsymbol{\omega}_{\alpha}\boldsymbol{\gamma}_{\beta}\right)\right\}\\
&+ \mathrm{Tr}\left\{\tilde{\boldsymbol{Q}}^{(\beta\alpha)}\frac{\partial \boldsymbol{\Omega}_{\beta}}{\partial X^{\alpha}} - \tilde{\boldsymbol{Q}}^{[\beta\alpha]}\boldsymbol{\Omega}_{\alpha}\boldsymbol{\Omega}_{\beta}\right\}\left|\frac{\partial X}{\partial x}\right| - \mathrm{Tr}\left\{\tilde{\boldsymbol{q}}^{(\beta\alpha)}\frac{\partial \boldsymbol{\omega}_{\beta}}{\partial x^{\alpha}} - \tilde{\boldsymbol{q}}^{[\beta\alpha]}\boldsymbol{\omega}_{\alpha}\boldsymbol{\omega}_{\beta}\right\}.
\end{aligned} \tag{5.38}$$

The partial derivative terms will again be merged with the corresponding terms in the action functional (5.15), whereas the remaining terms of Eq. (5.38) define the amended gauge Hamiltonian $\tilde{\mathcal{H}}_{\text{Gauge}_2}$

$$\begin{aligned}
\tilde{\mathcal{H}}_{\text{Gauge}_2} = \bar{\psi}\,\boldsymbol{\omega}_{\alpha}\,\tilde{\kappa}^{\alpha} - \tilde{\bar{\kappa}}^{\alpha}\boldsymbol{\omega}_{\alpha}\,\psi + \tfrac{1}{2}\,\mathrm{Tr}\left\{\tilde{\boldsymbol{k}}^{\beta\alpha}\left(\boldsymbol{\gamma}_{\beta}\,\boldsymbol{\omega}_{\alpha} - \boldsymbol{\omega}_{\alpha}\boldsymbol{\gamma}_{\beta} + \boldsymbol{\omega}_{\beta}\,\boldsymbol{\gamma}_{\alpha} - \boldsymbol{\gamma}_{\alpha}\,\boldsymbol{\omega}_{\beta}\right)\right\}\\
+ \tfrac{1}{2}\,\mathrm{Tr}\left\{\tilde{\boldsymbol{q}}^{\beta\alpha}\left(\boldsymbol{\omega}_{\beta}\,\boldsymbol{\omega}_{\alpha} - \boldsymbol{\omega}_{\alpha}\,\boldsymbol{\omega}_{\beta}\right)\right\}
\end{aligned} \tag{5.39}$$

is mapped into the transformed gauge Hamiltonian $\tilde{\mathcal{H}}'_{\text{Gauge}_2}$ of exactly the same form in the transformed fields. With the explicitly written spinor indices, Eq. (5.39) is then

$$\begin{aligned}
\tilde{\mathcal{H}}_{\text{Gauge}_2} = \bar{\psi}_a\,\omega^{a}{}_{b\alpha}\,\tilde{\kappa}^{b\alpha} - \tilde{\bar{\kappa}}_a{}^{\alpha}\,\omega^{a}{}_{b\alpha}\,\psi^{b} + \tfrac{1}{2}\tilde{q}^{a}{}_{b\beta\alpha}\left(\omega^{b}{}_{c\beta}\,\omega^{c}{}_{a\alpha} - \omega^{b}{}_{c\alpha}\,\omega^{c}{}_{a\beta}\right)\\
+ \tfrac{1}{2}\tilde{k}^{a}{}_{b\beta\alpha}\left(\gamma^{b}{}_{c\beta}\,\omega^{c}{}_{a\alpha} - \omega^{b}{}_{c\alpha}\,\gamma^{c}{}_{a\beta} + \omega^{b}{}_{c\beta}\,\gamma^{c}{}_{a\alpha} - \gamma^{b}{}_{c\alpha}\,\omega^{c}{}_{a\beta}\right).
\end{aligned} \tag{5.40}$$

With the Hamiltonian $\tilde{\mathcal{H}}_{\text{D}}$ of the spinor fields in flat spacetime, the gauge Hamiltonian $\tilde{\mathcal{H}}_{\text{Gauge}_2}$ then yields the total Hamiltonian

$$\tilde{\mathcal{H}}_2 = \tilde{\mathcal{H}}_{\text{D}} + \tilde{\mathcal{H}}_{\text{Gauge}_2}, \tag{5.41}$$

which describes the dynamics in a curved spacetime in the action functional:

$$\begin{aligned}
S_1 = \int_V \Bigg[& \tilde{\bar{\kappa}}^{\alpha}\frac{\partial \psi}{\partial x^{\alpha}} + \frac{\partial \bar{\psi}}{\partial x^{\alpha}}\tilde{\kappa}^{\alpha}\\
& + \tfrac{1}{2}\,\mathrm{Tr}\left\{\tilde{\boldsymbol{k}}^{\beta\alpha}\left(\frac{\partial \boldsymbol{\gamma}_{\beta}}{\partial x^{\alpha}} - \frac{\partial \boldsymbol{\gamma}_{\alpha}}{\partial x^{\beta}}\right) + \tilde{\boldsymbol{q}}^{\beta\alpha}\left(\frac{\partial \boldsymbol{\omega}_{\beta}}{\partial x^{\alpha}} - \frac{\partial \boldsymbol{\omega}_{\alpha}}{\partial x^{\beta}}\right)\right\} - \tilde{\mathcal{H}}_2\Bigg]\,\mathrm{d}^4x.
\end{aligned}$$

The full action now writes in explicit form inserting the final gauge Hamiltonian (5.39):

$$
\begin{aligned}
S_1 = \int_V \Bigg[& \tilde{\bar{\kappa}}^{\alpha} \left(\frac{\partial \psi}{\partial x^{\alpha}} + \boldsymbol{\omega}_{\alpha}\, \psi \right) + \left(\frac{\partial \bar{\psi}}{\partial x^{\alpha}} - \bar{\psi}\, \boldsymbol{\omega}_{\alpha} \right) \tilde{\kappa}^{\alpha} \\
& + \mathrm{Tr} \left\{ \tfrac{1}{2} \tilde{\boldsymbol{k}}^{\beta\alpha} \left(\frac{\partial \boldsymbol{\gamma}_{\beta}}{\partial x^{\alpha}} - \frac{\partial \boldsymbol{\gamma}_{\alpha}}{\partial x^{\beta}} + \boldsymbol{\omega}_{\alpha}\, \boldsymbol{\gamma}_{\beta} - \boldsymbol{\gamma}_{\beta}\, \boldsymbol{\omega}_{\alpha} - \boldsymbol{\omega}_{\beta}\, \boldsymbol{\gamma}_{\alpha} + \boldsymbol{\gamma}_{\alpha}\, \boldsymbol{\omega}_{\beta} \right) \right\} \\
& + \mathrm{Tr} \left\{ \tfrac{1}{2} \tilde{\boldsymbol{q}}^{\beta\alpha} \left(\frac{\partial \boldsymbol{\omega}_{\beta}}{\partial x^{\alpha}} - \frac{\partial \boldsymbol{\omega}_{\alpha}}{\partial x^{\beta}} + \boldsymbol{\omega}_{\alpha}\, \boldsymbol{\omega}_{\beta} - \boldsymbol{\omega}_{\beta}\, \boldsymbol{\omega}_{\alpha} \right) \right\} - \tilde{\mathcal{H}}_{\mathrm{D}} \Bigg] \mathrm{d}^4 x. \qquad (5.42)
\end{aligned}
$$

This action functional now constitutes a closed system, hence no external fields are involved anymore. It has the identical form if expressed in the transformed fields—and thus complies with the Principle of General Relativity. As before, $\tilde{\mathcal{H}}_2$ is the modified, form-invariant Hamiltonian density, while $\tilde{\mathcal{H}}_{\mathrm{D}}$ describes the dynamics of the spinor fields in absence of gravitational coupling.

We must finally add to $\tilde{\mathcal{H}}_{\mathrm{D}}$ the Hamiltonian $\mathcal{H}_{\mathrm{Gr}} = \mathcal{H}_{\mathrm{Gr}}(\tilde{\boldsymbol{k}}, \tilde{\boldsymbol{q}}, \boldsymbol{\gamma})$ of the free gravitational field, hence the description of its dynamics in regions void of any sources of gravitation. This will be discussed in the next section.

5.2.4 Step 3: Adding the "Free Gravitation Hamiltonian" $\tilde{\mathcal{H}}_{\mathrm{Gr}}$

As the system Hamiltonian $\tilde{\mathcal{H}}_{\mathrm{D}}$ does not depend on the gauge field, i.e. the spinor connection $\boldsymbol{\omega}_{\beta}$—which was introduced in the course of the gauge procedure—the field equation for the partial x^{α}-derivative of the gauge field follows from the final amended gauge Hamiltonian $\tilde{\mathcal{H}}_{\mathrm{Gauge}_2}$, Eq. (5.39), setting up the respective canonical equation:

$$
\frac{\partial \boldsymbol{\omega}_{\beta}}{\partial x^{\alpha}} - \frac{\partial \boldsymbol{\omega}_{\alpha}}{\partial x^{\beta}} = 2 \frac{\partial \tilde{\mathcal{H}}_{\mathrm{Gauge}_2}}{\partial \tilde{\boldsymbol{q}}^{\beta\alpha}} = \boldsymbol{\omega}_{\beta}\, \boldsymbol{\omega}_{\alpha} - \boldsymbol{\omega}_{\alpha}\, \boldsymbol{\omega}_{\beta}. \qquad (5.43)
$$

With the definition of the curvature spinor,

$$
\boldsymbol{R}_{\alpha\beta} \equiv \frac{\partial \boldsymbol{\omega}_{\beta}}{\partial x^{\alpha}} - \frac{\partial \boldsymbol{\omega}_{\alpha}}{\partial x^{\beta}} + \boldsymbol{\omega}_{\alpha}\, \boldsymbol{\omega}_{\beta} - \boldsymbol{\omega}_{\beta}\, \boldsymbol{\omega}_{\alpha}, \qquad (5.44)
$$

this obviously gives $\boldsymbol{R}_{\alpha\beta} \equiv 0$. The curvature spinor vanishes identically for the gauged system described by the Hamiltonian $\tilde{\mathcal{H}}_{\mathrm{D}} + \tilde{\mathcal{H}}_{\mathrm{Gauge}_2}$. This is a common feature of all gauge theories: a Hamiltonian, resp. Lagrangian, which describes the dynamics of the "free" gauge fields must be added "by hand" to the gauge-invariant action, based on physical reasoning in order to allow for dynamical gauge fields. Here, a Hamiltonian $\tilde{\mathcal{H}}_{\mathrm{Gr}}$ must describe the dynamics of the "free" gauge fields, i.e., the dynamics of the gauge fields $\boldsymbol{\omega}_\mu(x)$ in the absence of any coupling to the spinor fields and their conjugates. Moreover, $\tilde{\mathcal{H}}_{\mathrm{Gr}}(\tilde{\boldsymbol{k}}, \tilde{\boldsymbol{q}}, \boldsymbol{\gamma})$ must be form-invariant under the canonical transformation rules of the fields $\tilde{\boldsymbol{k}}$, $\tilde{\boldsymbol{q}}$, and $\boldsymbol{\gamma}$ in order not to spoil the form-invariance of the final system

$$\tilde{\mathcal{H}}_3 = \tilde{\mathcal{H}}_{\mathrm{D}} + \tilde{\mathcal{H}}_{\mathrm{Gauge}_2} + \tilde{\mathcal{H}}_{\mathrm{Gr}}. \tag{5.45}$$

Consequently, the Hamiltonian $\tilde{\mathcal{H}}_{\mathrm{Gr}}$ can only be a function of the momenta $\tilde{\boldsymbol{k}}^{\beta\alpha}$, $\tilde{\boldsymbol{q}}^{\beta\alpha}$, and the metric spinors $\boldsymbol{\gamma}_\xi$. The canonical equation for the derivatives of the gauge field is then

$$\frac{\partial \boldsymbol{\omega}_\beta}{\partial x^\alpha} - \frac{\partial \boldsymbol{\omega}_\alpha}{\partial x^\beta} = 2\frac{\partial\left(\tilde{\mathcal{H}}_{\mathrm{Gauge}_2} + \tilde{\mathcal{H}}_{\mathrm{Gr}}\right)}{\partial \tilde{\boldsymbol{q}}^{\beta\alpha}} = \boldsymbol{\omega}_\beta\,\boldsymbol{\omega}_\alpha - \boldsymbol{\omega}_\alpha\,\boldsymbol{\omega}_\beta + 2\frac{\partial \tilde{\mathcal{H}}_{\mathrm{Gr}}}{\partial \tilde{\boldsymbol{q}}^{\beta\alpha}},$$

and the curvature spinor (5.44) is now a dynamic quantity:

$$\frac{1}{2}\boldsymbol{R}_{\beta\alpha} = -\frac{\partial \tilde{\mathcal{H}}_{\mathrm{Gr}}}{\partial \tilde{\boldsymbol{q}}^{\beta\alpha}}.$$

In analogy, a dependence of $\tilde{\mathcal{H}}_{\mathrm{Gr}}$ on $\tilde{\boldsymbol{k}}^{\beta\alpha}$ yields the canonical equation

$$\frac{\partial \boldsymbol{\gamma}_\beta}{\partial x^\alpha} - \frac{\partial \boldsymbol{\gamma}_\alpha}{\partial x^\beta} = 2\frac{\partial\left(\tilde{\mathcal{H}}_{\mathrm{Gauge}_2} + \tilde{\mathcal{H}}_{\mathrm{Gr}}\right)}{\partial \tilde{\boldsymbol{k}}^{\beta\alpha}} = -\boldsymbol{\omega}_\alpha\boldsymbol{\gamma}_\beta + \boldsymbol{\gamma}_\beta\,\boldsymbol{\omega}_\alpha + \boldsymbol{\omega}_\beta\,\boldsymbol{\gamma}_\alpha - \boldsymbol{\gamma}_\alpha\,\boldsymbol{\omega}_\beta + 2\frac{\partial \tilde{\mathcal{H}}_{\mathrm{Gr}}}{\partial \tilde{\boldsymbol{k}}^{\beta\alpha}},$$

hence

$$\frac{\partial \boldsymbol{\gamma}_\beta}{\partial x^\alpha} - \frac{\partial \boldsymbol{\gamma}_\alpha}{\partial x^\beta} + \boldsymbol{\omega}_\alpha\boldsymbol{\gamma}_\beta - \boldsymbol{\gamma}_\beta\boldsymbol{\omega}_\alpha - \boldsymbol{\omega}_\beta\boldsymbol{\gamma}_\alpha + \boldsymbol{\gamma}_\alpha\boldsymbol{\omega}_\beta = 2\frac{\partial \tilde{\mathcal{H}}_{\mathrm{Gr}}}{\partial \tilde{\boldsymbol{k}}^{\beta\alpha}}.$$

The dynamics of $\boldsymbol{\gamma}_\beta$ is thus determined by the dependence of $\tilde{\mathcal{H}}_{\mathrm{Gr}}$ on $\tilde{\boldsymbol{k}}^{\beta\alpha}$. It follows that the $\tilde{\boldsymbol{k}}^{\beta\alpha}$ is skew-symmetric in its tensor indices, $\tilde{\boldsymbol{k}}^{\beta\alpha} = -\tilde{\boldsymbol{k}}^{\alpha\beta}$.

5.2.5 *Canonical Equations for the Final Gauged Hamiltonian $\tilde{\mathcal{H}}_3$*

The canonical equations emerging from the action functional (5.42) form a closed system of eight coupled field equations for the interaction of the spin- 1/2 fields, described by $\tilde{\mathcal{H}}_{\mathrm{D}}$, and the gravitational field, described by $\tilde{\mathcal{H}}_{\mathrm{Gr}}$:

$$\frac{\partial \bar{\psi}_c}{\partial x^\nu} = \frac{\partial \tilde{\mathcal{H}}_3}{\partial \tilde{\kappa}^{c\nu}} = \frac{\partial \tilde{\mathcal{H}}_{\mathrm{D}}}{\partial \tilde{\kappa}^{c\nu}} + \bar{\psi}_b \, \omega^b{}_{c\nu} \tag{5.46a}$$

$$\frac{\partial \psi^c}{\partial x^\nu} = \frac{\partial \tilde{\mathcal{H}}_3}{\partial \tilde{\bar{\kappa}}_c{}^{\nu}} = \frac{\partial \tilde{\mathcal{H}}_{\mathrm{D}}}{\partial \tilde{\bar{\kappa}}_c{}^{\nu}} - \omega^c{}_{b\nu} \, \psi^b \tag{5.46b}$$

$$\frac{\partial \tilde{\bar{\kappa}}_c{}^{\beta}}{\partial x^\beta} = -\frac{\partial \tilde{\mathcal{H}}_3}{\partial \psi^c} = -\frac{\partial \tilde{\mathcal{H}}_{\mathrm{D}}}{\partial \psi^c} + \tilde{\bar{\kappa}}_b{}^{\beta} \, \omega^b{}_{c\beta} \tag{5.46c}$$

$$\frac{\partial \tilde{\kappa}^{c\beta}}{\partial x^\beta} = -\frac{\partial \tilde{\mathcal{H}}_3}{\partial \bar{\psi}_c} = -\frac{\partial \tilde{\mathcal{H}}_{\mathrm{D}}}{\partial \bar{\psi}_c} - \omega^c{}_{b\beta} \, \tilde{\kappa}^{b\beta} \tag{5.46d}$$

$$\frac{\partial \gamma^a{}_{b\mu}}{\partial x^\nu} - \frac{\partial \gamma^a{}_{b\nu}}{\partial x^\mu} = 2\frac{\partial \tilde{\mathcal{H}}_3}{\partial \tilde{k}^b{}_{a\mu\nu}} = 2\frac{\partial \tilde{\mathcal{H}}_{\mathrm{Gr}}}{\partial \tilde{k}^b{}_{a\mu\nu}} - \omega^a{}_{c\nu} \, \gamma^c{}_{b\mu} + \gamma^a{}_{c\mu} \, \omega^c{}_{b\nu} + \omega^a{}_{c\mu} \, \gamma^c{}_{b\nu} - \gamma^a{}_{c\nu} \, \omega^c{}_{b\mu} \tag{5.46e}$$

$$\frac{\partial \omega^a{}_{b\mu}}{\partial x^\nu} - \frac{\partial \omega^a{}_{b\nu}}{\partial x^\mu} = 2\frac{\partial \tilde{\mathcal{H}}_3}{\partial \tilde{q}^b{}_{a\mu\nu}} = 2\frac{\partial \tilde{\mathcal{H}}_{\mathrm{Gr}}}{\partial \tilde{q}^b{}_{a\mu\nu}} - \omega^a{}_{c\nu} \, \omega^c{}_{b\mu} + \omega^a{}_{c\mu} \, \omega^c{}_{b\nu} \tag{5.46f}$$

$$\frac{\partial \tilde{k}^a{}_{b\mu\beta}}{\partial x^\beta} = -\frac{\partial \tilde{\mathcal{H}}_3}{\partial \gamma^b{}_{a\mu}} = -\frac{\partial \tilde{\mathcal{H}}_{\mathrm{D}}}{\partial \gamma^b{}_{a\mu}} - \frac{\partial \tilde{\mathcal{H}}_{\mathrm{Gr}}}{\partial \gamma^b{}_{a\mu}} + \tilde{k}^a{}_{c\mu\beta} \, \omega^c{}_{b\beta} - \omega^a{}_{c\beta} \, \tilde{k}^c{}_{b\mu\beta} \tag{5.46g}$$

$$\begin{aligned}\frac{\partial \tilde{q}^a{}_{b\mu\beta}}{\partial x^\beta} &= -\frac{\partial \tilde{\mathcal{H}}_3}{\partial \omega^b{}_{a\mu}} = -\frac{\partial \tilde{\mathcal{H}}_{\mathrm{Gauge}}}{\partial \omega^b{}_{a\mu}} \\ &= \psi^a \, \tilde{\bar{\kappa}}_b{}^{\mu} - \tilde{\kappa}^{a\mu} \, \bar{\psi}_b + \omega^a{}_{c\beta} \, \tilde{q}^c{}_{b\beta\mu} - \tilde{q}^a{}_{c\beta mu} \, \omega^c{}_{b\beta} + \gamma^a{}_{c\beta} \, \tilde{k}^c{}_{b\beta\mu} - \tilde{k}^a{}_{c\beta\mu} \, \gamma^c{}_{b\beta} \,.\end{aligned} \tag{5.46h}$$

Hiding the spinor indices yields a concise notation of the above equations:

$$\frac{\partial \bar{\psi}}{\partial x^\nu} = \frac{\partial \tilde{\mathcal{H}}_3}{\partial \tilde{\kappa}^{\nu}} = \frac{\partial \tilde{\mathcal{H}}_{\mathrm{D}}}{\partial \tilde{\kappa}^{\nu}} + \bar{\psi} \, \boldsymbol{\omega}_\nu \tag{5.47a}$$

$$\frac{\partial \psi}{\partial x^\nu} = \frac{\partial \tilde{\mathcal{H}}_3}{\partial \tilde{\bar{\kappa}}^{\nu}} = \frac{\partial \tilde{\mathcal{H}}_{\mathrm{D}}}{\partial \tilde{\bar{\kappa}}^{\nu}} - \boldsymbol{\omega}_\nu \, \psi \tag{5.47b}$$

$$\frac{\partial \tilde{\bar{\kappa}}^{\beta}}{\partial x^\beta} = -\frac{\partial \tilde{\mathcal{H}}_3}{\partial \psi} = -\frac{\partial \tilde{\mathcal{H}}_{\mathrm{D}}}{\partial \psi} + \tilde{\bar{\kappa}}^{\beta} \, \boldsymbol{\omega}_\beta \tag{5.47c}$$

$$\frac{\partial \tilde{\kappa}^{\beta}}{\partial x^\beta} = -\frac{\partial \tilde{\mathcal{H}}_3}{\partial \bar{\psi}} = -\frac{\partial \tilde{\mathcal{H}}_{\mathrm{D}}}{\partial \bar{\psi}} - \boldsymbol{\omega}_\beta \, \tilde{\kappa}^{\beta} \tag{5.47d}$$

$$\frac{\partial \boldsymbol{\gamma}_\mu}{\partial x^\nu} - \frac{\partial \boldsymbol{\gamma}_\nu}{\partial x^\mu} = 2\frac{\partial \tilde{\mathcal{H}}_3}{\partial \tilde{\boldsymbol{k}}^{\mu\nu}} = 2\frac{\partial \tilde{\mathcal{H}}_{\mathrm{Gr}}}{\partial \tilde{\boldsymbol{k}}^{\mu\nu}} - \boldsymbol{\omega}_\nu \, \boldsymbol{\gamma}_\mu + \boldsymbol{\gamma}_\mu \, \boldsymbol{\omega}_\nu + \boldsymbol{\omega}_\mu \, \boldsymbol{\gamma}_\nu - \boldsymbol{\gamma}_\nu \, \boldsymbol{\omega}_\mu \tag{5.47e}$$

$$\frac{\partial \boldsymbol{\omega}_\mu}{\partial x^\nu} - \frac{\partial \boldsymbol{\omega}_\nu}{\partial x^\mu} = 2\frac{\partial \tilde{\mathcal{H}}_3}{\partial \tilde{\boldsymbol{q}}^{\mu\nu}} = 2\frac{\partial \tilde{\mathcal{H}}_{\mathrm{Gr}}}{\partial \tilde{\boldsymbol{q}}^{\mu\nu}} - \boldsymbol{\omega}_\nu \, \boldsymbol{\omega}_\mu + \boldsymbol{\omega}_\mu \, \boldsymbol{\omega}_\nu \tag{5.47f}$$

$$\frac{\partial \tilde{\boldsymbol{k}}^{\mu\beta}}{\partial x^\beta} = -\frac{\partial \tilde{\mathcal{H}}_3}{\partial \boldsymbol{\gamma}_\mu} = -\frac{\partial \tilde{\mathcal{H}}_{\mathrm{D}}}{\partial \boldsymbol{\gamma}_\mu} - \frac{\partial \tilde{\mathcal{H}}_{\mathrm{Gr}}}{\partial \boldsymbol{\gamma}_\mu} + \tilde{\boldsymbol{k}}^{\mu\beta} \, \boldsymbol{\omega}_\beta - \boldsymbol{\omega}_\beta \, \tilde{\boldsymbol{k}}^{\mu\beta} \tag{5.47g}$$

$$\begin{aligned}\frac{\partial \tilde{\boldsymbol{q}}^{\mu\beta}}{\partial x^\beta} &= -\frac{\partial \tilde{\mathcal{H}}_3}{\partial \boldsymbol{\omega}_\mu} = -\frac{\partial \tilde{\mathcal{H}}_{\mathrm{Gauge}}}{\partial \boldsymbol{\omega}_\mu} \\ &= \psi \, \tilde{\bar{\kappa}}^{\mu} - \tilde{\kappa}^{\mu} \, \bar{\psi} + \boldsymbol{\omega}_\beta \, \tilde{\boldsymbol{q}}^{\beta\mu} - \tilde{\boldsymbol{q}}^{\beta\mu} \, \boldsymbol{\omega}_\beta + \boldsymbol{\gamma}_\beta \, \tilde{\boldsymbol{k}}^{\beta\mu} - \tilde{\boldsymbol{k}}^{\beta\mu} \, \boldsymbol{\gamma}_\beta .\end{aligned} \tag{5.47h}$$

All equations actually constitute tensor equations, and are equivalent to those found in the vierbein formulation of Chap. 4, see the explicit proof below.

✓ Explicit Calculation

To verify that the obtained field equations (5.47) agree with those of the vierbein formulation (4.79), we translate the field equations from the spinor into the tetrad representation. With the definitions

$$\boldsymbol{\gamma}_\mu \equiv \boldsymbol{\gamma}_j\, e^j{}_\mu \quad \Rightarrow \quad e^j{}_\mu \equiv \tfrac{1}{4}\,\mathrm{Tr}\left\{\boldsymbol{\gamma}^j\,\boldsymbol{\gamma}_\mu\right\} \tag{5.48a}$$

$$\tilde{\boldsymbol{k}}^{\mu\nu} \equiv \tfrac{1}{4}\boldsymbol{\gamma}^j\,\tilde{k}_j{}^{\mu\nu} \quad \Rightarrow \quad \tilde{k}_j{}^{\mu\nu} \equiv \mathrm{Tr}\left\{\boldsymbol{\gamma}_j\,\tilde{\boldsymbol{k}}^{\mu\nu}\right\} \tag{5.48b}$$

$$\boldsymbol{\omega}_\mu \equiv -\tfrac{\mathrm{i}}{4}\boldsymbol{\sigma}_i{}^j\,\omega^i{}_{j\mu} \qquad \omega_{[ij]\mu} \equiv \tfrac{\mathrm{i}}{2}\,\mathrm{Tr}\left\{\boldsymbol{\sigma}_{ij}\,\boldsymbol{\omega}_\mu\right\} \tag{5.48c}$$

$$\tilde{\boldsymbol{q}}^{\mu\nu} \equiv -\tfrac{\mathrm{i}}{2}\boldsymbol{\sigma}_i{}^j\,\tilde{q}_j{}^{i\mu\nu} \qquad \tilde{q}^{[ij]\mu\nu} \equiv -\tfrac{\mathrm{i}}{4}\,\mathrm{Tr}\left\{\boldsymbol{\sigma}^{ij}\,\tilde{\boldsymbol{q}}^{\mu\nu}\right\} \tag{5.48d}$$

one gets by virtue of the Dirac algebra identities

$$\boldsymbol{\gamma}_n\,\boldsymbol{\sigma}_i{}^j - \boldsymbol{\sigma}_i{}^j\,\boldsymbol{\gamma}_n \equiv -2\mathrm{i}\left(\delta^j_n\,\boldsymbol{\gamma}_i - \eta_{in}\,\boldsymbol{\gamma}^j\right)$$

$$\boldsymbol{\sigma}_n{}^m\,\boldsymbol{\sigma}_i{}^j - \boldsymbol{\sigma}_i{}^j\,\boldsymbol{\sigma}_n{}^m \equiv -2\mathrm{i}\left(\delta^j_n\,\boldsymbol{\sigma}_i{}^m - \eta_{in}\,\boldsymbol{\sigma}^{jm} - \delta^m_i\,\boldsymbol{\sigma}_n{}^j + \eta^{mj}\,\boldsymbol{\sigma}_{ni}\right)$$

$$\mathrm{Tr}\left\{\boldsymbol{\sigma}_n{}^m\,\boldsymbol{\sigma}_i{}^j\right\} \equiv 4\left(\eta_{ni}\,\eta^{mj} - \delta^j_n\,\delta^m_i\right)$$

the following correspondences of the dynamical quantities:

$$\begin{aligned}
\boldsymbol{\gamma}_\mu\,\boldsymbol{\omega}_\nu - \boldsymbol{\omega}_\nu\,\boldsymbol{\gamma}_\mu &= -\tfrac{\mathrm{i}}{4}\left(\boldsymbol{\gamma}_n\,\boldsymbol{\sigma}_i{}^j - \boldsymbol{\sigma}_i{}^j\,\boldsymbol{\gamma}_n\right)e^n{}_\mu\,\omega^i{}_{j\nu}\\
&= -\tfrac{1}{2}\left(\delta^j_n\,\boldsymbol{\gamma}_i - \eta_{in}\,\boldsymbol{\gamma}^j\right)e^n{}_\mu\,\omega^i{}_{j\nu}\\
&= -\boldsymbol{\gamma}_i\,\omega^i{}_{j\nu}\,e^j{}_\mu\\
\boldsymbol{\omega}_\mu\,\boldsymbol{\omega}_\nu - \boldsymbol{\omega}_\nu\,\boldsymbol{\omega}_\mu &= -\frac{1}{16}\omega^n{}_{m\mu}\,\omega^i{}_{j\nu}\left(\boldsymbol{\sigma}_n{}^m\,\boldsymbol{\sigma}_i{}^j - \boldsymbol{\sigma}_i{}^j\,\boldsymbol{\sigma}_n{}^m\right)\\
&= \tfrac{\mathrm{i}}{8}\omega^n{}_{m\mu}\,\omega^i{}_{j\nu}\left(\delta^j_n\,\boldsymbol{\sigma}_i{}^m - \eta_{in}\,\boldsymbol{\sigma}^{jm} - \delta^m_i\,\boldsymbol{\sigma}_n{}^j + \eta^{mj}\,\boldsymbol{\sigma}_{ni}\right)\\
&= -\tfrac{\mathrm{i}}{4}\left(\omega^i{}_{n\mu}\,\omega^n{}_{j\nu} - \omega^i{}_{n\nu}\,\omega^n{}_{j\mu}\right)\boldsymbol{\sigma}_i{}^j\\
\tilde{\boldsymbol{k}}^{\mu\alpha}\,\boldsymbol{\omega}_\alpha - \boldsymbol{\omega}_\alpha\,\tilde{\boldsymbol{k}}^{\mu\alpha} &= -\frac{\mathrm{i}}{16}\tilde{k}_n{}^{\mu\alpha}\,\omega^i{}_{j\alpha}\left(\boldsymbol{\gamma}^n\,\boldsymbol{\sigma}_i{}^j - \boldsymbol{\sigma}_i{}^j\,\boldsymbol{\gamma}^n\right) = -\tfrac{1}{8}\tilde{k}_n{}^{\mu\alpha}\,\omega^i{}_{j\alpha}\left(\eta^{nj}\,\boldsymbol{\gamma}_i - \delta^n_i\,\boldsymbol{\gamma}^j\right)\\
&= \tfrac{1}{4}\tilde{k}_i{}^{\mu\alpha}\,\omega^i{}_{j\alpha}\,\boldsymbol{\gamma}^j\\
\tilde{\boldsymbol{k}}^{\mu\beta}\,\boldsymbol{\gamma}_\beta - \boldsymbol{\gamma}_\beta\,\tilde{\boldsymbol{k}}^{\mu\beta} &= \tfrac{1}{4}\tilde{k}_j{}^{\mu\beta}\,e^i{}_\beta\left(\boldsymbol{\gamma}^j\,\boldsymbol{\gamma}_i - \boldsymbol{\gamma}_i\,\boldsymbol{\gamma}^j\right)\\
&= \tfrac{\mathrm{i}}{2}\,\tilde{k}_j{}^{\mu\beta}\,e^i{}_\beta\,\boldsymbol{\sigma}_i{}^j\\
\tilde{\boldsymbol{q}}^{\mu\beta}\,\boldsymbol{\omega}_\beta - \boldsymbol{\omega}_\beta\,\tilde{\boldsymbol{q}}^{\mu\beta} &= \tfrac{1}{8}\tilde{q}_m{}^{n\mu\beta}\,\omega^i{}_{j\beta}\left(\boldsymbol{\sigma}^m{}_n\,\boldsymbol{\sigma}_i{}^j - \boldsymbol{\sigma}_i{}^j\,\boldsymbol{\sigma}^m{}_n\right) = -\tfrac{1}{8}\tilde{q}_m{}^{n\mu\beta}\,\omega^i{}_{j\beta}\left(\boldsymbol{\sigma}_n{}^m\,\boldsymbol{\sigma}_i{}^j - \boldsymbol{\sigma}_i{}^j\,\boldsymbol{\sigma}_n{}^m\right)\\
&= \tfrac{\mathrm{i}}{4}\tilde{q}_m{}^{n\mu\beta}\,\omega^i{}_{j\beta}\left(\delta^j_n\,\boldsymbol{\sigma}_i{}^m - \eta_{in}\,\boldsymbol{\sigma}^{jm} - \delta^m_i\,\boldsymbol{\sigma}_n{}^j + \eta^{mj}\,\boldsymbol{\sigma}_{ni}\right)
\end{aligned}$$

$$= \tfrac{i}{2}\left(\tilde{q}_j^{\ n\mu\beta}\,\omega^i_{\ n\beta} - \tilde{q}_n^{\ i\mu\beta}\,\omega^n_{\ j\beta}\right)\boldsymbol{\sigma}_i^{\ j}.$$

The derivatives of the Hamiltonian are equivalently expressed as:

$$\frac{\partial\tilde{\mathcal{H}}_3}{\partial\boldsymbol{\gamma}_\mu} = \frac{\partial\tilde{\mathcal{H}}_3}{\partial e^i_{\ \alpha}}\frac{\partial e^i_{\ \alpha}}{\partial\boldsymbol{\gamma}_\mu} = \tfrac{1}{4}\frac{\partial\tilde{\mathcal{H}}_3}{\partial e^i_{\ \alpha}}\frac{\partial}{\partial\boldsymbol{\gamma}_\mu}\operatorname{Tr}\left\{\boldsymbol{\gamma}^i\boldsymbol{\gamma}_\alpha\right\} = \tfrac{1}{4}\boldsymbol{\gamma}^i\frac{\partial\tilde{\mathcal{H}}_3}{\partial e^i_{\ \mu}} \quad\Leftrightarrow\quad \frac{\partial\tilde{\mathcal{H}}_3}{\partial e^i_{\ \mu}}e^i_{\ \nu} = \operatorname{Tr}\left\{\frac{\partial\tilde{\mathcal{H}}_3}{\partial\boldsymbol{\gamma}_\mu}\boldsymbol{\gamma}_\nu\right\}$$

$$\frac{\partial\tilde{\mathcal{H}}_3}{\partial\tilde{\boldsymbol{k}}^{\mu\nu}} = \frac{\partial\tilde{\mathcal{H}}_3}{\partial\tilde{k}_i^{\ \alpha\beta}}\frac{\partial\tilde{k}_i^{\ \alpha\beta}}{\partial\tilde{\boldsymbol{k}}^{\mu\nu}} = \frac{\partial\tilde{\mathcal{H}}_3}{\partial\tilde{k}_i^{\ \alpha\beta}}\frac{\partial}{\partial\tilde{\boldsymbol{k}}^{\mu\nu}}\operatorname{Tr}\left\{\boldsymbol{\gamma}_i\tilde{\boldsymbol{k}}^{\alpha\beta}\right\} = \boldsymbol{\gamma}_i\frac{\partial\tilde{\mathcal{H}}_3}{\partial\tilde{k}_i^{\ \mu\nu}}$$

$$\frac{\partial\tilde{\mathcal{H}}_3}{\partial\tilde{\boldsymbol{q}}^{\mu\nu}} = \frac{\partial\tilde{\mathcal{H}}_3}{\partial\tilde{q}_i^{\ j\alpha\beta}}\frac{\partial\tilde{q}_i^{\ j\alpha\beta}}{\partial\tilde{\boldsymbol{q}}^{\mu\nu}} = -\tfrac{i}{4}\frac{\partial\tilde{\mathcal{H}}_3}{\partial\tilde{q}_i^{\ j\alpha\beta}}\frac{\partial}{\partial\tilde{\boldsymbol{q}}^{\mu\nu}}\operatorname{Tr}\left\{\boldsymbol{\sigma}_i^{\ j}\tilde{\boldsymbol{q}}^{\alpha\beta}\right\} = -\tfrac{i}{4}\frac{\partial\tilde{\mathcal{H}}_3}{\partial\tilde{q}_i^{\ j\mu\nu}}\boldsymbol{\sigma}_i^{\ j}$$

$$\frac{\partial\tilde{\mathcal{H}}_3}{\partial\boldsymbol{\omega}_\mu} = \frac{\partial\tilde{\mathcal{H}}_3}{\partial\omega^i_{\ j\alpha}}\frac{\partial\omega^i_{\ j\alpha}}{\partial\boldsymbol{\omega}_\mu} = \tfrac{i}{2}\frac{\partial\tilde{\mathcal{H}}_3}{\partial\omega^i_{\ j\alpha}}\frac{\partial}{\partial\boldsymbol{\omega}_\mu}\operatorname{Tr}\left\{\boldsymbol{\sigma}^i_{\ j}\boldsymbol{\omega}_\alpha\right\} = \tfrac{i}{2}\frac{\partial\tilde{\mathcal{H}}_3}{\partial\omega^i_{\ j\mu}}\boldsymbol{\sigma}^i_{\ j}.$$

The field equations (5.46) are converted by means of Eqs. (5.48) into

$$\frac{\partial\psi}{\partial x^\nu} = \frac{\partial\tilde{\mathcal{H}}_3}{\partial\tilde{\kappa}^\nu} = \frac{\partial\tilde{\mathcal{H}}_\mathrm{D}}{\partial\tilde{\kappa}^\nu} - \tfrac{i}{4}\bar{\psi}\,\omega^i_{\ j\nu}\boldsymbol{\sigma}_i^{\ j} \tag{5.49a}$$

$$\frac{\partial\psi}{\partial x^\nu} = \frac{\partial\tilde{\mathcal{H}}_3}{\partial\bar{\tilde{\kappa}}^\nu} = \frac{\partial\tilde{\mathcal{H}}_\mathrm{D}}{\partial\bar{\tilde{\kappa}}^\nu} + \tfrac{i}{4}\omega^i_{\ j\nu}\boldsymbol{\sigma}_i^{\ j}\,\psi \tag{5.49b}$$

$$\frac{\partial\bar{\tilde{\kappa}}^\beta}{\partial x^\beta} = -\frac{\partial\tilde{\mathcal{H}}_3}{\partial\psi} = -\frac{\partial\tilde{\mathcal{H}}_\mathrm{D}}{\partial\psi} - \tfrac{i}{4}\bar{\tilde{\kappa}}^\beta\,\omega^i_{\ j\beta}\boldsymbol{\sigma}_i^{\ j} \tag{5.49c}$$

$$\frac{\partial\tilde{\kappa}^\beta}{\partial x^\beta} = -\frac{\partial\tilde{\mathcal{H}}_3}{\partial\bar{\psi}} = -\frac{\partial\tilde{\mathcal{H}}_\mathrm{D}}{\partial\bar{\psi}} + \tfrac{i}{4}\omega^i_{\ j\beta}\boldsymbol{\sigma}_i^{\ j}\,\tilde{\kappa}^\beta \tag{5.49d}$$

$$\frac{\partial\tilde{k}_j^{\ [\mu\beta]}}{\partial x^\beta} = -\frac{\partial\tilde{\mathcal{H}}_3}{\partial e^j_{\ \mu}} = -\frac{\partial\tilde{\mathcal{H}}_\mathrm{D}}{\partial e^j_{\ \mu}} - \frac{\partial\tilde{\mathcal{H}}_\mathrm{Gr}}{\partial e^j_{\ \mu}} + \tilde{k}_i^{\ [\mu\beta]}\omega^i_{\ j\beta} \tag{5.49e}$$

$$\frac{\partial\tilde{q}_i^{\ j\mu\beta}}{\partial x^\beta} = -\frac{\partial\tilde{\mathcal{H}}_3}{\partial\omega^i_{\ j\mu}} = -\frac{\partial\tilde{\mathcal{H}}_{\mathrm{Gauge}_2}}{\partial\omega^i_{\ j\mu}}$$

$$= \tfrac{i}{4}\bar{\psi}\,\boldsymbol{\sigma}_i^{\ j}\tilde{\kappa}^\mu - \tfrac{i}{4}\bar{\tilde{\kappa}}^\mu\boldsymbol{\sigma}_i^{\ j}\psi + \tfrac{1}{2}\tilde{k}_i^{\ \beta\mu}e_\beta^{\ j} - \tfrac{1}{2}e_{i_\beta}\,\tilde{k}^{j\beta\mu} + \tilde{q}_i^{\ n\beta\mu}\omega^j_{\ n\beta} - \omega_{in\beta}\,\tilde{q}^{jn\beta\mu} \tag{5.49f}$$

$$\frac{\partial e^i_{\ \mu}}{\partial x^\nu} - \frac{\partial e^i_{\ \nu}}{\partial x^\mu} = 2\frac{\partial\tilde{\mathcal{H}}_3}{\partial\tilde{k}_i^{\ \mu\nu}} = 2\frac{\partial\tilde{\mathcal{H}}_\mathrm{Gr}}{\partial\tilde{k}_i^{\ \mu\nu}} + \omega^i_{\ j\mu}e^j_{\ \nu} - \omega^i_{\ j\nu}e^j_{\ \mu} \tag{5.49g}$$

$$\frac{\partial\omega^i_{\ j\mu}}{\partial x^\nu} - \frac{\partial\omega^i_{\ j\nu}}{\partial x^\mu} = 2\frac{\partial\tilde{\mathcal{H}}_3}{\partial\tilde{q}_i^{\ j\mu\nu}} = 2\frac{\partial\tilde{\mathcal{H}}_\mathrm{Gr}}{\partial\tilde{q}_i^{\ j\mu\nu}} + \omega^i_{\ n\mu}\,\omega^n_{\ j\nu} - \omega^i_{\ n\nu}\,\omega^n_{\ j\mu} \tag{5.49h}$$

and thus indeed coincide with Eq. (4.79) of the vierbein description.

5.2.6 Consistency Equation

Calculating the x^{μ}-derivative of the canonical equation (5.47h), the second derivative of $\tilde{\boldsymbol{q}}^{\mu\beta}$ vanishes by virtue of its skew-symmetry in μ and β:

$$\frac{\partial}{\partial x^{\mu}}\left(\psi\, b\tilde{\bar{p}}i^{\mu}-\tilde{\kappa}^{\mu}\,\bar{\psi}-\tilde{\boldsymbol{k}}^{\beta\mu}\,\boldsymbol{\gamma}_{\beta}+\boldsymbol{\gamma}_{\beta}\,\tilde{\boldsymbol{k}}^{\beta\mu}-\tilde{\boldsymbol{q}}^{\beta\mu}\,\boldsymbol{\omega}_{\beta}+\boldsymbol{\omega}_{\beta}\,\tilde{\boldsymbol{q}}^{\beta\mu}\right)=0.$$

Inserting the canonical field equations (5.47) yields

$$\begin{aligned}
0=&\left(\frac{\partial\tilde{\mathcal{H}}_{\mathrm{D}}}{\partial\tilde{\bar{\kappa}}^{\mu}}-\boldsymbol{\omega}_{\mu}\,\psi\right)\tilde{\bar{\kappa}}^{\mu}-\psi\left(\frac{\partial\tilde{\mathcal{H}}_{\mathrm{D}}}{\partial\psi}-\tilde{\bar{\kappa}}^{\mu}\boldsymbol{\omega}_{\mu}\right)+\left(\frac{\partial\tilde{\mathcal{H}}_{\mathrm{D}}}{\partial\bar{\psi}}+\boldsymbol{\omega}_{\mu}\,\tilde{\kappa}^{\mu}\right)\bar{\psi}-\tilde{\kappa}^{\mu}\left(\frac{\partial\tilde{\mathcal{H}}_{\mathrm{D}}}{\partial\tilde{\kappa}^{\mu}}+\bar{\psi}\,\boldsymbol{\omega}_{\mu}\right)\\
&-\boldsymbol{\omega}_{\mu}\,\tilde{\boldsymbol{k}}^{\beta\mu}\boldsymbol{\gamma}_{\beta}+\tilde{\boldsymbol{k}}^{\beta\mu}\boldsymbol{\omega}_{\mu}\,\boldsymbol{\gamma}_{\beta}+\frac{\partial\tilde{\mathcal{H}}_{\mathrm{D}}}{\partial\boldsymbol{\gamma}_{\beta}}\boldsymbol{\gamma}_{\beta}+\frac{\partial\tilde{\mathcal{H}}_{\mathrm{Gr}}}{\partial\boldsymbol{\gamma}_{\beta}}\boldsymbol{\gamma}_{\beta}-\tilde{\boldsymbol{k}}^{\beta\mu}\boldsymbol{\omega}_{\mu}\,\boldsymbol{\gamma}_{\beta}+\tilde{\boldsymbol{k}}^{\beta\mu}\boldsymbol{\gamma}_{\beta}\,\boldsymbol{\omega}_{\mu}-\tilde{\boldsymbol{k}}^{\beta\mu}\frac{\partial\tilde{\mathcal{H}}_{\mathrm{Gr}}}{\partial\tilde{\boldsymbol{k}}^{\beta\mu}}\\
&+\boldsymbol{\gamma}_{\beta}\,\boldsymbol{\omega}_{\mu}\,\tilde{\boldsymbol{k}}^{\beta\mu}-\boldsymbol{\gamma}_{\beta}\,\tilde{\boldsymbol{k}}^{\beta\mu}\boldsymbol{\omega}_{\mu}-\boldsymbol{\gamma}_{\beta}\frac{\partial\tilde{\mathcal{H}}_{\mathrm{D}}}{\partial\boldsymbol{\gamma}_{\beta}}-\boldsymbol{\gamma}_{\beta}\frac{\partial\tilde{\mathcal{H}}_{\mathrm{Gr}}}{\partial\boldsymbol{\gamma}_{\beta}}+\boldsymbol{\omega}_{\mu}\,\boldsymbol{\gamma}_{\beta}\,\tilde{\boldsymbol{k}}^{\beta\mu}-\boldsymbol{\gamma}_{\beta}\,\boldsymbol{\omega}_{\mu}\,\tilde{\boldsymbol{k}}^{\beta\mu}+\frac{\partial\tilde{\mathcal{H}}_{\mathrm{Gr}}}{\partial\tilde{\boldsymbol{k}}^{\beta\mu}}\tilde{\boldsymbol{k}}^{\beta\mu}\\
&+\left(\tilde{\kappa}^{\beta}\,\bar{\psi}-\psi\,\tilde{\bar{\kappa}}^{\beta}+\tilde{\boldsymbol{q}}^{\beta\mu}\boldsymbol{\omega}_{\mu}-\boldsymbol{\omega}_{\mu}\,\tilde{\boldsymbol{q}}^{\beta\mu}+\tilde{\boldsymbol{k}}^{\beta\mu}\boldsymbol{\gamma}_{\mu}-\boldsymbol{\gamma}_{\mu}\,\tilde{\boldsymbol{k}}^{\beta\mu}\right)\boldsymbol{\omega}_{\beta}-\tilde{\boldsymbol{q}}^{\beta\mu}\boldsymbol{\omega}_{\mu}\,\boldsymbol{\omega}_{\beta}-\tilde{\boldsymbol{q}}^{\beta\mu}\frac{\partial\tilde{\mathcal{H}}_{\mathrm{Gr}}}{\partial\tilde{\boldsymbol{q}}^{\beta\mu}}\\
&+\boldsymbol{\omega}_{\beta}\left(\psi\,\tilde{\bar{\kappa}}^{\beta}-\tilde{\kappa}^{\beta}\,\bar{\psi}-\tilde{\boldsymbol{q}}^{\beta\mu}\boldsymbol{\omega}_{\mu}+\boldsymbol{\omega}_{\mu}\,\tilde{\boldsymbol{q}}^{\beta\mu}-\tilde{\boldsymbol{k}}^{\beta\mu}\boldsymbol{\gamma}_{\mu}+\boldsymbol{\gamma}_{\mu}\,\tilde{\boldsymbol{k}}^{\beta\mu}\right)+\boldsymbol{\omega}_{\mu}\,\boldsymbol{\omega}_{\beta}\,\tilde{\boldsymbol{q}}^{\beta\mu}+\frac{\partial\tilde{\mathcal{H}}_{\mathrm{Gr}}}{\partial\tilde{\boldsymbol{q}}^{\beta\mu}}\tilde{\boldsymbol{q}}^{\beta\mu}.
\end{aligned}$$

The terms which do not cancel are:

$$\begin{aligned}
0=&\frac{\partial\tilde{\mathcal{H}}_{\mathrm{D}}}{\partial\tilde{\bar{\kappa}}^{\beta}}\tilde{\bar{\kappa}}^{\beta}-\psi\frac{\partial\tilde{\mathcal{H}}_{\mathrm{D}}}{\partial\psi}+\frac{\partial\tilde{\mathcal{H}}_{\mathrm{D}}}{\partial\bar{\psi}}\bar{\psi}-\tilde{\kappa}^{\beta}\frac{\partial\tilde{\mathcal{H}}_{\mathrm{D}}}{\partial\tilde{\kappa}^{\beta}}+\frac{\partial\tilde{\mathcal{H}}_{\mathrm{D}}}{\partial\boldsymbol{\gamma}_{\beta}}\boldsymbol{\gamma}_{\beta}-\boldsymbol{\gamma}_{\beta}\frac{\partial\tilde{\mathcal{H}}_{\mathrm{D}}}{\partial\boldsymbol{\gamma}_{\beta}}\\
&+\frac{\partial\tilde{\mathcal{H}}_{\mathrm{Gr}}}{\partial\boldsymbol{\gamma}_{\beta}}\boldsymbol{\gamma}_{\beta}-\boldsymbol{\gamma}_{\beta}\frac{\partial\tilde{\mathcal{H}}_{\mathrm{Gr}}}{\partial\boldsymbol{\gamma}_{\beta}}+\frac{\partial\tilde{\mathcal{H}}_{\mathrm{Gr}}}{\partial\tilde{\boldsymbol{k}}^{\beta\mu}}\tilde{\boldsymbol{k}}^{\beta\mu}-\tilde{\boldsymbol{k}}^{\beta\mu}\frac{\partial\tilde{\mathcal{H}}_{\mathrm{Gr}}}{\partial\tilde{\boldsymbol{k}}^{\beta\mu}}+\frac{\partial\tilde{\mathcal{H}}_{\mathrm{Gr}}}{\partial\tilde{\boldsymbol{q}}^{\beta\mu}}\tilde{\boldsymbol{q}}^{\beta\mu}-\tilde{\boldsymbol{q}}^{\beta\mu}\frac{\partial\tilde{\mathcal{H}}_{\mathrm{Gr}}}{\partial\tilde{\boldsymbol{q}}^{\beta\mu}},
\end{aligned}\tag{5.50}$$

or, writing explicitly the spinor and spacetime indices:

$$\begin{aligned}
0=&\frac{\partial\tilde{\mathcal{H}}_{\mathrm{D}}}{\partial\tilde{\bar{\kappa}}_{a}{}^{\beta}}\tilde{\bar{\kappa}}_{b}{}^{\beta}-\psi^{a}\frac{\partial\tilde{\mathcal{H}}_{\mathrm{D}}}{\partial\psi^{b}}+\frac{\partial\tilde{\mathcal{H}}_{\mathrm{D}}}{\partial\bar{\psi}_{a}}\bar{\psi}_{b}-\tilde{\kappa}^{a\beta}\frac{\partial\tilde{\mathcal{H}}_{\mathrm{D}}}{\partial\tilde{\kappa}^{b\beta}}+\frac{\partial\tilde{\mathcal{H}}_{\mathrm{D}}}{\partial\gamma^{c}{}_{a\beta}}\gamma^{c}{}_{b\beta}-\gamma^{a}{}_{c\beta}\frac{\partial\tilde{\mathcal{H}}_{\mathrm{D}}}{\partial\gamma^{b}{}_{c\beta}}\\
&+\frac{\partial\tilde{\mathcal{H}}_{\mathrm{Gr}}}{\partial\gamma^{c}{}_{a\beta}}\gamma^{c}{}_{b\beta}-\gamma^{a}{}_{c\beta}\frac{\partial\tilde{\mathcal{H}}_{\mathrm{Gr}}}{\partial\gamma^{b}{}_{c\beta}}+\frac{\partial\tilde{\mathcal{H}}_{\mathrm{Gr}}}{\partial\tilde{k}^{c}{}_{a\beta\mu}}\tilde{k}^{c}{}_{b\beta\mu}-\tilde{k}^{a}{}_{c\beta\mu}\frac{\partial\tilde{\mathcal{H}}_{\mathrm{Gr}}}{\partial\tilde{k}^{b}{}_{c\beta\mu}}+\frac{\partial\tilde{\mathcal{H}}_{\mathrm{Gr}}}{\partial\tilde{q}^{c}{}_{a\beta\mu}}\tilde{q}^{c}{}_{b\beta\mu}-\tilde{q}^{a}{}_{c\beta\mu}\frac{\partial\tilde{\mathcal{H}}_{\mathrm{Gr}}}{\partial\tilde{q}^{b}{}_{c\beta\mu}}.
\end{aligned}$$

This Einstein-type equation relates the derivatives of the Hamiltonian $\tilde{\mathcal{H}}_0$ of the matter fields with the derivatives emerging from the model Hamiltonian $\tilde{\mathcal{H}}_{\mathrm{Gr}}$ for the free gravitational field. It holds for any given Hamiltonian densities $\tilde{\mathcal{H}}_0$ and $\tilde{\mathcal{H}}_{\mathrm{Gr}}$. The particular equation for the spinor and gravitational fields can be set up only after specifying those Hamiltonians, which will be done in the following sections.

This also applies for the canonical field equation (5.46g):

$$\frac{\partial \tilde{\boldsymbol{k}}^{\mu\beta}}{\partial x^{\beta}} - \tilde{\boldsymbol{k}}^{\mu\beta}\boldsymbol{\omega}_{\beta} + \boldsymbol{\omega}_{\beta}\tilde{\boldsymbol{k}}^{\mu\beta} = -\frac{\partial \tilde{\mathcal{H}}_{\mathrm{D}}}{\partial \boldsymbol{\gamma}_{\mu}} - \frac{\partial \tilde{\mathcal{H}}_{\mathrm{Gr}}}{\partial \boldsymbol{\gamma}_{\mu}},$$

which reduces to the spinor representation of the generic Einstein equation for the case $\tilde{\boldsymbol{k}}^{\mu\beta}(x) \equiv 0$, hence for zero torsion. With

$$\frac{\partial \tilde{\boldsymbol{k}}^{\mu\beta}}{\partial x^{\beta}} - \tilde{\boldsymbol{k}}^{\mu\beta}\boldsymbol{\omega}_{\beta} + \boldsymbol{\omega}_{\beta}\tilde{\boldsymbol{k}}^{\mu\beta} \equiv \tilde{\boldsymbol{k}}^{\mu\beta}{}_{;\beta} - \tilde{\boldsymbol{k}}^{\xi\beta} S^{\mu}{}_{\xi\beta} - 2\tilde{\boldsymbol{k}}^{\mu\xi} S^{\beta}{}_{\xi\beta},$$

we get the tensor equation (5.46g) in the form:

$$\tilde{\boldsymbol{k}}^{\mu\beta}{}_{;\beta} - \tilde{\boldsymbol{k}}^{\xi\beta} S^{\mu}{}_{\xi\beta} - 2\tilde{\boldsymbol{k}}^{\mu\xi} S^{\beta}{}_{\xi\beta} = -\frac{\partial \tilde{\mathcal{H}}_{\mathrm{D}}}{\partial \boldsymbol{\gamma}_{\mu}} - \frac{\partial \tilde{\mathcal{H}}_{\mathrm{Gr}}}{\partial \boldsymbol{\gamma}_{\mu}}.$$

5.2.7 *Generalized Dirac Equation of the "Gauged" System $\tilde{\mathcal{H}}_3 = \tilde{\mathcal{H}}_{\mathrm{D}} + \tilde{\mathcal{H}}_{\mathrm{Gauge}} + \tilde{\mathcal{H}}_{\mathrm{Gr}}$*

The Dirac Hamiltonian (5.10) can now be expressed equivalently in a dynamic space-time background:

$$\tilde{\mathcal{H}}_{\mathrm{D}}(\psi, \bar{\psi}, \tilde{\kappa}^{\mu}, \tilde{\bar{\kappa}}^{\mu}, \boldsymbol{\gamma}_{\mu}) = \frac{\mathrm{i}M}{2}\left(\bar{\psi}\boldsymbol{\gamma}_{\beta}\tilde{\kappa}^{\beta} - \frac{6}{\sqrt{-g}}\tilde{\bar{\kappa}}^{\alpha}\boldsymbol{\tau}_{\alpha\beta}\tilde{\kappa}^{\beta} - \tilde{\bar{\kappa}}^{\alpha}\boldsymbol{\gamma}_{\alpha}\psi\right) + (m - M)\,\bar{\psi}\psi\sqrt{-g}, \tag{5.51}$$

with the independent sets of fields ψ, $\bar{\psi}$, $\tilde{\kappa}^{\mu}$, $\tilde{\bar{\kappa}}^{\mu}$, and $\boldsymbol{\gamma}_{\mu}$, while the factor $\sqrt{-g}$—with g the determinant of the covariant metric $g_{\mu\nu}$—represents merely a function of the $\boldsymbol{\gamma}_{\mu}$. The $\boldsymbol{\gamma}_{\nu}$-derivatives of $g_{\alpha\beta}$ and $\sqrt{-g}$ can be derived from the $\boldsymbol{\gamma}$-matrix representation (5.2) of the metric $g_{\alpha\beta}$

$$\frac{\partial g_{\alpha\beta}}{\partial \boldsymbol{\gamma}_{\nu}} = \tfrac{1}{4}\frac{\partial}{\partial \boldsymbol{\gamma}_{\nu}}\mathrm{Tr}\left\{\boldsymbol{\gamma}_{\alpha}\boldsymbol{\gamma}_{\beta}\right\} = \tfrac{1}{4}\left(\boldsymbol{\gamma}_{\alpha}\,\delta^{\nu}_{\beta} + \boldsymbol{\gamma}_{\beta}\,\delta^{\nu}_{\alpha}\right) \tag{5.52a}$$

$$\frac{\partial g^{\alpha\beta}}{\partial \boldsymbol{\gamma}_{\nu}} = \tfrac{1}{4}\frac{\partial}{\partial \boldsymbol{\gamma}_{\nu}}\mathrm{Tr}\left\{\boldsymbol{\gamma}^{\alpha}\boldsymbol{\gamma}^{\beta}\right\} = -\tfrac{1}{4}\left(\boldsymbol{\gamma}^{\alpha} g^{\beta\nu} + \boldsymbol{\gamma}^{\beta} g^{\alpha\nu}\right) \tag{5.52b}$$

$$\frac{\partial\sqrt{-g}}{\partial \boldsymbol{\gamma}_{\nu}} = \frac{\partial\sqrt{-g}}{\partial g_{\alpha\beta}}\frac{\partial g_{\alpha\beta}}{\partial \boldsymbol{\gamma}_{\nu}} = \left(\tfrac{1}{2}g^{\beta\alpha}\sqrt{-g}\right)\tfrac{1}{4}\left(\boldsymbol{\gamma}_{\alpha}\,\delta^{\nu}_{\beta} + \boldsymbol{\gamma}_{\beta}\,\delta^{\nu}_{\alpha}\right) = \tfrac{1}{4}\boldsymbol{\gamma}^{\nu}\sqrt{-g} \tag{5.52c}$$

$$\frac{\partial\sqrt{-g}}{\partial \boldsymbol{\gamma}^{\nu}} = -\tfrac{1}{4}\boldsymbol{\gamma}_{\nu}\sqrt{-g} \tag{5.52d}$$

$$\frac{\partial}{\partial \boldsymbol{\gamma}_\nu} \frac{1}{\sqrt{-g}} = -\tfrac{1}{4} \boldsymbol{\gamma}^\nu \frac{1}{\sqrt{-g}} \tag{5.52e}$$

$$\frac{1}{\sqrt{-g}} \frac{\partial \sqrt{-g}}{\partial x^\nu} \mathbb{1} = \frac{1}{\sqrt{-g}} \frac{\partial \sqrt{-g}}{\partial g_{\alpha\beta}} \frac{\partial g_{\alpha\beta}}{\partial x^\nu} \mathbb{1} = \tfrac{1}{2} \left(\boldsymbol{\gamma}^\beta \frac{\partial \boldsymbol{\gamma}_\beta}{\partial x^\nu} + \frac{\partial \boldsymbol{\gamma}_\beta}{\partial x^\nu} \boldsymbol{\gamma}^\beta \right) = -\tfrac{1}{2} \left(\boldsymbol{\gamma}_\beta \frac{\partial \boldsymbol{\gamma}^\beta}{\partial x^\nu} + \frac{\partial \boldsymbol{\gamma}^\beta}{\partial x^\nu} \boldsymbol{\gamma}_\beta \right). \tag{5.52f}$$

The metric (1, 1)-spinor-(0, 1)-tensor $\boldsymbol{\gamma}_\nu$ is the dynamical variable which describes the geometry of the coordinate space and simultaneously mediates between the coordinate space and the spinor space.

With $\tilde{\mathcal{H}}_3$ the gauged system for the Dirac Hamiltonian $\tilde{\mathcal{H}}_{\mathrm{D}}$ from Eq. (5.51), the spinor-covariant canonical equations follow from Eqs. (5.46a) to (5.46d) as

$$\frac{\partial \bar{\psi}}{\partial x^\nu} = \frac{\partial \tilde{\mathcal{H}}_3}{\partial \tilde{\kappa}^\nu} = \bar{\psi} \left(\frac{\mathrm{i}M}{2} \boldsymbol{\gamma}_\nu + \boldsymbol{\omega}_\nu \right) - 3\mathrm{i}M \, \bar{\tilde{\kappa}}^\beta \boldsymbol{\tau}_{\beta\nu} \frac{1}{\sqrt{-g}} \tag{5.53a}$$

$$\frac{\partial \psi}{\partial x^\nu} = \frac{\partial \tilde{\mathcal{H}}_3}{\partial \bar{\tilde{\kappa}}^\nu} = -\left(\frac{\mathrm{i}M}{2} \boldsymbol{\gamma}_\nu + \boldsymbol{\omega}_\nu \right) \psi - 3\mathrm{i}M \boldsymbol{\tau}_{\nu\beta} \tilde{\kappa}^\beta \frac{1}{\sqrt{-g}} \tag{5.53b}$$

$$\frac{\partial \bar{\tilde{\kappa}}^\alpha}{\partial x^\alpha} = -\frac{\partial \tilde{\mathcal{H}}_3}{\partial \psi} = \bar{\tilde{\kappa}}^\alpha \left(\frac{\mathrm{i}M}{2} \boldsymbol{\gamma}_\alpha + \boldsymbol{\omega}_\alpha \right) - (m - M) \, \bar{\psi} \sqrt{-g} \tag{5.53c}$$

$$\frac{\partial \tilde{\kappa}^\alpha}{\partial x^\alpha} = -\frac{\partial \tilde{\mathcal{H}}_3}{\partial \bar{\psi}} = -\left(\frac{\mathrm{i}M}{2} \boldsymbol{\gamma}_\alpha + \boldsymbol{\omega}_\alpha \right) \tilde{\kappa}^\alpha - (m - M) \, \psi \sqrt{-g}. \tag{5.53d}$$

Equations (5.53a) and (5.53b) are solved for the momentum spinors by contraction with $\boldsymbol{\sigma}^{\nu\alpha}$ from the right and with $\boldsymbol{\sigma}^{\alpha\nu}$ from the left, respectively:

$$\bar{\tilde{\kappa}}^\alpha = \left[\frac{\mathrm{i}}{2} \bar{\psi} \boldsymbol{\gamma}^\alpha + \frac{\mathrm{i}}{3M} \left(\frac{\partial \bar{\psi}}{\partial x^\beta} - \bar{\psi} \, \boldsymbol{\omega}_\beta \right) \boldsymbol{\sigma}^{\beta\alpha} \right] \sqrt{-g} \tag{5.53a'}$$

$$\tilde{\kappa}^\alpha = \left[-\frac{\mathrm{i}}{2} \boldsymbol{\gamma}^\alpha \psi + \frac{\mathrm{i}}{3M} \boldsymbol{\sigma}^{\alpha\beta} \left(\frac{\partial \psi}{\partial x^\beta} + \boldsymbol{\omega}_\beta \, \psi \right) \right] \sqrt{-g}. \tag{5.53b'}$$

The momentum tensors can now be eliminated by inserting Eqs. (5.53a′) and (5.53b′) into the canonical equations (5.53c) and (5.53d), respectively, as shown in the explicit calculation below.

✓ Explicit Calculation

Inserting Eq. (5.53b′) into (5.53d) yields:

$$\frac{\partial}{\partial x^\alpha} \left\{ \left[-\frac{\mathrm{i}}{2} \boldsymbol{\gamma}^\alpha \psi + \frac{\mathrm{i}}{3M} \boldsymbol{\sigma}^{\alpha\beta} \left(\frac{\partial \psi}{\partial x^\beta} + \boldsymbol{\omega}_\beta \psi \right) \right] \sqrt{-g} \right\}$$

$$= \left(\frac{\mathrm{i}M}{2}\boldsymbol{\gamma}_\alpha + \boldsymbol{\omega}_\alpha\right)\left[\frac{\mathrm{i}}{2}\boldsymbol{\gamma}^\alpha\psi - \frac{\mathrm{i}}{3M}\boldsymbol{\sigma}^{\alpha\beta}\left(\frac{\partial\psi}{\partial x^\beta} + \boldsymbol{\omega}_\beta\psi\right)\right]\sqrt{-g} - (m - M)\,\psi\sqrt{-g}$$

Expanding

$$\left[-\frac{\mathrm{i}}{2}\boldsymbol{\gamma}^\alpha\psi + \frac{\mathrm{i}}{3M}\boldsymbol{\sigma}^{\alpha\beta}\left(\frac{\partial\psi}{\partial x^\beta} + \boldsymbol{\omega}_\beta\psi\right)\right]\frac{1}{\sqrt{-g}}\frac{\partial\sqrt{-g}}{\partial x^\alpha}$$
$$-\frac{\mathrm{i}}{2}\frac{\partial\boldsymbol{\gamma}^\alpha}{\partial x^\alpha}\psi - \frac{\mathrm{i}}{2}\boldsymbol{\gamma}^\alpha\frac{\partial\psi}{\partial x^\alpha} + \frac{\mathrm{i}}{3M}\frac{\partial\boldsymbol{\sigma}^{\alpha\beta}}{\partial x^\alpha}\left(\frac{\partial\psi}{\partial x^\beta} + \boldsymbol{\omega}_\beta\psi\right) + \frac{\mathrm{i}}{3M}\boldsymbol{\sigma}^{\alpha\beta}\left(\cancel{\frac{\partial^2\psi}{\partial x^\beta\partial x^\alpha}} + \frac{\partial}{\partial x^\alpha}(\boldsymbol{\omega}_\beta\psi)\right)$$
$$= \frac{\mathrm{i}M}{2}\boldsymbol{\gamma}_\alpha\left[\cancel{\frac{\mathrm{i}}{2}\boldsymbol{\gamma}^\alpha\psi} - \frac{\mathrm{i}}{3M}\boldsymbol{\sigma}^{\alpha\beta}\left(\frac{\partial\psi}{\partial x^\beta} + \boldsymbol{\omega}_\beta\psi\right)\right]$$
$$+\boldsymbol{\omega}_\alpha\left[\frac{\mathrm{i}}{2}\boldsymbol{\gamma}^\alpha\psi - \frac{\mathrm{i}}{3M}\boldsymbol{\sigma}^{\alpha\beta}\left(\frac{\partial\psi}{\partial x^\beta} + \boldsymbol{\omega}_\beta\psi\right)\right] - (m - \cancel{M})\,\psi,$$

hence

$$\left[-\frac{\mathrm{i}}{2}\boldsymbol{\gamma}^\alpha\psi + \frac{\mathrm{i}}{3M}\boldsymbol{\sigma}^{\alpha\beta}\left(\frac{\partial\psi}{\partial x^\beta} + \boldsymbol{\omega}_\beta\psi\right)\right]\frac{1}{\sqrt{-g}}\frac{\partial\sqrt{-g}}{\partial x^\alpha}$$
$$-\frac{\mathrm{i}}{2}\frac{\partial\boldsymbol{\gamma}^\alpha}{\partial x^\alpha}\psi - \frac{\mathrm{i}}{2}\boldsymbol{\gamma}^\alpha\frac{\partial\psi}{\partial x^\alpha} + \frac{\mathrm{i}}{3M}\frac{\partial\boldsymbol{\sigma}^{\alpha\beta}}{\partial x^\alpha}\left(\frac{\partial\psi}{\partial x^\beta} + \boldsymbol{\omega}_\beta\psi\right) + \frac{\mathrm{i}}{3M}\boldsymbol{\sigma}^{\alpha\beta}\frac{\partial}{\partial x^\alpha}(\boldsymbol{\omega}_\beta\psi)$$
$$= \frac{\mathrm{i}}{2}\boldsymbol{\gamma}^\beta\left(\frac{\partial\psi}{\partial x^\beta} + \boldsymbol{\omega}_\beta\psi\right) + \frac{\mathrm{i}}{2}\boldsymbol{\omega}_\alpha\boldsymbol{\gamma}^\alpha\psi - \frac{\mathrm{i}}{3M}\boldsymbol{\omega}_\alpha\boldsymbol{\sigma}^{\alpha\beta}\left(\frac{\partial\psi}{\partial x^\beta} + \boldsymbol{\omega}_\beta\psi\right) - m\psi,$$

and thus

$$\frac{\mathrm{i}}{3M}\left[\boldsymbol{\sigma}^{\alpha\beta}\left(\frac{\partial\psi}{\partial x^\beta} + \boldsymbol{\omega}_\beta\psi\right)\frac{1}{\sqrt{-g}}\frac{\partial\sqrt{-g}}{\partial x^\alpha} + \boldsymbol{\omega}_\alpha\boldsymbol{\sigma}^{\alpha\beta}\left(\frac{\partial\psi}{\partial x^\beta} + \boldsymbol{\omega}_\beta\psi\right)\right.$$
$$\left.+\frac{\partial\boldsymbol{\sigma}^{\alpha\beta}}{\partial x^\alpha}\left(\frac{\partial\psi}{\partial x^\beta} + \boldsymbol{\omega}_\beta\psi\right) + \boldsymbol{\sigma}^{\alpha\beta}\frac{\partial\boldsymbol{\omega}_\beta}{\partial x^\alpha}\psi + \boldsymbol{\sigma}^{\alpha\beta}\boldsymbol{\omega}_\beta\frac{\partial\psi}{\partial x^\alpha}\right]$$
$$= \frac{\mathrm{i}}{2}\boldsymbol{\gamma}^\alpha\left(\frac{\partial\psi}{\partial x^\alpha} + \boldsymbol{\omega}_\alpha\psi\right) + \frac{\mathrm{i}}{2}\boldsymbol{\omega}_\alpha\boldsymbol{\gamma}^\alpha\psi - m\psi + \frac{\mathrm{i}}{2}\boldsymbol{\gamma}^\alpha\psi\frac{1}{\sqrt{-g}}\frac{\partial\sqrt{-g}}{\partial x^\alpha} + \frac{\mathrm{i}}{2}\frac{\partial\boldsymbol{\gamma}^\alpha}{\partial x^\alpha}\psi + \frac{\mathrm{i}}{2}\boldsymbol{\gamma}^\alpha\frac{\partial\psi}{\partial x^\alpha}.$$

Finally

$$\frac{\mathrm{i}}{3M}\left[\boldsymbol{\sigma}^{\alpha\beta}\left(\frac{\partial\boldsymbol{\omega}_\beta}{\partial x^\alpha} + \boldsymbol{\omega}_\alpha\boldsymbol{\omega}_\beta\right)\psi + \left(\frac{\partial\boldsymbol{\sigma}^{\alpha\beta}}{\partial x^\alpha} + \boldsymbol{\omega}_\alpha\boldsymbol{\sigma}^{\alpha\beta} - \boldsymbol{\sigma}^{\alpha\beta}\boldsymbol{\omega}_\alpha + \boldsymbol{\sigma}^{\alpha\beta}\frac{\partial\ln\sqrt{-g}}{\partial x^\alpha}\right)\left(\frac{\partial\psi}{\partial x^\beta} + \boldsymbol{\omega}_\beta\psi\right)\right]$$
$$= \mathrm{i}\boldsymbol{\gamma}^\alpha\left(\frac{\partial\psi}{\partial x^\alpha} + \boldsymbol{\omega}_\alpha\psi\right) + \frac{\mathrm{i}}{2}\left(\frac{\partial\boldsymbol{\gamma}^\alpha}{\partial x^\alpha} + \boldsymbol{\omega}_\alpha\boldsymbol{\gamma}^\alpha - \boldsymbol{\gamma}^\alpha\boldsymbol{\omega}_\alpha + \boldsymbol{\gamma}^\alpha\frac{\partial\ln\sqrt{-g}}{\partial x^\alpha}\right)\psi - m\psi$$

Collecting the terms, this yields the final form of the generalized Dirac equation for the spinor ψ in a general, not necessarily metric compatible spacetime including torsion, which includes the anomalous gravitational coupling effect for a "fermion length" parameter $\ell \equiv 1/M$:

$$\begin{aligned} &\mathrm{i}\gamma^\alpha\left(\frac{\partial\psi}{\partial x^\alpha}+\omega_\alpha\psi\right)+\frac{\mathrm{i}}{2}\left[\frac{\partial\gamma^\alpha}{\partial x^\alpha}+\omega_\alpha\gamma^\alpha-\gamma^\alpha\omega_\alpha-\tfrac{1}{2}\gamma^\alpha\left(\gamma_\beta\frac{\partial\gamma^\beta}{\partial x^\alpha}+\frac{\partial\gamma^\beta}{\partial x^\alpha}\gamma_\beta\right)\right]\psi-m\psi\\ &=\frac{\mathrm{i}\ell}{3}\left\{\sigma^{\alpha\xi}\left(\frac{\partial\omega_\xi}{\partial x^\alpha}+\omega_\alpha\omega_\xi\right)\psi\right.\\ &\quad\left.+\left[\frac{\partial\sigma^{\alpha\xi}}{\partial x^\alpha}+\omega_\alpha\sigma^{\alpha\xi}-\sigma^{\alpha\xi}\omega_\alpha-\tfrac{1}{2}\sigma^{\alpha\xi}\left(\gamma_\beta\frac{\partial\gamma^\beta}{\partial x^\alpha}+\frac{\partial\gamma^\beta}{\partial x^\alpha}\gamma_\beta\right)\right]\left(\frac{\partial\psi}{\partial x^\xi}+\omega_\xi\psi\right)\right\}. \end{aligned} \quad (5.54)$$

This is the most general form of the Dirac equation with "", which emerges from the Gasiorowicz term in the Dirac Lagrangian (5.6).

Neglecting the anomalous gravitational coupling effect by setting $\ell = 0$, Eq. (5.54) reduces to

$$\mathrm{i}\gamma^\alpha\frac{\partial\psi}{\partial x^\alpha}+\frac{\mathrm{i}}{2}\left[\frac{\partial\gamma^\alpha}{\partial x^\alpha}+\omega_\alpha\gamma^\alpha+\gamma^\alpha\omega_\alpha-\tfrac{1}{2}\gamma^\alpha\left(\gamma_\beta\frac{\partial\gamma^\beta}{\partial x^\alpha}+\frac{\partial\gamma^\beta}{\partial x^\alpha}\gamma_\beta\right)\right]\psi-m\psi=0.$$

The term in brackets thus induces an effective (1, 1)-"mass matrix" $\boldsymbol{M} \equiv M^a{}_b$, which originates from the coupling of the spinor to the dynamic and curved spacetime:

$$\mathrm{i}\gamma^\alpha\frac{\partial\psi}{\partial x^\alpha}-(m+\boldsymbol{M})\,\psi=0,\quad \boldsymbol{M}=-\frac{\mathrm{i}}{2}\left[\frac{\partial\gamma^\alpha}{\partial x^\alpha}+\omega_\alpha\gamma^\alpha+\gamma^\alpha\omega_\alpha-\tfrac{1}{2}\gamma^\alpha\left(\gamma_\beta\frac{\partial\gamma^\beta}{\partial x^\alpha}+\frac{\partial\gamma^\beta}{\partial x^\alpha}\gamma_\beta\right)\right].$$

So, even if the fermion's rest mass is zero ($m = 0$), the spinor's spacetime interaction induces $\boldsymbol{M}\psi \equiv M^a{}_b\,\psi^b$ to be non-zero.

5.2.8 Affine Connection

The derivative of the transformation rule of the metric (1, 1)-spinor-(0, 1)-tensor γ_ξ from Eq. (5.18c)

$$\gamma_\beta = \boldsymbol{S}^{-1}\boldsymbol{\Gamma}_\xi\boldsymbol{S}\frac{\partial X^\xi}{\partial x^\beta}$$

can be expressed in terms of the affine connection $\gamma^\lambda{}_{\xi\beta}$, which is defined by its transformation rule as

$$\frac{\partial^2 X^\xi}{\partial x^\beta\partial x^\alpha}=\gamma^\lambda{}_{\beta\alpha}\frac{\partial X^\xi}{\partial x^\lambda}-\Gamma^\xi{}_{\tau\lambda}\frac{\partial X^\tau}{\partial x^\beta}\frac{\partial X^\lambda}{\partial x^\alpha}. \quad (5.55)$$

Then

$$\frac{\partial \boldsymbol{\gamma}_\beta}{\partial x^\alpha} = \left(\frac{\partial \boldsymbol{S}^{-1}}{\partial x^\alpha}\boldsymbol{\Gamma}_\xi \boldsymbol{S} + \boldsymbol{S}^{-1}\frac{\partial \boldsymbol{\Gamma}_\xi}{\partial X^\eta}\frac{\partial X^\eta}{\partial x^\alpha}\boldsymbol{S} + \boldsymbol{S}^{-1}\boldsymbol{\Gamma}_\xi \frac{\partial \boldsymbol{S}}{\partial x^\alpha}\right)\frac{\partial X^\xi}{\partial x^\beta} + \boldsymbol{S}^{-1}\boldsymbol{\Gamma}_\xi \boldsymbol{S}\frac{\partial^2 X^\xi}{\partial x^\beta \partial x^\alpha}.$$

With the following representations of the rule (5.27) for the spinor connection

$$\frac{\partial \boldsymbol{S}^{-1}}{\partial x^\alpha} = \boldsymbol{S}^{-1}\boldsymbol{\Omega}_\eta \frac{\partial X^\eta}{\partial x^\alpha} - \boldsymbol{\omega}_\alpha \boldsymbol{S}^{-1}, \qquad \frac{\partial \boldsymbol{S}}{\partial x^\alpha} = \boldsymbol{S}\boldsymbol{\omega}_\alpha - \boldsymbol{\Omega}_\eta \boldsymbol{S}\frac{\partial X^\eta}{\partial x^\alpha},$$

one finds

$$\frac{\partial \boldsymbol{\Gamma}_\xi}{\partial X^\eta} + \boldsymbol{\Omega}_\eta \boldsymbol{\Gamma}_\xi - \boldsymbol{\Gamma}_\xi \boldsymbol{\Omega}_\eta - \boldsymbol{\Gamma}_\lambda \Gamma^\lambda{}_{\xi\eta} = \boldsymbol{S}\left(\frac{\partial \boldsymbol{\gamma}_\beta}{\partial x^\alpha} + \boldsymbol{\omega}_\alpha \boldsymbol{\gamma}_\beta - \boldsymbol{\gamma}_\beta \boldsymbol{\omega}_\alpha - \boldsymbol{\gamma}_\lambda \gamma^\lambda{}_{\beta\alpha}\right)\boldsymbol{S}^{-1}\frac{\partial x^\beta}{\partial X^\xi}\frac{\partial x^\alpha}{\partial X^\eta}.$$

The sum in parentheses thus transforms as a (1, 1)-spinor-(1, 1)-tensor and thus defines the covariant x^α-derivative of the metric (1, 1)-spinor-(0, 1)-vector $\boldsymbol{\gamma}_\beta$:

$$\boldsymbol{\gamma}_{\beta;\alpha} = \frac{\partial \boldsymbol{\gamma}_\beta}{\partial x^\alpha} + \boldsymbol{\omega}_\alpha \boldsymbol{\gamma}_\beta - \boldsymbol{\gamma}_\beta \boldsymbol{\omega}_\alpha - \boldsymbol{\gamma}_\lambda \gamma^\lambda{}_{\beta\alpha}, \tag{5.56}$$

which thus transforms as a (1, 1)-spinor-(0, 2)-tensor:

$$\boldsymbol{\Gamma}_{\xi;\eta} = \boldsymbol{S}\boldsymbol{\gamma}_{\beta;\alpha}\boldsymbol{S}^{-1}\frac{\partial x^\beta}{\partial X^\xi}\frac{\partial x^\alpha}{\partial X^\eta}.$$

The canonical equation (5.46f) for $\boldsymbol{\gamma}_\mu$ can identically be rewritten as

$$\begin{aligned} 2\frac{\partial \tilde{\mathcal{H}}_{\mathrm{Gr}}}{\partial \tilde{\boldsymbol{k}}^{\mu\nu}} &= \frac{\partial \boldsymbol{\gamma}_\mu}{\partial x^\nu} + \boldsymbol{\omega}_\nu \boldsymbol{\gamma}_\mu - \boldsymbol{\gamma}_\mu \boldsymbol{\omega}_\nu - \left(\frac{\partial \boldsymbol{\gamma}_\nu}{\partial x^\mu} + \boldsymbol{\omega}_\mu \boldsymbol{\gamma}_\nu - \boldsymbol{\gamma}_\nu \boldsymbol{\omega}_\mu\right) \\ &= \boldsymbol{\gamma}_{\mu;\nu} - \boldsymbol{\gamma}_{\nu;\mu} + 2\boldsymbol{\gamma}_\beta S^\beta{}_{\mu\nu}. \end{aligned}$$

5.2.9 *Restriction to Metric Compatibility* ($\boldsymbol{\gamma}^\mu{}_{;\nu} \equiv 0$)

A vanishing covariant derivative $\boldsymbol{\gamma}^\mu{}_{;\nu} = 0$ of the metric spinor $\boldsymbol{\gamma}^\mu$ implies a covariantly constant metric tensor $g_{\mu\nu}$, hence the condition. We can also regard this as the definition of the affine connection $\gamma^\xi{}_{\alpha\beta}$:

$$\boldsymbol{\gamma}^\mu{}_{;\nu} = \frac{\partial \boldsymbol{\gamma}^\mu}{\partial x^\nu} + \boldsymbol{\omega}_\nu \boldsymbol{\gamma}^\mu - \boldsymbol{\gamma}^\mu \boldsymbol{\omega}_\nu + \boldsymbol{\gamma}^\beta \gamma^\mu{}_{\beta\nu} = 0, \tag{5.57}$$

hence

$$\frac{\partial \boldsymbol{\gamma}^\mu}{\partial x^\nu} + \boldsymbol{\omega}_\nu \boldsymbol{\gamma}^\mu - \boldsymbol{\gamma}^\mu \boldsymbol{\omega}_\nu = -\boldsymbol{\gamma}^\beta \gamma^\mu{}_{\beta\nu}. \tag{5.58}$$

Consequently, one gets for a vanishing covariant derivative of the commutator (5.7) of the metric spinor:

$$\frac{\partial \boldsymbol{\sigma}^{\mu\xi}}{\partial x^{\nu}} + \boldsymbol{\omega}_{\nu}\boldsymbol{\sigma}^{\mu\xi} - \boldsymbol{\sigma}^{\mu\xi}\boldsymbol{\omega}_{\nu} = -\boldsymbol{\sigma}^{\beta\xi}\,\gamma^{\mu}_{\ \beta\nu} - \boldsymbol{\sigma}^{\mu\beta}\,\gamma^{\xi}_{\ \beta\nu} \tag{5.59}$$

and for

$$\begin{aligned}\boldsymbol{\gamma}_{\beta}\frac{\partial \boldsymbol{\gamma}^{\beta}}{\partial x^{\alpha}} + \frac{\partial \boldsymbol{\gamma}^{\beta}}{\partial x^{\alpha}}\boldsymbol{\gamma}_{\beta} &= \boldsymbol{\gamma}_{\beta}\left(\cancel{\boldsymbol{\gamma}^{\beta}\boldsymbol{\omega}_{\alpha}} - \cancel{\boldsymbol{\omega}_{\alpha}\boldsymbol{\gamma}^{\beta}} - \boldsymbol{\gamma}^{\xi}\,\gamma^{\beta}_{\ \xi\alpha}\right) + \left(\cancel{\boldsymbol{\gamma}^{\beta}\boldsymbol{\omega}_{\alpha}} - \cancel{\boldsymbol{\omega}_{\alpha}\boldsymbol{\gamma}^{\beta}} - \boldsymbol{\gamma}^{\xi}\,\gamma^{\beta}_{\ \xi\alpha}\right)\boldsymbol{\gamma}_{\beta} \\ &= -\left(\boldsymbol{\gamma}_{\beta}\boldsymbol{\gamma}^{\xi} + \boldsymbol{\gamma}^{\xi}\boldsymbol{\gamma}_{\beta}\right)\gamma^{\beta}_{\ \xi\alpha} = -2\delta^{\xi}_{\beta}\,\gamma^{\beta}_{\ \xi\alpha}\,\mathbb{1} \\ &= -2\,\gamma^{\beta}_{\ \beta\alpha}\,\mathbb{1}. \end{aligned}\tag{5.60}$$

With the Cartan torsion tensor $S^{\beta}_{\ \xi\alpha}$ defined as the skew-symmetric part of the affine connection:

$$S^{\beta}_{\ \xi\alpha} = \tfrac{1}{2}\left(\gamma^{\beta}_{\ \xi\alpha} - \gamma^{\beta}_{\ \alpha\xi}\right), \tag{5.61}$$

the generalized Dirac equation (5.54) then simplifies for metric compatibility:

$$\begin{aligned}0 = \mathrm{i}\boldsymbol{\gamma}^{\beta}&\left(\frac{\partial \psi}{\partial x^{\beta}} + \boldsymbol{\omega}_{\beta}\,\psi\right) - \left(m + \mathrm{i}\boldsymbol{\gamma}^{\beta}S^{\alpha}_{\ \alpha\beta}\right)\psi \\ &- \frac{\mathrm{i}\ell}{3}\left[\boldsymbol{\sigma}^{\alpha\beta}\left(\frac{\partial \boldsymbol{\omega}_{\beta}}{\partial x^{\alpha}} + \boldsymbol{\omega}_{\alpha}\,\boldsymbol{\omega}_{\beta}\right)\psi + \left(\boldsymbol{\sigma}^{\xi\alpha}S^{\beta}_{\ \xi\alpha} + 2\boldsymbol{\sigma}^{\alpha\beta}S^{\xi}_{\ \xi\alpha}\right)\left(\frac{\partial \psi}{\partial x^{\beta}} + \boldsymbol{\omega}_{\beta}\psi\right)\right].\end{aligned}\tag{5.62}$$

✓ Explicit Calculation

For the spinor gauge theory, it is necessary to work out the correlation $\boldsymbol{R}_{\alpha\beta} \leftrightarrow R^{\eta}_{\ \xi\alpha\beta}$. Actually, both are physically the same quantities in different representations, with the Latin indices referring to the spinor indices and the Greek indices referring to the coordinate space.

The covariant derivative of the metric spinor $\boldsymbol{\gamma}_{\mu}$ vanishes for the case of metric compatibility ($\boldsymbol{\gamma}_{\mu;\alpha} \equiv 0$):

$$\frac{\partial \boldsymbol{\gamma}_{\mu}}{\partial x^{\alpha}} = \boldsymbol{\gamma}_{\mu}\boldsymbol{\omega}_{\alpha} - \boldsymbol{\omega}_{\alpha}\boldsymbol{\gamma}_{\mu} + \boldsymbol{\gamma}_{\xi}\,\gamma^{\xi}_{\ \mu\alpha}\,. \tag{5.63}$$

In order to work out the correlation of the Riemann tensor and the curvature spinor, defined in Eq. (5.44), we calculate the derivative of Eq. (5.63)

$$\frac{\partial^2 \boldsymbol{\gamma}_{\mu}}{\partial x^{\alpha}\partial x^{\beta}} = \frac{\partial \boldsymbol{\gamma}_{\mu}}{\partial x^{\beta}}\boldsymbol{\omega}_{\alpha} + \boldsymbol{\gamma}_{\mu}\frac{\partial \boldsymbol{\omega}_{\alpha}}{\partial x^{\beta}} - \frac{\partial \boldsymbol{\omega}_{\alpha}}{\partial x^{\beta}}\boldsymbol{\gamma}_{\mu} - \boldsymbol{\omega}_{\alpha}\frac{\partial \boldsymbol{\gamma}_{\mu}}{\partial x^{\beta}} + \frac{\partial \boldsymbol{\gamma}_{\xi}}{\partial x^{\beta}}\gamma^{\xi}_{\ \mu\alpha} + \boldsymbol{\gamma}_{\xi}\frac{\partial \gamma^{\xi}_{\ \mu\alpha}}{\partial x^{\beta}}.$$

Swapping the sequence of differentiation, the difference is encountered as

$$0 = \boldsymbol{\gamma}_\mu \left(\frac{\partial \boldsymbol{\omega}_\alpha}{\partial x^\beta} - \frac{\partial \boldsymbol{\omega}_\beta}{\partial x^\alpha} \right) - \left(\frac{\partial \boldsymbol{\omega}_\alpha}{\partial x^\beta} - \frac{\partial \boldsymbol{\omega}_\beta}{\partial x^\alpha} \right) \boldsymbol{\gamma}_\mu + \boldsymbol{\gamma}_\xi \left(\frac{\partial \gamma^\xi{}_{\mu\alpha}}{\partial x^\beta} - \frac{\partial \gamma^\xi{}_{\mu\beta}}{\partial x^\alpha} \right)$$
$$+ \frac{\partial \boldsymbol{\gamma}_\mu}{\partial x^\beta} \boldsymbol{\omega}_\alpha - \frac{\partial \boldsymbol{\gamma}_\mu}{\partial x^\alpha} \boldsymbol{\omega}_\beta - \boldsymbol{\omega}_\alpha \frac{\partial \boldsymbol{\gamma}_\mu}{\partial x^\beta} + \boldsymbol{\omega}_\beta \frac{\partial \boldsymbol{\gamma}_\mu}{\partial x^\alpha} + \frac{\partial \boldsymbol{\gamma}_\xi}{\partial x^\beta} \gamma^\xi{}_{\mu\alpha} - \frac{\partial \boldsymbol{\gamma}_\xi}{\partial x^\alpha} \gamma^\xi{}_{\mu\beta}.$$

Inserting Eq. (5.63) for the first derivatives of $\boldsymbol{\gamma}_\mu$ yields

$$0 = \boldsymbol{\gamma}_\mu \left(\frac{\partial \boldsymbol{\omega}_\alpha}{\partial x^\beta} - \frac{\partial \boldsymbol{\omega}_\beta}{\partial x^\alpha} \right) - \left(\frac{\partial \boldsymbol{\omega}_\alpha}{\partial x^\beta} - \frac{\partial \boldsymbol{\omega}_\beta}{\partial x^\alpha} \right) \boldsymbol{\gamma}_\mu + \boldsymbol{\gamma}_\xi \left(\frac{\partial \gamma^\xi{}_{\mu\alpha}}{\partial x^\beta} - \frac{\partial \gamma^\xi{}_{\mu\beta}}{\partial x^\alpha} \right)$$
$$+ \left(\boldsymbol{\gamma}_\mu \boldsymbol{\omega}_\beta - \cancel{\boldsymbol{\omega}_\beta \boldsymbol{\gamma}_\mu} + \cancel{\boldsymbol{\gamma}_\xi \gamma^\xi{}_{\mu\beta}} \right) \boldsymbol{\omega}_\alpha - \left(\boldsymbol{\gamma}_\mu \boldsymbol{\omega}_\alpha - \cancel{\boldsymbol{\omega}_\alpha \boldsymbol{\gamma}_\mu} + \cancel{\boldsymbol{\gamma}_\xi \gamma^\xi{}_{\mu\alpha}} \right) \boldsymbol{\omega}_\beta$$
$$- \boldsymbol{\omega}_\alpha \left(\cancel{\boldsymbol{\gamma}_\mu \boldsymbol{\omega}_\beta} - \boldsymbol{\omega}_\beta \boldsymbol{\gamma}_\mu + \cancel{\boldsymbol{\gamma}_\xi \gamma^\xi{}_{\mu\beta}} \right) + \boldsymbol{\omega}_\beta \left(\cancel{\boldsymbol{\gamma}_\mu \boldsymbol{\omega}_\alpha} - \boldsymbol{\omega}_\alpha \boldsymbol{\gamma}_\mu + \cancel{\boldsymbol{\gamma}_\xi \gamma^\xi{}_{\mu\alpha}} \right)$$
$$+ \left(\cancel{\boldsymbol{\gamma}_\xi \boldsymbol{\omega}_\beta} - \cancel{\boldsymbol{\omega}_\beta \boldsymbol{\gamma}_\xi} + \boldsymbol{\gamma}_\eta \gamma^\eta{}_{\xi\beta} \right) \gamma^\xi{}_{\mu\alpha} - \left(\cancel{\boldsymbol{\gamma}_\xi \boldsymbol{\omega}_\alpha} - \cancel{\boldsymbol{\omega}_\alpha \boldsymbol{\gamma}_\xi} + \boldsymbol{\gamma}_\eta \gamma^\eta{}_{\xi\alpha} \right) \gamma^\xi{}_{\mu\beta}.$$

Rearranging the terms gives

$$0 = \boldsymbol{\gamma}_\mu \left(\frac{\partial \boldsymbol{\omega}_\alpha}{\partial x^\beta} - \frac{\partial \boldsymbol{\omega}_\beta}{\partial x^\alpha} + \boldsymbol{\omega}_\beta \boldsymbol{\omega}_\alpha - \boldsymbol{\omega}_\alpha \boldsymbol{\omega}_\beta \right) - \left(\frac{\partial \boldsymbol{\omega}_\alpha}{\partial x^\beta} - \frac{\partial \boldsymbol{\omega}_\beta}{\partial x^\alpha} + \boldsymbol{\omega}_\beta \boldsymbol{\omega}_\alpha - \boldsymbol{\omega}_\alpha \boldsymbol{\omega}_\beta \right) \boldsymbol{\gamma}_\mu$$
$$+ \boldsymbol{\gamma}_\xi \left(\frac{\partial \gamma^\xi{}_{\mu\alpha}}{\partial x^\beta} - \frac{\partial \gamma^\xi{}_{\mu\beta}}{\partial x^\alpha} + \gamma^\xi{}_{\eta\beta} \gamma^\eta{}_{\mu\alpha} - \gamma^\xi{}_{\eta\alpha} \gamma^\eta{}_{\mu\beta} \right),$$

while all other terms cancel. With the Riemann-Cartan tensor $R^\xi{}_{\mu\beta\alpha}$, defined by the sum proportional to $\boldsymbol{\gamma}_\xi$, and the curvature spinor (5.44), the equation writes concisely

$$\boldsymbol{\gamma}_\mu \boldsymbol{R}_{\beta\alpha} - \boldsymbol{R}_{\beta\alpha} \boldsymbol{\gamma}_\mu + \boldsymbol{\gamma}_\xi R^\xi{}_{\mu\beta\alpha} = 0. \tag{5.64}$$

Equation (5.64) is identically satisfied by

$$\boldsymbol{R}_{\beta\alpha} = -\tfrac{\mathrm{i}}{4} \boldsymbol{\sigma}_\xi{}^\eta R^\xi{}_{\eta\beta\alpha} \quad \Leftrightarrow \quad R^\xi{}_{\mu\beta\alpha} = \tfrac{\mathrm{i}}{2} \operatorname{Tr} \left\{ \boldsymbol{\sigma}^\xi{}_\mu \boldsymbol{R}_{\beta\alpha} \right\}, \tag{5.65}$$

as can be seen by contracting Eq. (5.64) with $\boldsymbol{\gamma}^\eta$ and taking the trace of the spinor indices:

$$0 = \operatorname{Tr} \left\{ \boldsymbol{\gamma}_\mu \boldsymbol{R}_{\beta\alpha} \boldsymbol{\gamma}^\eta - \boldsymbol{R}_{\beta\alpha} \boldsymbol{\gamma}_\mu \boldsymbol{\gamma}^\eta \right\} + \operatorname{Tr} \left\{ \boldsymbol{\gamma}_\xi \boldsymbol{\gamma}^\eta \right\} R^\xi{}_{\mu\beta\alpha}$$
$$= \operatorname{Tr} \left\{ \left(\boldsymbol{\gamma}^\eta \boldsymbol{\gamma}_\mu - \boldsymbol{\gamma}_\mu \boldsymbol{\gamma}^\eta \right) \boldsymbol{R}_{\beta\alpha} \right\} + \operatorname{Tr} \left\{ \boldsymbol{\gamma}_\xi \boldsymbol{\gamma}^\eta \right\} R^\xi{}_{\mu\beta\alpha} = -2\mathrm{i} \operatorname{Tr} \left\{ \boldsymbol{\sigma}^\eta{}_\mu \boldsymbol{R}_{\beta\alpha} \right\} + 4 \delta^\eta_\xi R^\xi{}_{\mu\beta\alpha}.$$

The last step follows from the skew-symmetry of the Riemann tensor in its first index pair.

We may finally express the leftmost term in the second line of (5.62) in terms of the curvature spinor (5.44)

$$\begin{aligned} & \mathrm{i}\gamma^{\beta}\left(\frac{\partial\psi}{\partial x^{\beta}}+\boldsymbol{\omega}_{\beta}\,\psi+S^{\alpha}{}_{\beta\alpha}\,\psi\right)-m\psi \\ & -\frac{\mathrm{i}\ell}{3}\left[\tfrac{1}{2}\boldsymbol{\sigma}^{\alpha\beta}\boldsymbol{R}_{\alpha\beta}\,\psi+\left(\boldsymbol{\sigma}^{\xi\alpha}S^{\beta}{}_{\xi\alpha}+2\boldsymbol{\sigma}^{\alpha\beta}S^{\xi}{}_{\xi\alpha}\right)\left(\frac{\partial\psi}{\partial x^{\beta}}+\boldsymbol{\omega}_{\beta}\,\psi\right)\right]=0. \end{aligned} \tag{5.66}$$

This shows that a gravitational Pauli coupling exists even if the spacetime geometry is torsion-free and furnishes an effective mass term

$$\mathrm{i}\gamma^{\beta}\left(\frac{\partial\psi}{\partial x^{\beta}}+\boldsymbol{\omega}_{\beta}\,\psi\right)-\left(m+\frac{\ell}{3}\gamma^{\alpha}\gamma^{\beta}\boldsymbol{R}_{\alpha\beta}\right)\psi=0, \tag{5.67}$$

which can be expressed equivalently in terms of the Ricci scalar R according to Eq. (5.70)

$$\mathrm{i}\gamma^{\beta}\left(\frac{\partial\psi}{\partial x^{\beta}}+\boldsymbol{\omega}_{\beta}\,\psi\right)-\left(m+\frac{\ell}{6}R\right)\psi=0, \tag{5.68}$$

as worked out in the following explicit calculation.

✓ Explict Calculation

Full contraction of the Riemann tensor with metric spinors (Dirac matrices). From the definition of the Dirac algebra for a general contravariant metric $g^{\eta\alpha}(x)=g^{\alpha\eta}(x)$,

$$\tfrac{1}{2}\left(\gamma^{a}{}_{c^{\eta}}\,\gamma^{c}{}_{b^{\alpha}}+\gamma^{a}{}_{c^{\alpha}}\,\gamma^{c}{}_{b^{\eta}}\right)=g^{\eta\alpha}\,\delta^{a}_{b}\quad\Longleftrightarrow\quad\tfrac{1}{2}\left(\gamma^{\eta}\gamma^{\alpha}+\gamma^{\alpha}\gamma^{\eta}\right)=g^{\eta\alpha}\,\mathbb{1},$$

where the upper case Latin indices refer to the spinor space and $\mathbb{1}$ denotes the unit matrix in spinor space, one concludes

$$\begin{aligned} \gamma^{\eta}\gamma^{\alpha}\gamma^{\beta} &= \left(\gamma^{\eta}\gamma^{\alpha}+\gamma^{\alpha}\gamma^{\eta}\right)\gamma^{\beta}-\gamma^{\alpha}\gamma^{\eta}\gamma^{\beta} \\ &= 2g^{\eta\alpha}\gamma^{\beta}-\gamma^{\alpha}\gamma^{\eta}\gamma^{\beta} \\ &= 2g^{\eta\alpha}\gamma^{\beta}-\gamma^{\alpha}\left(\gamma^{\eta}\gamma^{\beta}+\gamma^{\beta}\gamma^{\eta}\right)+\gamma^{\alpha}\gamma^{\beta}\gamma^{\eta} \\ &= 2g^{\eta\alpha}\gamma^{\beta}-2g^{\beta\eta}\gamma^{\alpha}+\left(\gamma^{\alpha}\gamma^{\beta}+\gamma^{\beta}\gamma^{\alpha}\right)\gamma^{\eta}-\gamma^{\beta}\gamma^{\alpha}\gamma^{\eta} \\ &= 2g^{\eta\alpha}\gamma^{\beta}-2g^{\beta\eta}\gamma^{\alpha}+2g^{\alpha\beta}\gamma^{\eta}-\gamma^{\beta}\gamma^{\alpha}\gamma^{\eta}, \end{aligned}$$

hence

$$\boldsymbol{\gamma}^{\eta}\boldsymbol{\gamma}^{\alpha}\boldsymbol{\gamma}^{\beta}+\boldsymbol{\gamma}^{\beta}\boldsymbol{\gamma}^{\alpha}\boldsymbol{\gamma}^{\eta}=2\left(g^{\eta\alpha}\boldsymbol{\gamma}^{\beta}-g^{\beta\eta}\boldsymbol{\gamma}^{\alpha}+g^{\alpha\beta}\boldsymbol{\gamma}^{\eta}\right). \tag{5.69}$$

By virtue of the skew-symmetry of the Riemann tensor in both its first and last index pair

$$R_{\xi\eta\alpha\beta}=-R_{\eta\xi\alpha\beta}, \qquad R_{\xi\eta\alpha\beta}=-R_{\xi\eta\beta\alpha},$$

the contraction $R_{\xi\eta\alpha\beta}\,\boldsymbol{\gamma}^{\xi}\boldsymbol{\gamma}^{\eta}\boldsymbol{\gamma}^{\alpha}\boldsymbol{\gamma}^{\beta}$ yields

$$\begin{aligned}
R_{\xi\eta\alpha\beta}\,\boldsymbol{\gamma}^{\xi}\boldsymbol{\gamma}^{\eta}\boldsymbol{\gamma}^{\alpha}\boldsymbol{\gamma}^{\beta} &= \tfrac{1}{2}R_{\xi\eta\alpha\beta}\left[\boldsymbol{\gamma}^{\xi}\left(\boldsymbol{\gamma}^{\eta}\boldsymbol{\gamma}^{\alpha}\boldsymbol{\gamma}^{\beta}\right)+\left(\boldsymbol{\gamma}^{\xi}\boldsymbol{\gamma}^{\eta}\boldsymbol{\gamma}^{\alpha}\right)\boldsymbol{\gamma}^{\beta}\right]\\
&= R_{\xi\eta\alpha\beta}\left[\boldsymbol{\gamma}^{\xi}\left(g^{\eta\alpha}\boldsymbol{\gamma}^{\beta}-g^{\beta\eta}\boldsymbol{\gamma}^{\alpha}+\cancel{g^{\alpha\beta}\boldsymbol{\gamma}^{\eta}}-\tfrac{1}{2}\boldsymbol{\gamma}^{\beta}\boldsymbol{\gamma}^{\alpha}\boldsymbol{\gamma}^{\eta}\right)\right.\\
&\qquad\left.+\left(\cancel{g^{\xi\eta}\boldsymbol{\gamma}^{\alpha}}-g^{\alpha\xi}\boldsymbol{\gamma}^{\eta}+g^{\eta\alpha}\boldsymbol{\gamma}^{\xi}-\tfrac{1}{2}\boldsymbol{\gamma}^{\alpha}\boldsymbol{\gamma}^{\eta}\boldsymbol{\gamma}^{\xi}\right)\boldsymbol{\gamma}^{\beta}\right]\\
&= R_{\xi\eta\alpha\beta}\left(2g^{\eta\alpha}\boldsymbol{\gamma}^{\xi}\boldsymbol{\gamma}^{\beta}-g^{\beta\eta}\boldsymbol{\gamma}^{\xi}\boldsymbol{\gamma}^{\alpha}-g^{\alpha\xi}\boldsymbol{\gamma}^{\eta}\boldsymbol{\gamma}^{\beta}-\tfrac{1}{2}\boldsymbol{\gamma}^{\xi}\boldsymbol{\gamma}^{\beta}\boldsymbol{\gamma}^{\alpha}\boldsymbol{\gamma}^{\eta}-\tfrac{1}{2}\boldsymbol{\gamma}^{\alpha}\boldsymbol{\gamma}^{\eta}\boldsymbol{\gamma}^{\xi}\boldsymbol{\gamma}^{\beta}\right)\\
&= -4R_{\eta\beta}\,\boldsymbol{\gamma}^{\eta}\boldsymbol{\gamma}^{\beta}-\tfrac{1}{2}R_{\xi\eta\alpha\beta}\left(\boldsymbol{\gamma}^{\xi}\boldsymbol{\gamma}^{\beta}\boldsymbol{\gamma}^{\alpha}\boldsymbol{\gamma}^{\eta}+\boldsymbol{\gamma}^{\alpha}\boldsymbol{\gamma}^{\eta}\boldsymbol{\gamma}^{\xi}\boldsymbol{\gamma}^{\beta}\right).
\end{aligned}$$

Making use of further Riemann tensor identities, which hold for zero torsion,

$$R_{\xi\eta\alpha\beta}=R_{\alpha\beta\xi\eta}, \qquad R_{\eta\beta}=R_{\beta\eta}, \qquad R_{\xi\eta\alpha\beta}+R_{\xi\alpha\beta\eta}+R_{\xi\beta\eta\alpha}=0,$$

hence

$$\left(R_{\xi\eta\alpha\beta}+R_{\xi\alpha\beta\eta}+R_{\xi\beta\eta\alpha}\right)\boldsymbol{\gamma}^{\xi}\boldsymbol{\gamma}^{\beta}\boldsymbol{\gamma}^{\alpha}\boldsymbol{\gamma}^{\eta}=0 \quad\Leftrightarrow\quad R_{\xi\eta\alpha\beta}\left(\boldsymbol{\gamma}^{\xi}\boldsymbol{\gamma}^{\beta}\boldsymbol{\gamma}^{\alpha}\boldsymbol{\gamma}^{\eta}+\boldsymbol{\gamma}^{\xi}\boldsymbol{\gamma}^{\alpha}\boldsymbol{\gamma}^{\eta}\boldsymbol{\gamma}^{\beta}+\boldsymbol{\gamma}^{\xi}\boldsymbol{\gamma}^{\eta}\boldsymbol{\gamma}^{\beta}\boldsymbol{\gamma}^{\alpha}\right)=0,$$

yields

$$-\tfrac{1}{2}R_{\xi\eta\alpha\beta}\left(\boldsymbol{\gamma}^{\xi}\boldsymbol{\gamma}^{\beta}\boldsymbol{\gamma}^{\alpha}\boldsymbol{\gamma}^{\eta}+\boldsymbol{\gamma}^{\alpha}\boldsymbol{\gamma}^{\eta}\boldsymbol{\gamma}^{\xi}\boldsymbol{\gamma}^{\beta}\right)=\tfrac{1}{2}R_{\xi\eta\alpha\beta}\left(\boldsymbol{\gamma}^{\xi}\boldsymbol{\gamma}^{\eta}\boldsymbol{\gamma}^{\beta}\boldsymbol{\gamma}^{\alpha}+\boldsymbol{\gamma}^{\xi}\boldsymbol{\gamma}^{\alpha}\boldsymbol{\gamma}^{\eta}\boldsymbol{\gamma}^{\beta}-\boldsymbol{\gamma}^{\alpha}\boldsymbol{\gamma}^{\eta}\boldsymbol{\gamma}^{\xi}\boldsymbol{\gamma}^{\beta}\right).$$

The full contraction of the Riemann tensor with the Dirac matrices now simplifies to

$$\begin{aligned}
R_{\xi\eta\alpha\beta}\,\boldsymbol{\gamma}^{\xi}\boldsymbol{\gamma}^{\eta}\boldsymbol{\gamma}^{\alpha}\boldsymbol{\gamma}^{\beta} &= -2R_{\eta\beta}\left(\boldsymbol{\gamma}^{\eta}\boldsymbol{\gamma}^{\beta}+\boldsymbol{\gamma}^{\beta}\boldsymbol{\gamma}^{\eta}\right)+\tfrac{1}{2}R_{\xi\eta\alpha\beta}\left(\boldsymbol{\gamma}^{\xi}\boldsymbol{\gamma}^{\eta}\boldsymbol{\gamma}^{\beta}\boldsymbol{\gamma}^{\alpha}+\boldsymbol{\gamma}^{\xi}\boldsymbol{\gamma}^{\alpha}\boldsymbol{\gamma}^{\eta}\boldsymbol{\gamma}^{\beta}+\boldsymbol{\gamma}^{\alpha}\boldsymbol{\gamma}^{\xi}\boldsymbol{\gamma}^{\eta}\boldsymbol{\gamma}^{\beta}\right)\\
&= -4R_{\eta\beta}\,g^{\eta\beta}\mathbb{1}+\tfrac{1}{2}R_{\xi\eta\alpha\beta}\left[-\boldsymbol{\gamma}^{\xi}\boldsymbol{\gamma}^{\eta}\boldsymbol{\gamma}^{\alpha}\boldsymbol{\gamma}^{\beta}+\left(\boldsymbol{\gamma}^{\xi}\boldsymbol{\gamma}^{\alpha}+\boldsymbol{\gamma}^{\alpha}\boldsymbol{\gamma}^{\xi}\right)\boldsymbol{\gamma}^{\eta}\boldsymbol{\gamma}^{\beta}\right]\\
&= -4R\,\mathbb{1}-\tfrac{1}{2}R_{\xi\eta\alpha\beta}\,\boldsymbol{\gamma}^{\xi}\boldsymbol{\gamma}^{\eta}\boldsymbol{\gamma}^{\alpha}\boldsymbol{\gamma}^{\beta}+R_{\xi\eta\alpha\beta}\,g^{\xi\alpha}\boldsymbol{\gamma}^{\eta}\boldsymbol{\gamma}^{\beta}\\
&= -3R\,\mathbb{1}-\tfrac{1}{2}R_{\xi\eta\alpha\beta}\,\boldsymbol{\gamma}^{\xi}\boldsymbol{\gamma}^{\eta}\boldsymbol{\gamma}^{\alpha}\boldsymbol{\gamma}^{\beta},
\end{aligned}$$

and finally,

$$R_{\xi\eta\alpha\beta}\,\boldsymbol{\gamma}^{\xi}\boldsymbol{\gamma}^{\eta}\boldsymbol{\gamma}^{\alpha}\boldsymbol{\gamma}^{\beta}=-2R\,\mathbb{1} \quad\Leftrightarrow\quad R_{\xi\eta\alpha\beta}\,\gamma^{a}{}_{c\xi}\,\gamma^{c}{}_{d\eta}\,\gamma^{d}{}_{e\alpha}\,\gamma^{e}{}_{b\beta}=-2R_{\xi\eta\alpha\beta}\,g^{\xi\alpha}\,g^{\eta\beta}\,\delta^{a}_{b}. \tag{5.70}$$

5.3 SO(1, 3) × Diff(*M*) Symmetry Group for Spin-3/2 Fields

The uncoupled action for a spin-3/2 field and the spacetime-dependent Dirac matrix $\boldsymbol{\gamma}_\mu$, the spinor-connection field $\boldsymbol{\omega}_\mu$, and their corresponding momentum fields is given by:

$$S_0 = \int_V \left(\tilde{\bar{\kappa}}^{\mu\nu} \frac{\partial \psi_\mu}{\partial x^\nu} + \frac{\partial \bar{\psi}_\mu}{\partial x^\nu} \tilde{\kappa}^{\mu\nu} + \mathrm{Tr}\left\{ \tilde{\boldsymbol{k}}^{\mu\nu} \frac{\partial \boldsymbol{\gamma}_\mu}{\partial x^\nu} \right\} + \mathrm{Tr}\left\{ \boldsymbol{q}^{\mu\nu} \frac{\partial \boldsymbol{\omega}_\mu}{\partial x^\nu} \right\} - \tilde{\mathcal{H}}_0 - \tilde{\mathcal{H}}_{\mathrm{Gr}} \right) \mathrm{d}^4 x. \tag{5.71}$$

The generating function for the pertaining gauge theory has the spinor representation

$$\begin{aligned} \tilde{\mathcal{F}}_3^\mu \Big|_x = &- \left(\tilde{\bar{\kappa}}^{\beta\mu} \boldsymbol{S}^{-1} \Psi_\alpha + \bar{\Psi}_\alpha \boldsymbol{S} \tilde{\kappa}^{\beta\mu} \right) \frac{\partial X^\alpha}{\partial x^\beta} \\ &- \mathrm{Tr}\left\{ \tilde{\boldsymbol{k}}^{\beta\mu} \boldsymbol{S}^{-1} \boldsymbol{\Gamma}_\alpha \boldsymbol{S} \frac{\partial X^\alpha}{\partial x^\beta} \right\} - \mathrm{Tr}\left\{ \tilde{\boldsymbol{q}}^{\beta\mu} \left(\boldsymbol{S}^{-1} \boldsymbol{\Omega}_\alpha \boldsymbol{S} \frac{\partial X^\alpha}{\partial x^\beta} + \boldsymbol{S}^{-1} \frac{\partial \boldsymbol{S}}{\partial x^\beta} \right) \right\}. \end{aligned} \tag{5.72}$$

The canonical transformation rules provide the required transformations of the fields:

$$\tilde{\bar{\mathcal{K}}}^{\alpha\mu}(X) = -\frac{\partial \tilde{\mathcal{F}}_3^\xi(x)}{\partial \Psi_\alpha(X)} \frac{\partial X^\mu}{\partial x^\xi} \left| \frac{\partial x}{\partial X} \right| = \tilde{\bar{\kappa}}^{\beta\xi}(x) \boldsymbol{S}^{-1} \frac{\partial X^\alpha}{\partial x^\beta} \frac{\partial X^\mu}{\partial x^\xi} \left| \frac{\partial x}{\partial X} \right| \tag{5.73a}$$

$$\tilde{\mathcal{K}}^{\alpha\mu}(X) = -\frac{\partial \tilde{\mathcal{F}}_3^\xi(x)}{\partial \bar{\Psi}_\alpha(X)} \frac{\partial X^\mu}{\partial x^\xi} \left| \frac{\partial x}{\partial X} \right| = \boldsymbol{S} \tilde{\kappa}^{\beta\xi}(x) \frac{\partial X^\alpha}{\partial x^\beta} \frac{\partial X^\mu}{\partial x^\xi} \left| \frac{\partial x}{\partial X} \right| \tag{5.73b}$$

$$\delta_\nu^\mu \bar{\psi}_\beta(x) = -\frac{\partial \tilde{\mathcal{F}}_3^\mu(x)}{\partial \tilde{\kappa}^{\beta\nu}(x)} = \delta_\nu^\mu \bar{\Psi}_\alpha(X) \boldsymbol{S} \frac{\partial X^\alpha}{\partial x^\beta} \tag{5.73c}$$

$$\delta_\nu^\mu \psi_\beta(x) = -\frac{\partial \tilde{\mathcal{F}}_3^\mu(x)}{\partial \tilde{\bar{\kappa}}^{\beta\nu}(x)} = \delta_\nu^\mu \boldsymbol{S}^{-1} \Psi_\alpha(X) \frac{\partial X^\alpha}{\partial x^\beta} \tag{5.73d}$$

$$\tilde{\boldsymbol{K}}^{\alpha\mu}(X) = -\frac{\partial \tilde{\mathcal{F}}_3^\xi(x)}{\partial \boldsymbol{\Gamma}_\alpha(X)} \frac{\partial X^\mu}{\partial x^\xi} \left| \frac{\partial x}{\partial X} \right| = \boldsymbol{S} \tilde{\boldsymbol{k}}^{\beta\xi}(x) \boldsymbol{S}^{-1} \frac{\partial X^\alpha}{\partial x^\beta} \frac{\partial X^\mu}{\partial x^\xi} \left| \frac{\partial x}{\partial X} \right| \tag{5.73e}$$

$$\delta_\nu^\mu \boldsymbol{\gamma}_\beta(x) = -\frac{\partial \tilde{\mathcal{F}}_3^\mu(x)}{\partial \tilde{\boldsymbol{k}}^{\beta\nu}(x)} = \delta_\nu^\mu \boldsymbol{S}^{-1} \boldsymbol{\Gamma}_\alpha(X) \boldsymbol{S} \frac{\partial X^\alpha}{\partial x^\beta} \tag{5.73f}$$

$$\tilde{\boldsymbol{Q}}^{\alpha\mu}(X) = -\frac{\partial \tilde{\mathcal{F}}_3^\xi(x)}{\partial \boldsymbol{\Omega}_\alpha(X)} \frac{\partial X^\mu}{\partial x^\xi} \left| \frac{\partial x}{\partial X} \right| = \boldsymbol{S} \tilde{\boldsymbol{q}}^{\beta\xi}(x) \boldsymbol{S}^{-1} \frac{\partial X^\alpha}{\partial x^\beta} \frac{\partial X^\mu}{\partial x^\xi} \left| \frac{\partial x}{\partial X} \right| \tag{5.73g}$$

$$\delta_\nu^\mu \boldsymbol{\omega}_\beta(x) = -\frac{\partial \tilde{\mathcal{F}}_3^\mu(x)}{\partial \tilde{\boldsymbol{q}}^{\beta\nu}(x)} = \delta_\nu^\mu \left(\boldsymbol{S}^{-1} \boldsymbol{\Omega}_\alpha(X) \boldsymbol{S} \frac{\partial X^\alpha}{\partial x^\beta} + \boldsymbol{S}^{-1} \frac{\partial \boldsymbol{S}}{\partial x^\beta} \right). \tag{5.73h}$$

The transformation rule for the Hamiltonian is obtained from the divergence of the explicit x-dependence of the generating function $\tilde{\mathcal{F}}_3^\mu$:

$$\tilde{\mathcal{H}}' \left|\frac{\partial X}{\partial x}\right| - \tilde{\mathcal{H}} = \left.\frac{\partial \tilde{\mathcal{F}}_3^{\mu}}{\partial x^{\mu}}\right|_{\text{expl}}. \tag{5.73i}$$

Thus, the correlation of the Hamiltonians is as always obtained by differentiating all coefficients (not the fields!) of $\tilde{\mathcal{F}}_3^{\mu}$ with respect to x^{μ} and then expressing all those derivatives in terms of the physical fields according to the canonical transformation rules (5.73). The strategy to replace the parameter dependencies of the divergence (5.73i) of the generating function (5.72) by the involved physical fields of the action (5.71) will be applied in the following to the spin-3/2 fields, while the other parts of the action (5.71) and their transformation rules were already discussed in Sect. 5.2.3.

5.3.1 *Terms Related to the Spinor Fields*

The divergence of $\tilde{\mathcal{F}}_3^{\mu}$ with regard to the explicit spacetime-dependent coefficients concerning the spinor fields $\bar{\Psi}_{\alpha}$ and Ψ_{α} gives:

$$\left.\frac{\partial \tilde{\mathcal{F}}_3^{\mu}}{\partial x^{\mu}}\right|_{\bar{\Psi}_{\alpha},\Psi_{\alpha}} = -\left(\tilde{\bar{\kappa}}^{\beta\mu}\frac{\partial \boldsymbol{S}^{-1}}{\partial x^{\mu}}\Psi_{\alpha} + \bar{\Psi}_{\alpha}\frac{\partial \boldsymbol{S}}{\partial x^{\mu}}\tilde{\kappa}^{\beta\mu}\right)\frac{\partial X^{\alpha}}{\partial x^{\beta}} - \left(\tilde{\bar{\kappa}}^{\beta\mu}\boldsymbol{S}^{-1}\Psi_{\alpha} + \bar{\Psi}_{\alpha}\boldsymbol{S}\tilde{\kappa}^{\beta\mu}\right)\frac{\partial^2 X^{\alpha}}{\partial x^{\beta}\partial x^{\mu}}. \tag{5.74}$$

At this point, we have two options. The second derivative term in X^{β} can be replaced by the transformation rule of the affine connections according to Eq. (5.55). The second option, which will be pursued here, is to replace the second derivative term by the x^{μ}-derivatives of the spinor fields $\bar{\psi}_{\beta}$ and ψ_{β} according to their transformation rules (5.73c) and (5.73d), respectively. This will be worked out in detail in the following explicit calculation.

✓ Explict Calculation

We have from Eq. (5.73d)

$$\psi_{\beta}(x) = \boldsymbol{S}^{-1}(x)\,\Psi_{\alpha}(X)\frac{\partial X^{\alpha}}{\partial x^{\beta}},$$

and hence

$$\frac{\partial \psi_{\beta}}{\partial x^{\mu}} = \frac{\partial \boldsymbol{S}^{-1}}{\partial x^{\mu}}\Psi_{\alpha}\frac{\partial X^{\alpha}}{\partial x^{\beta}} + \boldsymbol{S}^{-1}\frac{\partial \Psi_{\alpha}}{\partial X^{\xi}}\frac{\partial X^{\alpha}}{\partial x^{\beta}}\frac{\partial X^{\xi}}{\partial x^{\mu}} + \boldsymbol{S}^{-1}\Psi_{\alpha}\frac{\partial^2 X^{\alpha}}{\partial x^{\beta}\partial x^{\mu}}.$$

Thus

$$\tilde{\bar{\kappa}}^{(\beta\mu)}\boldsymbol{S}^{-1}\Psi_{\alpha}\frac{\partial^2 X^{\alpha}}{\partial x^{\beta}\partial x^{\mu}} = \tilde{\bar{\kappa}}^{(\beta\mu)}\frac{\partial \psi_{\beta}}{\partial x^{\mu}} - \tilde{\bar{\kappa}}^{(\beta\mu)}\boldsymbol{S}^{-1}\frac{\partial \Psi_{\alpha}}{\partial X^{\xi}}\frac{\partial X^{\alpha}}{\partial x^{\beta}}\frac{\partial X^{\xi}}{\partial x^{\mu}} - \tilde{\bar{\kappa}}^{(\beta\mu)}\frac{\partial \boldsymbol{S}^{-1}}{\partial x^{\mu}}\Psi_{\alpha}\frac{\partial X^{\alpha}}{\partial x^{\beta}}$$

$$= \tilde{\bar{\kappa}}^{(\beta\mu)} \frac{\partial \psi_\beta}{\partial x^\mu} - \tilde{\bar{\mathcal{K}}}^{(\beta\mu)} \frac{\partial \Psi_\beta}{\partial X^\mu} \left| \frac{\partial X}{\partial x} \right| + \tilde{\bar{\kappa}}^{(\beta\mu)} \boldsymbol{S}^{-1} \frac{\partial \boldsymbol{S}}{\partial x^\mu} \psi_\beta .$$

Similarly, from Eq. (5.73c):

$$\bar{\psi}_\beta(x) = \bar{\Psi}_\alpha(X)\, \boldsymbol{S}(x) \frac{\partial X^\alpha}{\partial x^\beta},$$

the partial derivative is

$$\frac{\partial \bar{\psi}_\beta}{\partial x^\mu} = \frac{\partial \bar{\Psi}_\alpha}{\partial X^\xi} \boldsymbol{S} \frac{\partial X^\alpha}{\partial x^\beta} \frac{\partial X^\xi}{\partial x^\mu} + \bar{\Psi}_\alpha \frac{\partial \boldsymbol{S}}{\partial x^\mu} \frac{\partial X^\alpha}{\partial x^\beta} + \bar{\Psi}_\alpha \boldsymbol{S} \frac{\partial^2 X^\alpha}{\partial x^\beta \partial x^\mu}.$$

Thus

$$\begin{aligned}
\bar{\Psi}_\alpha \boldsymbol{S}\, \tilde{\kappa}^{(\beta\mu)} \frac{\partial^2 X^\alpha}{\partial x^\beta \partial x^\mu} &= \frac{\partial \bar{\psi}_\beta}{\partial x^\mu} \tilde{\kappa}^{(\beta\mu)} - \frac{\partial \bar{\Psi}_\alpha}{\partial X^\xi} \boldsymbol{S}\, \tilde{\kappa}^{(\beta\mu)} \frac{\partial X^\alpha}{\partial x^\beta} \frac{\partial X^\xi}{\partial x^\mu} - \bar{\Psi}_\alpha \frac{\partial X^\alpha}{\partial x^\beta} \frac{\partial \boldsymbol{S}}{\partial x^\mu} \tilde{\kappa}^{(\beta\mu)} \\
&= \frac{\partial \bar{\psi}_\beta}{\partial x^\mu} \tilde{\kappa}^{(\beta\mu)} - \frac{\partial \bar{\Psi}_\beta}{\partial X^\mu} \tilde{\mathcal{K}}^{(\beta\mu)} \left| \frac{\partial X}{\partial x} \right| - \bar{\psi}_\beta \boldsymbol{S}^{-1} \frac{\partial \boldsymbol{S}}{\partial x^\mu} \tilde{\kappa}^{(\beta\mu)}
\end{aligned}$$

Equation (5.74) thus has the equivalent form:

$$\begin{aligned}
\left.\frac{\partial \tilde{\mathcal{F}}_3^\mu}{\partial x^\mu}\right|_{\bar{\Psi}_\alpha, \Psi_\alpha} &= -\tilde{\bar{\kappa}}^{(\beta\mu)} \frac{\partial \psi_\beta}{\partial x^\mu} - \frac{\partial \bar{\psi}_\beta}{\partial x^\mu} \tilde{\kappa}^{(\beta\mu)} + \left(\tilde{\bar{\mathcal{K}}}^{(\beta\mu)} \frac{\partial \Psi_\beta}{\partial X^\mu} + \frac{\partial \bar{\Psi}_\beta}{\partial X^\mu} \tilde{\mathcal{K}}^{(\beta\mu)} \right) \left| \frac{\partial X}{\partial x} \right| \\
&\quad + \tilde{\bar{\kappa}}^{\beta\mu} \boldsymbol{S}^{-1} \frac{\partial \boldsymbol{S}}{\partial x^\mu} \psi_\beta - \bar{\psi}_\beta \boldsymbol{S}^{-1} \frac{\partial \boldsymbol{S}}{\partial x^\mu} \tilde{\kappa}^{\beta\mu} - \tilde{\bar{\kappa}}^{(\beta\mu)} \boldsymbol{S}^{-1} \frac{\partial \boldsymbol{S}}{\partial x^\mu} \psi_\beta + \bar{\psi}_\beta \boldsymbol{S}^{-1} \frac{\partial \boldsymbol{S}}{\partial x^\mu} \tilde{\kappa}^{(\beta\mu)} \\
&= -\tilde{\bar{\kappa}}^{(\beta\mu)} \frac{\partial \psi_\beta}{\partial x^\mu} - \frac{\partial \bar{\psi}_\beta}{\partial x^\mu} \tilde{\kappa}^{(\beta\mu)} + \left(\tilde{\bar{\mathcal{K}}}^{(\beta\mu)} \frac{\partial \Psi_\beta}{\partial X^\mu} + \frac{\partial \bar{\Psi}_\beta}{\partial X^\mu} \tilde{\mathcal{K}}^{(\beta\mu)} \right) \left| \frac{\partial X}{\partial x} \right| \\
&\quad + \tilde{\bar{\kappa}}^{[\beta\mu]} \left(\boldsymbol{\omega}_\mu - \boldsymbol{S}^{-1} \boldsymbol{\Omega}_\alpha \boldsymbol{S} \frac{\partial X^\alpha}{\partial x^\mu} \right) \psi_\beta - \bar{\psi}_\beta \left(\boldsymbol{\omega}_\mu - \boldsymbol{S}^{-1} \boldsymbol{\Omega}_\alpha \boldsymbol{S} \frac{\partial X^\alpha}{\partial x^\mu} \right) \tilde{\kappa}^{[\beta\mu]} \\
&= -\tilde{\bar{\kappa}}^{(\beta\mu)} \frac{\partial \psi_\beta}{\partial x^\mu} - \frac{\partial \bar{\psi}_\beta}{\partial x^\mu} \tilde{\kappa}^{(\beta\mu)} + \tilde{\bar{\kappa}}^{[\beta\mu]} \boldsymbol{\omega}_\mu \psi_\beta - \bar{\psi}_\beta \boldsymbol{\omega}_\mu \tilde{\kappa}^{[\beta\mu]} \\
&\quad + \left(\tilde{\bar{\mathcal{K}}}^{(\beta\mu)} \frac{\partial \Psi_\beta}{\partial X^\mu} + \frac{\partial \bar{\Psi}_\beta}{\partial X^\mu} \tilde{\mathcal{K}}^{(\beta\mu)} - \tilde{\bar{\mathcal{K}}}^{[\beta\mu]} \boldsymbol{\Omega}_\mu \Psi_\beta + \bar{\Psi}_\beta \boldsymbol{\Omega}_\mu \tilde{\mathcal{K}}^{[\beta\mu]} \right) \left| \frac{\partial X}{\partial x} \right|, \quad (5.75)
\end{aligned}$$

where the gauge field transformation (5.26) was inserted for $S^{-1}\partial_\mu S$. We observe that all parameter dependencies of Eq. (5.74) were replaced by dependencies on the physical fields. Moreover, the physical fields occur in a symmetric form in the original and the transformed fields, which is required to finally derive a form-invariant action functional. The gauge Hamiltonian $\tilde{\mathcal{H}}_{\text{Gauge}}$ is thus given by

$$\tilde{\mathcal{H}}_{\text{Gauge}} = \tilde{\bar{\kappa}}^{(\beta\alpha)} \frac{\partial \psi_\beta}{\partial x^\alpha} + \frac{\partial \bar{\psi}_\beta}{\partial x^\alpha} \tilde{\kappa}^{(\beta\alpha)} - \tilde{\bar{\kappa}}^{[\beta\alpha]} \boldsymbol{\omega}_\mu \psi_\beta + \bar{\psi}_\beta \boldsymbol{\omega}_\alpha \tilde{\kappa}^{[\beta\alpha]},$$

which yields, inserted into the original action (5.71), the final form-invariant action functional

$$S_1 = \int_V \left[\tfrac{1}{2}\tilde{\tilde{\kappa}}^{\beta\alpha} \left(\frac{\partial \psi_\beta}{\partial x^\alpha} - \frac{\partial \psi_\alpha}{\partial x^\beta} + \boldsymbol{\omega}_\alpha \psi_\beta - \boldsymbol{\omega}_\beta \psi_\alpha \right) + \tfrac{1}{2} \left(\frac{\partial \bar{\psi}_\beta}{\partial x^\alpha} - \frac{\partial \bar{\psi}_\alpha}{\partial x^\beta} - \bar{\psi}_\beta \boldsymbol{\omega}_\alpha + \bar{\psi}_\alpha \boldsymbol{\omega}_\beta \right) \tilde{\kappa}^{\beta\alpha} \right.$$
$$+ \mathrm{Tr}\left\{ \tfrac{1}{2}\tilde{\boldsymbol{k}}^{\beta\alpha} \left(\frac{\partial \boldsymbol{\gamma}_\beta}{\partial x^\alpha} - \frac{\partial \boldsymbol{\gamma}_\alpha}{\partial x^\beta} + \boldsymbol{\omega}_\alpha \boldsymbol{\gamma}_\beta - \boldsymbol{\gamma}_\beta \boldsymbol{\omega}_\alpha - \boldsymbol{\omega}_\beta \boldsymbol{\gamma}_\alpha + \boldsymbol{\gamma}_\alpha \boldsymbol{\omega}_\beta \right) \right\}$$
$$\left. + \mathrm{Tr}\left\{ \tfrac{1}{2}\tilde{\boldsymbol{q}}^{\beta\alpha} \left(\frac{\partial \boldsymbol{\omega}_\beta}{\partial x^\alpha} - \frac{\partial \boldsymbol{\omega}_\alpha}{\partial x^\beta} + \boldsymbol{\omega}_\alpha \boldsymbol{\omega}_\beta - \boldsymbol{\omega}_\beta \boldsymbol{\omega}_\alpha \right) \right\} - \tilde{\mathcal{H}}_{\mathrm{D}} - \tilde{\mathcal{H}}_{\mathrm{Gr}} \right] \mathrm{d}^4 x. \tag{5.76}$$

This is the expected result: the spin-3/2 field ψ_β in the "gauged" action (5.76) shows up as the field tensor of a spin-1 field with respect to its spacetime index β and as a spin-1/2 field that directly couples to the gauge field $\boldsymbol{\omega}_\alpha$ by means of its spinor index.

References

1. S. Gasiorowicz, *Elementary Particle Physics* (Wiley, New York, 1966)
2. J. Struckmeier, A. Redelbach, Covariant Hamiltonian field theory. Int. J. Mod. Phys. E **17**, 435–491 (2008). https://doi.org/10.1142/s0218301308009458. arXiv:0811.0508
3. J. Struckmeier et al., Canonical transformation path to gauge theories of gravity. Phys. Rev. D **95**, 124048 (2017). https://doi.org/10.1103/PhysRevD.95.124048. arXiv:1704.07246

Chapter 6
Noether's Theorem

In this Chapter the infinitesimal version of the canonical transformation framework is applied [1] in the spirit of Noether's approach. As known from [2], each symmetry of a given dynamical system is associated with a pertaining conserved Noether current j_{N}^{μ}.

Dropping here the spinor field from the consideration enables us to formulate the theory completely in the coordinate frame [3]. The vierbein fields then appear in quadratic configurations only and can be replaced by the metric tensor. This recovers, along an alternative path though, the findings discussed in Chap. 4, but restricted to scalar and vector matter for which again a general form of an Einstein-type field equation is derived.

However, in the metric tensor formulation, the transformation properties of its conjugate momentum field evidently differ from those of the conjugate momentum field of the vierbein. The dynamics of gravity as compared to the results of Chap. 4 are reproduced. Regarding the vector field, we here do not assume a skew-symmetric field tensor, i.e. we chose the Proca rather than the Maxwell-Proca version of the free vector field Lagrangian. This results in a *direct* coupling of vector and gauge field—the affine connection—and thereby abolishes a U(1) symmetry of the system.

In the generating function (4.36), the identity

$$\tilde{k}_i^{\;\mu\nu}\,\Lambda^i_{\;I}\,E^I_{\;\alpha}\frac{\partial X^\alpha}{\partial x^\mu}=\tilde{k}_\beta^{\;\mu\nu}\,e_i^{\;\beta}\,\Lambda^i_{\;I}\,E^I_{\;\alpha}\frac{\partial X^\alpha}{\partial x^\mu}=\tilde{k}_\beta^{\;\mu\nu}\,e_i^{\;\beta}\,e^i_{\;\mu}=\tilde{k}^{\beta\mu\nu}\,g_{\beta\mu}$$
$$=\tilde{k}^{\beta\mu\nu}G_{\xi\alpha}\frac{\partial X^\xi}{\partial x^\beta}\frac{\partial X^\alpha}{\partial x^\mu}$$

can be used to transfer this term into a coordinate formulation. As we shall see below, this transformation from the Lorentz to the coordinate system will—after variation—change the physical interpretation of the momentum field $\tilde{k}_i^{\;\mu\nu}$. While in the vierbein formalism the canonical equation (4.79i) links it to the torsion field, the coordinate formalism will relate it to non-metricity. Moreover, as the canonical transformation

D. Vasak et al., *Covariant Canonical Gauge Gravity*, FIAS Interdisciplinary Science Series, https://doi.org/10.1007/978-3-031-43717-5_6

process identifies the affine rather than the spin connection as the appropriate gauge field, we must also modify the transformation property of its conjugate momentum field $\tilde{q}_{\eta}^{\;\alpha\beta\mu}$. This results (see also Refs. [4, 5]) in the modified generating function of the finite Diff(M) symmetry transformation

$$\begin{aligned}\tilde{\mathcal{F}}_3^{\mu}\Big|_x = &-\tilde{\pi}^{\mu}\,\Phi - \tilde{p}^{\alpha\mu}\,A_{\beta}\frac{\partial X^{\beta}}{\partial x^{\alpha}} - \tilde{k}^{\alpha\beta\mu}G_{\xi\lambda}\frac{\partial X^{\xi}}{\partial x^{\alpha}}\frac{\partial X^{\lambda}}{\partial x^{\beta}}\\ &-\tilde{q}_{\eta}^{\;\alpha\beta\mu}\left(\Gamma^{\tau}_{\;\xi\lambda}\frac{\partial x^{\eta}}{\partial X^{\tau}}\frac{\partial X^{\xi}}{\partial x^{\alpha}}\frac{\partial X^{\lambda}}{\partial x^{\beta}} + \frac{\partial x^{\eta}}{\partial X^{\tau}}\frac{\partial^2 X^{\tau}}{\partial x^{\alpha}\partial x^{\beta}}\right). \qquad (6.1)\end{aligned}$$

Notice that the transformed metric tensor $G_{\xi\lambda}(X)$ is *symmetric*, which induces the tensor $\tilde{k}^{\alpha\beta\mu}$ to be symmetric in its first index pair, α, β.

This chapter will thus illuminate how the absence of fermions influences the description of gravity. We review in Sect. 6.1 the canonical transformation representation of finite Diff(M) symmetry transformations in the coordinate formulation, pointing out the differences in the description of spacetime. In Sect. 6.2 Noether's theorem [2] is formulated in the realm of covariant Hamiltonian field theory. In order to work out the conserved Noether current for the Diff(M) symmetry transformation, the finite transformation is reformulated as the pertaining infinitesimal transformation, followed by a detailed discussion of the conserved Noether current.

The most general Einstein-type field equation is presented in the Lagrangian formulation in Sect. 6.3. This equation is shown to be equivalent to the "consistency equation" of Ref. [4]. The identity also provides the correlation of the metric (Hilbert) and the canonical energy-momentum tensors of a given Lagrangian system by means of the identity for scalar-valued functions discussed in Sect. 3.1, Chap. 3. Finally, we discuss the correlation of the spin part of the energy-momentum tensor of the source system with the torsion of spacetime, described by the then skew-symmetric part of the Ricci tensor.

6.1 Diff(*M*) Transformation in Coordinate Formulation

As discussed above are the Hamiltonians in the coordinate description presumed to depend on the metric $g_{\alpha\lambda}$ and the—in general—non-symmetric connection $\gamma^{\alpha}_{\;\xi\eta}$, in conjunction with their respective conjugates, $\tilde{k}^{\alpha\lambda\beta}$ and $\tilde{q}_{\alpha}^{\;\xi\eta\beta}$. It is straightforward to show that the (4.34) is then modified to

$$
\begin{aligned}
&- \phi\, \delta^{\beta}_{\xi} \frac{\partial \tilde{\pi}^{\xi}}{\partial x^{\beta}} - a_{\alpha} \delta^{\beta}_{\xi} \frac{\partial \tilde{p}^{\,\alpha\xi}}{\partial x^{\beta}} - g_{\alpha\lambda} \delta^{\beta}_{\xi} \frac{\partial \tilde{k}^{\,\alpha\lambda\xi}}{\partial x^{\beta}} - \gamma^{\alpha}{}_{\lambda\eta} \delta^{\beta}_{\xi} \frac{\partial \tilde{q}_{\alpha}{}^{\lambda\eta\xi}}{\partial x^{\beta}} - \tilde{\mathcal{H}} \\
&- \left(\tilde{\Pi}^{\beta} \frac{\partial \Phi}{\partial X^{\beta}} + \tilde{P}^{\,\alpha\beta} \frac{\partial A_{\alpha}}{\partial X^{\beta}} + \tilde{K}^{\alpha\lambda\beta} \frac{\partial G_{\alpha\lambda}}{\partial X^{\beta}} + \tilde{Q}_{\alpha}{}^{\xi\eta\beta} \frac{\partial \Gamma^{\alpha}{}_{\xi\eta}}{\partial X^{\beta}} - \tilde{\mathcal{H}}' \right) \left| \frac{\partial X}{\partial x} \right| \\
&= \frac{\partial \tilde{\mathcal{F}}_3^{\eta}}{\partial \Phi} \frac{\partial X^{\beta}}{\partial x^{\eta}} \frac{\partial \Phi}{\partial X^{\beta}} + \frac{\partial \tilde{\mathcal{F}}_3^{\eta}}{\partial A_{\alpha}} \frac{\partial X^{\beta}}{\partial x^{\eta}} \frac{\partial A_{\alpha}}{\partial X^{\beta}} + \frac{\partial \tilde{\mathcal{F}}_3^{\eta}}{\partial G_{\alpha\lambda}} \frac{\partial X^{\beta}}{\partial x^{\eta}} \frac{\partial G_{\alpha\lambda}}{\partial X^{\beta}} + \frac{\partial \tilde{\mathcal{F}}_3^{\kappa}}{\partial \Gamma^{\alpha}{}_{\xi\eta}} \frac{\partial X^{\beta}}{\partial x^{\kappa}} \frac{\partial \Gamma^{\alpha}{}_{\xi\eta}}{\partial X^{\beta}} \\
&+ \frac{\partial \tilde{\mathcal{F}}_3^{\beta}}{\partial \tilde{\pi}^{\xi}} \frac{\partial \tilde{\pi}^{\xi}}{\partial x^{\beta}} + \frac{\partial \tilde{\mathcal{F}}_3^{\beta}}{\partial \tilde{p}^{\alpha\xi}} \frac{\partial \tilde{p}^{\alpha\xi}}{\partial x^{\beta}} + \frac{\partial \tilde{\mathcal{F}}_3^{\beta}}{\partial \tilde{k}^{\,\alpha\lambda\xi}} \frac{\partial \tilde{k}^{\,\alpha\lambda\xi}}{\partial x^{\beta}} + \frac{\partial \tilde{\mathcal{F}}_3^{\beta}}{\partial \tilde{q}_{\alpha}{}^{\lambda\eta\xi}} \frac{\partial \tilde{q}_{\alpha}{}^{\lambda\eta\xi}}{\partial x^{\beta}} + \left. \frac{\partial \tilde{\mathcal{F}}_3^{\beta}}{\partial x^{\beta}} \right|_{\text{expl}} .
\end{aligned}
\tag{6.2}
$$

While the canonical relations (4.35) are reproduced for the matter fields under consideration, they are modified accordingly for the metric tensor $g_{\alpha\beta}$, the affine connection $\gamma^{\eta}{}_{\alpha\beta}$, and their respective conjugates, $\tilde{k}^{\alpha\beta\mu}$ and $\tilde{q}_{\eta}{}^{\alpha\beta\mu}$:

$$
\tilde{K}^{\alpha\beta\mu}(X) = -\frac{\partial \tilde{\mathcal{F}}_3^{\kappa}}{\partial G_{\alpha\beta}} \frac{\partial X^{\mu}}{\partial x^{\kappa}} \left| \frac{\partial x}{\partial X} \right| \qquad \delta^{\mu}_{\nu}\, g_{\alpha\beta}(x) = -\frac{\partial \tilde{\mathcal{F}}_3^{\mu}}{\partial \tilde{k}^{\alpha\beta\nu}} \tag{6.3a}
$$

$$
\tilde{Q}_{\eta}{}^{\alpha\beta\mu}(X) = -\frac{\partial \tilde{\mathcal{F}}_3^{\kappa}}{\partial \Gamma^{\eta}{}_{\alpha\beta}} \frac{\partial X^{\mu}}{\partial x^{\kappa}} \left| \frac{\partial x}{\partial X} \right| \qquad \delta^{\mu}_{\nu}\, \gamma^{\eta}{}_{\alpha\beta}(x) = -\frac{\partial \tilde{\mathcal{F}}_3^{\mu}}{\partial \tilde{q}_{\eta}{}^{\alpha\beta\nu}} . \tag{6.3b}
$$

Then the general rules (4.21) with the specific generating function (6.1) reproduce the transformation rules (4.37) for the matter fields and their conjugates,

$$
\phi(x) = \Phi(X), \qquad a_{\alpha}(x) = A_{\beta}(X) \frac{\partial X^{\beta}}{\partial x^{\alpha}}, \tag{6.4a}
$$

$$
\tilde{\Pi}^{\mu}(X) = \tilde{\pi}^{\kappa}(x) \frac{\partial X^{\mu}}{\partial x^{\kappa}} \left| \frac{\partial x}{\partial X} \right|, \qquad \tilde{P}^{\nu\mu}(X) = \tilde{p}^{\alpha\kappa}(x) \frac{\partial X^{\nu}}{\partial x^{\alpha}} \frac{\partial X^{\mu}}{\partial x^{\kappa}} \left| \frac{\partial x}{\partial X} \right|, \tag{6.4b}
$$

but they give a modified version for the metric and connection fields and their respective conjugates:

$$
g_{\alpha\beta}(x) = G_{\xi\lambda}(X) \frac{\partial X^{\xi}}{\partial x^{\alpha}} \frac{\partial X^{\lambda}}{\partial x^{\beta}} \tag{6.4c}
$$

$$
\gamma^{\eta}{}_{\alpha\beta}(x) = \Gamma^{\tau}{}_{\xi\lambda}(X) \frac{\partial x^{\eta}}{\partial X^{\tau}} \frac{\partial X^{\xi}}{\partial x^{\alpha}} \frac{\partial X^{\lambda}}{\partial x^{\beta}} + \frac{\partial x^{\eta}}{\partial X^{\tau}} \frac{\partial^2 X^{\tau}}{\partial x^{\alpha} \partial x^{\beta}} \tag{6.4d}
$$

$$
\tilde{K}^{\xi\lambda\mu}(X) = \tilde{k}^{\alpha\beta\kappa}(x) \frac{\partial X^{\xi}}{\partial x^{\alpha}} \frac{\partial X^{\lambda}}{\partial x^{\beta}} \frac{\partial X^{\mu}}{\partial x^{\kappa}} \left| \frac{\partial x}{\partial X} \right| \tag{6.4e}
$$

$$
\tilde{Q}_{\tau}{}^{\xi\lambda\mu}(X) = \tilde{q}_{\eta}{}^{\alpha\beta\kappa}(x) \frac{\partial x^{\eta}}{\partial X^{\tau}} \frac{\partial X^{\xi}}{\partial x^{\alpha}} \frac{\partial X^{\lambda}}{\partial x^{\beta}} \frac{\partial X^{\mu}}{\partial x^{\kappa}} \left| \frac{\partial x}{\partial X} \right| . \tag{6.4f}
$$

Finally, the particular transformation rule for a covariant Hamiltonian follows from the general rule (4.19c) as:

$$
\begin{aligned}
\left.\frac{\partial \tilde{\mathcal{F}}_3^\nu}{\partial x^\nu}\right|_{\text{expl}} &= \left.\tilde{\mathcal{H}}'\right|_x - \left.\tilde{\mathcal{H}}\right|_x \\
&= -\tilde{p}^{\alpha\mu} a_\sigma \frac{\partial x^\sigma}{\partial X^\xi}\frac{\partial^2 X^\xi}{\partial x^\alpha \partial x^\mu} - \tilde{k}^{\alpha\beta\mu}\left(g_{\sigma\beta}\frac{\partial x^\sigma}{\partial X^\xi}\frac{\partial^2 X^\xi}{\partial x^\alpha \partial x^\mu} + g_{\alpha\sigma}\frac{\partial x^\sigma}{\partial X^\xi}\frac{\partial^2 X^\xi}{\partial x^\beta \partial x^\mu}\right) \\
&\quad + \tilde{q}_\eta{}^{\alpha\beta\mu}\left(\gamma^\sigma{}_{\alpha\beta}\frac{\partial x^\eta}{\partial X^\xi}\frac{\partial^2 X^\xi}{\partial x^\sigma \partial x^\mu} - \gamma^\eta{}_{\sigma\beta}\frac{\partial x^\sigma}{\partial X^\xi}\frac{\partial^2 X^\xi}{\partial x^\alpha \partial x^\mu} - \gamma^\eta{}_{\alpha\sigma}\frac{\partial x^\sigma}{\partial X^\xi}\frac{\partial^2 X^\xi}{\partial x^\beta \partial x^\mu}\right. \\
&\quad + \frac{\partial x^\eta}{\partial X^\tau}\frac{\partial^2 X^\tau}{\partial x^\sigma \partial x^\beta}\frac{\partial x^\sigma}{\partial X^\xi}\frac{\partial^2 X^\xi}{\partial x^\alpha \partial x^\mu} + \frac{\partial x^\eta}{\partial X^\tau}\frac{\partial^2 X^\tau}{\partial x^\sigma \partial x^\alpha}\frac{\partial x^\sigma}{\partial X^\xi}\frac{\partial^2 X^\xi}{\partial x^\beta \partial x^\mu} \\
&\quad \left. - \frac{\partial x^\eta}{\partial X^\xi}\frac{\partial^3 X^\xi}{\partial x^\alpha \partial x^\beta \partial x^\mu}\right).
\end{aligned} \tag{6.4g}
$$

Starting from these modifications we shall work out in the following section the infinitesimal transformation approach and then, using Noether's theorem for the continuous symmetry group Diff(M), construct the pertinent Noether current and derive the key equations for theories of gravity.

6.2 Gauge Theory of Gravity from Noether's Theorem

For a general *infinitesimal* canonical transformation the generating function (6.1) is replaced by the following ansatz:

$$
\left.\tilde{\mathcal{F}}_3^\mu\right|_x = \left.\left(-\tilde{\pi}^\mu \Phi - \tilde{p}^{\alpha\mu} A_\alpha - \tilde{k}^{\xi\lambda\mu} G_{\xi\lambda} - \tilde{q}_\eta{}^{\alpha\beta\mu}\Gamma^\eta{}_{\alpha\beta} - \epsilon\, \tilde{j}_{\mathrm{N}}^\mu\right)\right|_x . \tag{6.5}
$$

Therein $\tilde{j}_{\mathrm{N}}^\mu = \tilde{j}_{\mathrm{N}}^\mu\left(\tilde{\pi}, \phi, \tilde{p}, a, \tilde{k}, g, \tilde{q}, \gamma, x\right)$ denotes at first an arbitrary function of all dynamical fields in the original frame. Obviously, for $\epsilon = 0$, Eq. (6.5) generates the identity transformation for all dynamical quantities the autonomous Hamiltonian $\tilde{\mathcal{H}} = \tilde{\mathcal{H}}\left(\tilde{\pi}, \phi, \tilde{p}, a, \tilde{k}, g, \tilde{q}, \gamma\right)$ depends on. All contributions of the general transformation rules (4.34) and (6.3), that are associated with a non-identical mapping of fields and spacetime, are then encoded in the particular expression for $\tilde{j}_{\mathrm{N}}^\mu(x)$. With $\delta\phi(x) := \Phi(x) - \phi(x)$, and similarly for all other fields, this leads to

$$
\delta^\mu_\nu \delta\phi = -\epsilon\frac{\partial \tilde{j}_{\mathrm{N}}^\mu}{\partial \tilde{\pi}^\nu}, \qquad \delta\tilde{\pi}^\mu = \epsilon\frac{\partial \tilde{j}_{\mathrm{N}}^\mu}{\partial \phi}, \qquad \delta^\mu_\nu \delta a_\alpha = -\epsilon\frac{\partial \tilde{j}_{\mathrm{N}}^\mu}{\partial \tilde{p}^{\alpha\nu}}, \qquad \delta\tilde{p}^{\alpha\mu} = \epsilon\frac{\partial \tilde{j}_{\mathrm{N}}^\mu}{\partial a_\alpha} \tag{6.6a}
$$

$$
\delta^\mu_\nu \delta g_{\lambda\xi} = -\epsilon\frac{\partial \tilde{j}_{\mathrm{N}}^\mu}{\partial \tilde{k}^{\lambda\xi\nu}}, \quad \delta\tilde{k}^{\lambda\xi\mu} = \epsilon\frac{\partial \tilde{j}_{\mathrm{N}}^\mu}{\partial g_{\lambda\xi}}, \quad \delta^\mu_\nu \delta\gamma^\eta{}_{\alpha\beta} = -\epsilon\frac{\partial \tilde{j}_{\mathrm{N}}^\mu}{\partial \tilde{q}_\eta{}^{\alpha\beta\nu}}, \quad \delta\tilde{q}_\eta{}^{\alpha\beta\mu} = \epsilon\frac{\partial \tilde{j}_{\mathrm{N}}^\mu}{\partial \gamma^\eta{}_{\alpha\beta}} \tag{6.6b}
$$

and, in analogy to Eq. (4.35g),

$$\delta\tilde{\mathcal{H}}\Big|_{\mathrm{CT}} \equiv \tilde{\mathcal{H}}'\Big|_{x} - \tilde{\mathcal{H}}\Big|_{x} = \frac{\partial\tilde{\mathcal{F}}_3^{\alpha}}{\partial x^{\alpha}}\Bigg|_{\mathrm{expl}} = -\epsilon\,\frac{\partial\tilde{j}_{\mathrm{N}}^{\alpha}}{\partial x^{\alpha}}\Bigg|_{\mathrm{expl}}. \tag{6.6c}$$

The explicit representation of the infinitesimal transformation rules (6.6) for the generating function $\tilde{\mathcal{F}}_3^{\mu}$ of a chart transformation will be presented below. The subscript "CT" indicates a variation from a canonical transformation.

On the other hand, for a closed system, where the Hamiltonian does not explicitly depend on x, the variation of the Hamiltonian from field variations is

$$\begin{aligned}\delta\tilde{\mathcal{H}} = {}& \frac{\partial\tilde{\mathcal{H}}}{\partial\phi}\delta\phi + \frac{\partial\tilde{\mathcal{H}}}{\partial\tilde{\pi}^{\alpha}}\delta\tilde{\pi}^{\alpha} + \frac{\partial\tilde{\mathcal{H}}}{\partial a_{\alpha}}\delta a_{\alpha} + \frac{\partial\tilde{\mathcal{H}}}{\partial\tilde{p}^{\alpha\beta}}\delta\tilde{p}^{\alpha\beta} \\ & + \frac{\partial\tilde{\mathcal{H}}}{\partial g_{\lambda\xi}}\delta g_{\lambda\xi} + \frac{\partial\tilde{\mathcal{H}}}{\partial\tilde{k}^{\lambda\xi\beta}}\delta\tilde{k}^{\lambda\xi\beta} + \frac{\partial\tilde{\mathcal{H}}}{\partial\gamma^{\eta}{}_{\lambda\xi}}\delta\gamma^{\eta}{}_{\lambda\xi} + \frac{\partial\tilde{\mathcal{H}}}{\partial\tilde{q}_{\eta}{}^{\lambda\xi\beta}}\delta\tilde{q}_{\eta}{}^{\lambda\xi\beta}.\end{aligned}$$

Inserting the covariant canonical field equations as discussed in Sect. 4.2,

$$-\frac{\partial\tilde{\pi}^{\beta}}{\partial x^{\beta}} = \frac{\partial\tilde{\mathcal{H}}}{\partial\phi} \qquad \frac{\partial\phi}{\partial x^{\alpha}} = \frac{\partial\tilde{\mathcal{H}}}{\partial\tilde{\pi}^{\alpha}} \qquad -\frac{\partial\tilde{p}^{\alpha\beta}}{\partial x^{\beta}} = \frac{\partial\tilde{\mathcal{H}}}{\partial a_{\alpha}} \qquad \frac{\partial a_{\alpha}}{\partial x^{\beta}} = \frac{\partial\tilde{\mathcal{H}}}{\partial\tilde{p}^{\alpha\beta}} \tag{6.7a}$$

$$-\frac{\partial\tilde{k}^{\lambda\xi\beta}}{\partial x^{\beta}} = \frac{\partial\tilde{\mathcal{H}}}{\partial g_{\lambda\xi}} \quad \frac{\partial g_{\lambda\xi}}{\partial x^{\beta}} = \frac{\partial\tilde{\mathcal{H}}}{\partial\tilde{k}^{\lambda\xi\beta}} \quad -\frac{\partial\tilde{q}_{\eta}{}^{\lambda\xi\beta}}{\partial x^{\beta}} = \frac{\partial\tilde{\mathcal{H}}}{\partial\gamma^{\eta}{}_{\lambda\xi}} \quad \frac{\partial\gamma^{\eta}{}_{\lambda\xi}}{\partial x^{\beta}} = \frac{\partial\tilde{\mathcal{H}}}{\partial\tilde{q}_{\eta}{}^{\lambda\xi\beta}}, \tag{6.7b}$$

the variation of $\tilde{\mathcal{H}}$ is expressed along the system's spacetime evolution as

$$\begin{aligned}\delta\tilde{\mathcal{H}} = {}& -\frac{\partial\tilde{\pi}^{\beta}}{\partial x^{\alpha}}\delta^{\alpha}_{\beta}\,\delta\phi + \frac{\partial\phi}{\partial x^{\alpha}}\delta\tilde{\pi}^{\alpha} - \frac{\partial\tilde{p}^{\alpha\beta}}{\partial x^{\xi}}\delta^{\xi}_{\beta}\,\delta a_{\alpha} + \frac{\partial a_{\alpha}}{\partial x^{\beta}}\delta\tilde{p}^{\alpha\beta} \\ & - \frac{\partial\tilde{k}^{\lambda\xi\beta}}{\partial x^{\alpha}}\delta^{\alpha}_{\beta}\,\delta g_{\lambda\xi} + \frac{\partial g_{\lambda\xi}}{\partial x^{\beta}}\delta\tilde{k}^{\lambda\xi\beta} - \frac{\partial\tilde{q}_{\eta}{}^{\lambda\xi\beta}}{\partial x^{\alpha}}\delta^{\alpha}_{\beta}\,\delta\gamma^{\eta}{}_{\lambda\xi} + \frac{\partial\gamma^{\eta}{}_{\lambda\xi}}{\partial x^{\beta}}\delta\tilde{q}_{\eta}{}^{\lambda\xi\beta}.\end{aligned} \tag{6.8}$$

With the transformation rules (6.6), this gives in terms of the derivatives of $\tilde{j}_{\mathrm{N}}^{\mu}$:

$$\begin{aligned}\delta\tilde{\mathcal{H}} &= \epsilon\frac{\partial\tilde{\pi}^{\beta}}{\partial x^{\alpha}}\frac{\partial\tilde{j}_{\mathrm{N}}^{\alpha}}{\partial\tilde{\pi}^{\beta}} + \epsilon\frac{\partial\phi}{\partial x^{\alpha}}\frac{\partial\tilde{j}_{\mathrm{N}}^{\alpha}}{\partial\phi} + \epsilon\frac{\partial\tilde{p}^{\alpha\beta}}{\partial x^{\xi}}\frac{\partial\tilde{j}_{\mathrm{N}}^{\xi}}{\partial\tilde{p}^{\alpha\beta}} + \epsilon\frac{\partial a_{\alpha}}{\partial x^{\beta}}\frac{\partial\tilde{j}_{\mathrm{N}}^{\beta}}{\partial a_{\alpha}} \\ &\quad + \epsilon\frac{\partial\tilde{k}^{\lambda\xi\beta}}{\partial x^{\alpha}}\frac{\partial\tilde{j}_{\mathrm{N}}^{\alpha}}{\partial\tilde{k}^{\lambda\xi\beta}} + \epsilon\frac{\partial g_{\lambda\xi}}{\partial x^{\beta}}\frac{\partial\tilde{j}_{\mathrm{N}}^{\beta}}{\partial g_{\lambda\xi}} + \epsilon\frac{\partial\tilde{q}_{\eta}{}^{\lambda\xi\beta}}{\partial x^{\alpha}}\frac{\partial\tilde{j}_{\mathrm{N}}^{\alpha}}{\partial\tilde{q}_{\eta}{}^{\lambda\xi\beta}} + \epsilon\frac{\partial\gamma^{\eta}{}_{\lambda\xi}}{\partial x^{\beta}}\frac{\partial\tilde{j}_{\mathrm{N}}^{\beta}}{\partial\gamma^{\eta}{}_{\lambda\xi}} \\ &= \epsilon\frac{\partial\tilde{j}_{\mathrm{N}}^{\alpha}}{\partial x^{\alpha}} - \epsilon\frac{\partial\tilde{j}_{\mathrm{N}}^{\alpha}}{\partial x^{\alpha}}\Bigg|_{\mathrm{expl}} \\ &= \epsilon\frac{\partial\tilde{j}_{\mathrm{N}}^{\alpha}}{\partial x^{\alpha}} + \delta\tilde{\mathcal{H}}\Big|_{\mathrm{CT}}.\end{aligned}$$

The requirement that both variations, $\delta\tilde{\mathcal{H}}$ and $\delta\tilde{\mathcal{H}}\big|_{\mathrm{CT}}$, coincide ensures that the canonical transformation generated by (6.5) defines a symmetry transformation. This is exactly the case if the divergence of $\tilde{j}_{\mathrm{N}}^{\mu}$ vanishes:

$$\delta\tilde{\mathcal{H}} - \delta\tilde{\mathcal{H}}\Big|_{\mathrm{CT}} \stackrel{!}{=} 0 \quad \Leftrightarrow \quad \frac{\partial \tilde{j}_{\mathrm{N}}^{\alpha}}{\partial x^{\alpha}} \stackrel{!}{=} 0. \tag{6.9}$$

The vector $\tilde{j}_{\mathrm{N}}^{\mu}$ in the generating function (6.5) then defines the conserved Noether current.

6.2.1 *Infinitesimal Chart Transformations*

In order to work out the particular form of the Noether current $\tilde{j}_{\mathrm{N}}^{\mu}$ associated with an invariance of a given field theory under a specific form of (passive) diffeomorphisms, we first express an arbitrary but infinitesimal chart transition map $x^{\mu} \mapsto X^{\mu}$ on the manifold M as follows:

$$X^{\mu} = x^{\mu} + \epsilon\, h^{\mu}(x), \tag{6.10}$$

with $\epsilon \ll 1$. To first order in ϵ, the spacetime dependent coefficients of (6.1) are then expressed as

$$\frac{\partial X^{\mu}}{\partial x^{\nu}} = \delta^{\mu}_{\nu} + \epsilon\frac{\partial h^{\mu}}{\partial x^{\nu}}, \quad \frac{\partial x^{\mu}}{\partial X^{\nu}} = \delta^{\mu}_{\nu} - \epsilon\frac{\partial h^{\mu}}{\partial x^{\nu}}, \quad \frac{\partial^2 X^{\tau}}{\partial x^{\alpha}\partial x^{\beta}} = \epsilon\frac{\partial^2 h^{\tau}}{\partial x^{\alpha}\partial x^{\beta}}, \quad \left|\frac{\partial x}{\partial X}\right| = 1 - \epsilon\frac{\partial h^{\tau}}{\partial x^{\tau}}. \tag{6.11}$$

The transformation rules (6.4a) to (6.4f) of the fields give now up to first order in ϵ

$$\delta\phi(x) = \Phi(x) - \phi(x) \qquad = -\epsilon h^{\beta}\frac{\partial\phi}{\partial x^{\beta}} = -\epsilon\mathcal{L}_h\phi(x) \tag{6.12a}$$

$$\delta a_{\alpha}(x) = A_{\alpha}(x) - a_{\alpha}(x) \qquad = -\epsilon\left(h^{\beta}\frac{\partial a_{\alpha}}{\partial x^{\beta}} + a_{\beta}\frac{\partial h^{\beta}}{\partial x^{\alpha}}\right) = -\epsilon\mathcal{L}_h a_{\alpha}(x) \tag{6.12b}$$

$$\delta g_{\xi\lambda}(x) = G_{\xi\lambda}(x) - g_{\xi\lambda}(x) \qquad = -\epsilon\left(h^{\beta}\frac{\partial g_{\xi\lambda}}{\partial x^{\beta}} + g_{\beta\lambda}\frac{\partial h^{\beta}}{\partial x^{\xi}} + g_{\beta\xi}\frac{\partial h^{\beta}}{\partial x^{\lambda}}\right) = -\epsilon\mathcal{L}_h g_{\xi\lambda}(x) \tag{6.12c}$$

and

$$\begin{aligned}\delta\gamma^{\eta}{}_{\lambda\tau}(x) &= \Gamma^{\eta}{}_{\lambda\tau}(x) - \gamma^{\eta}{}_{\lambda\tau}(x)\\ &= -\epsilon\left(h^{\beta}\frac{\partial\gamma^{\eta}{}_{\lambda\tau}}{\partial x^{\beta}} - \gamma^{\beta}{}_{\lambda\tau}\frac{\partial h^{\eta}}{\partial x^{\beta}} + \gamma^{\eta}{}_{\beta\tau}\frac{\partial h^{\beta}}{\partial x^{\lambda}} + \gamma^{\eta}{}_{\lambda\beta}\frac{\partial h^{\beta}}{\partial x^{\tau}} + \frac{\partial^2 h^{\eta}}{\partial x^{\lambda}\partial x^{\tau}}\right)\\ &= -\epsilon\mathcal{L}_h\gamma^{\eta}{}_{\lambda\tau}(x).\end{aligned} \tag{6.12d}$$

The differences of the scalar, vector, and tensor fields are thus proportional to *Lie derivatives* along the vector field $h^\beta(x)$, denoted by $\mathscr{L}_h$, see also (2.42). This reflects the general fact that a continuous group of chart transitions which lie infinitesimally close to the identity define the Lie algebra of said group. The difference of the non-tensorial connection field $\gamma^\eta{}_{\lambda\tau}(x)$ is equally proportional to its Lie derivative which is a tensor (see Schouten [6], Eq. (II 10.34)).

✓ Explict Calculation

As an example, we derive explicitly the Lie derivative $\mathscr{L}_h$ of a covariant vector field $a_\mu(x)$ with respect to the contravariant vector field $h^\beta(x)$ occurring in Eq. (6.10). With Eq. (6.11) we have to first order in ϵ

$$a_\mu(x) = A_\beta(X)\frac{\partial X^\beta}{\partial x^\mu} = A_\beta(X)\left(\delta^\beta_\mu + \epsilon\frac{\partial h^\beta}{\partial x^\mu}\right) = A_\mu(X) + \epsilon\, a_\beta(x)\frac{\partial h^\beta}{\partial x^\mu} + \mathcal{O}(\epsilon^2).$$

On the other hand, the Taylor expansion of $A_\mu(X)$ truncated after the linear term in ϵ yields

$$A_\mu(X) = A_\mu(x) + \epsilon\, h^\beta\frac{\partial A_\mu(x)}{\partial x^\beta} = A_\mu(x) + \epsilon\, h^\beta\frac{\partial a_\mu(x)}{\partial x^\beta} + \mathcal{O}(\epsilon^2).$$

Inserting $A_\mu(X)$, we get

$$a_\mu(x) = A_\mu(x) + \epsilon\, h^\beta\frac{\partial a_\mu(x)}{\partial x^\beta} + \epsilon\, a_\beta(x)\frac{\partial h^\beta}{\partial x^\mu} + \mathcal{O}(\epsilon^2),$$

hence

$$\frac{a_\mu(x) - A_\mu(x)}{\epsilon} = h^\beta\frac{\partial a_\mu(x)}{\partial x^\beta} + a_\beta(x)\frac{\partial h^\beta}{\partial x^\mu} + \mathcal{O}(\epsilon).$$

Taking the limit $\epsilon \to 0$ defines the *Lie derivative* $\mathscr{L}_h$ with respect to the vector field $h^\beta(x)$ of $a_\mu(x)$

$$\mathscr{L}_h a_\mu := \lim_{\epsilon\to 0}\frac{a_\mu(x) - A_\mu(x)}{\epsilon} = h^\beta\frac{\partial a_\mu}{\partial x^\beta} + a_\beta\frac{\partial h^\beta}{\partial x^\mu}.$$

The corresponding differences of the conjugate momentum tensors are proportional to conventional Lie derivatives of the respective tensor densities,

$$
\begin{aligned}
\delta\tilde{\pi}^{\mu}(x) &= \tilde{\Pi}^{\mu}(x) - \tilde{\pi}^{\mu}(x) \\
&= \epsilon\left[-h^{\beta}\frac{\partial\tilde{\pi}^{\mu}}{\partial x^{\beta}} + \tilde{\pi}^{\beta}\left(\frac{\partial h^{\mu}}{\partial x^{\beta}} - \delta^{\mu}_{\beta}\frac{\partial h^{\tau}}{\partial x^{\tau}}\right)\right] = -\epsilon\mathscr{L}_h\tilde{\pi}^{\mu}(x) \qquad (6.13a)\\
\delta\tilde{p}^{\alpha\mu}(x) &= \tilde{P}^{\alpha\mu}(x) - \tilde{p}^{\alpha\mu}(x) \\
&= \epsilon\left[-h^{\beta}\frac{\partial\tilde{p}^{\alpha\mu}}{\partial x^{\beta}} + \tilde{p}^{\beta\mu}\frac{\partial h^{\alpha}}{\partial x^{\beta}} + \tilde{p}^{\alpha\beta}\left(\frac{\partial h^{\mu}}{\partial x^{\beta}} - \delta^{\mu}_{\beta}\frac{\partial h^{\tau}}{\partial x^{\tau}}\right)\right] = -\epsilon\mathscr{L}_h\tilde{p}^{\alpha\mu}(x) \\
& \qquad (6.13b)\\
\delta\tilde{k}^{\lambda\tau\mu}(x) &= \tilde{K}^{\lambda\tau\mu}(x) - \tilde{k}^{\lambda\tau\mu}(x) \\
&= \epsilon\left[-h^{\beta}\frac{\partial\tilde{k}^{\lambda\tau\mu}}{\partial x^{\beta}} + \tilde{k}^{\beta\tau\mu}\frac{\partial h^{\lambda}}{\partial x^{\beta}} + \tilde{k}^{\lambda\beta\mu}\frac{\partial h^{\tau}}{\partial x^{\beta}} + \tilde{k}^{\lambda\tau\beta}\left(\frac{\partial h^{\mu}}{\partial x^{\beta}} - \delta^{\mu}_{\beta}\frac{\partial h^{\tau}}{\partial x^{\tau}}\right)\right] \\
&= -\epsilon\mathscr{L}_h\tilde{k}^{\lambda\tau\mu}(x), \qquad (6.13c)
\end{aligned}
$$

and

$$
\begin{aligned}
\delta\tilde{q}_{\eta}{}^{\lambda\tau\mu}(x) &= \tilde{Q}_{\eta}{}^{\lambda\tau\mu}(x) - \tilde{q}_{\eta}{}^{\lambda\tau\mu}(x) \\
&= \epsilon\left[-h^{\beta}\frac{\partial\tilde{q}_{\eta}{}^{\lambda\tau\mu}}{\partial x^{\beta}} - \tilde{q}_{\beta}{}^{\lambda\tau\mu}\frac{\partial h^{\beta}}{\partial x^{\eta}} + \tilde{q}_{\eta}{}^{\beta\tau\mu}\frac{\partial h^{\lambda}}{\partial x^{\beta}} + \tilde{q}_{\eta}{}^{\lambda\beta\mu}\frac{\partial h^{\tau}}{\partial x^{\beta}} + \tilde{q}_{\eta}{}^{\lambda\tau\beta}\left(\frac{\partial h^{\mu}}{\partial x^{\beta}} - \delta^{\mu}_{\beta}\frac{\partial h^{\tau}}{\partial x^{\tau}}\right)\right] \\
&= -\epsilon\mathscr{L}_h\tilde{q}_{\eta}{}^{\lambda\tau\mu}(x). \qquad (6.13d)
\end{aligned}
$$

An important particular feature of the covariant Hamiltonian formalism of field theories is that merely the *divergences* of momentum fields are determined by the system's Hamiltonian rather than the individual components of the respective canonical momentum tensors. The canonical momentum tensors are thus determined only up to arbitrary divergence-free tensors, leaving the freedom to modify them by a divergence free expression. Taking advantage of this property, we first differentiate the field variation as given in Eq. (6.13a),

$$
\begin{aligned}
\frac{\partial(\delta\tilde{\pi}^{\mu}(x))}{\partial x^{\mu}} = \frac{\partial\tilde{\Pi}^{\mu}}{\partial x^{\mu}} - \frac{\partial\tilde{\pi}^{\mu}}{\partial x^{\mu}} &= \epsilon\left[\cancel{-\frac{\partial h^{\beta}}{\partial x^{\mu}}\frac{\partial\tilde{\pi}^{\mu}}{\partial x^{\beta}}} - h^{\beta}\frac{\partial^2\tilde{\pi}^{\mu}}{\partial x^{\beta}\partial x^{\mu}} + \frac{\partial\tilde{\pi}^{\beta}}{\partial x^{\mu}}\left(\cancel{\frac{\partial h^{\mu}}{\partial x^{\beta}}} - \delta^{\mu}_{\beta}\frac{\partial h^{\tau}}{\partial x^{\tau}}\right)\right] \\
&= -\epsilon\left(h^{\mu}\frac{\partial^2\tilde{\pi}^{\beta}}{\partial x^{\mu}\partial x^{\beta}} + \frac{\partial h^{\mu}}{\partial x^{\mu}}\frac{\partial\tilde{\pi}^{\beta}}{\partial x^{\beta}}\right) \\
&= -\epsilon\frac{\partial}{\partial x^{\mu}}\left(h^{\mu}\frac{\partial\tilde{\pi}^{\beta}}{\partial x^{\beta}}\right).
\end{aligned}
$$

Integrating this equation, we supersede the transformation rule (6.13a) by

$$
\delta\tilde{\pi}^{\mu} = -\epsilon h^{\mu}\frac{\partial\tilde{\pi}^{\beta}}{\partial x^{\beta}} = \epsilon h^{\mu}\frac{\partial\tilde{\mathcal{H}}}{\partial\phi}, \qquad (6.14a)
$$

which amounts to replacing the momentum tensor components $\delta\tilde{\pi}^{\mu}$ by modified components of an equivalent momentum tensor with the same divergence. Similarly, the divergence of the variation $\delta\tilde{p}^{\alpha\mu}$ give

$$\begin{aligned}
\frac{\partial\tilde{P}^{\alpha\mu}}{\partial x^{\mu}}-\frac{\partial\tilde{p}^{\alpha\mu}}{\partial x^{\mu}}&=\epsilon\left[\cancel{-\frac{\partial h^{\beta}}{\partial x^{\mu}}\frac{\partial\tilde{p}^{\alpha\mu}}{\partial x^{\beta}}}-h^{\beta}\frac{\partial^{2}\tilde{p}^{\alpha\mu}}{\partial x^{\beta}\partial x^{\mu}}+\frac{\partial\tilde{p}^{\beta\mu}}{\partial x^{\mu}}\frac{\partial h^{\alpha}}{\partial x^{\beta}}+\tilde{p}^{\beta\mu}\frac{\partial^{2}h^{\alpha}}{\partial x^{\beta}\partial x^{\mu}}\right.\\
&\quad\left.+\frac{\partial\tilde{p}^{\alpha\beta}}{\partial x^{\mu}}\left(\cancel{\frac{\partial h^{\mu}}{\partial x^{\beta}}}-\delta^{\mu}_{\beta}\frac{\partial h^{\tau}}{\partial x^{\tau}}\right)\right]\\
&=-\epsilon\left(h^{\mu}\frac{\partial^{2}\tilde{p}^{\alpha\beta}}{\partial x^{\mu}\partial x^{\beta}}-\frac{\partial\tilde{p}^{\beta\mu}}{\partial x^{\mu}}\frac{\partial h^{\alpha}}{\partial x^{\beta}}-\tilde{p}^{\beta\mu}\frac{\partial^{2}h^{\alpha}}{\partial x^{\beta}\partial x^{\mu}}+\frac{\partial\tilde{p}^{\alpha\beta}}{\partial x^{\beta}}\frac{\partial h^{\mu}}{\partial x^{\mu}}\right)\\
&=-\epsilon\frac{\partial}{\partial x^{\mu}}\left(h^{\mu}\frac{\partial\tilde{p}^{\alpha\beta}}{\partial x^{\beta}}-\tilde{p}^{\beta\mu}\frac{\partial h^{\alpha}}{\partial x^{\beta}}\right),
\end{aligned}$$

and the transformation rule for the momentum density $\tilde{p}^{\alpha\mu}$ from Eq. (6.13b) can be equivalently expressed as

$$\delta\tilde{p}^{\alpha\mu}=-\epsilon\left(h^{\mu}\frac{\partial\tilde{p}^{\alpha\beta}}{\partial x^{\beta}}-\tilde{p}^{\beta\mu}\frac{\partial h^{\alpha}}{\partial x^{\beta}}\right)=\epsilon\left(h^{\mu}\frac{\partial\tilde{\mathcal{H}}}{\partial a_{\alpha}}+\tilde{p}^{\beta\mu}\frac{\partial h^{\alpha}}{\partial x^{\beta}}\right). \tag{6.14b}$$

Along the same lines the corresponding modified transformation rules for the momenta $\tilde{k}^{\lambda\xi\mu}$ and $\tilde{q}_{\eta}{}^{\lambda\tau\mu}$ are constructed as:

$$\delta\tilde{k}^{\lambda\xi\mu}=\epsilon\left(h^{\mu}\frac{\partial\tilde{\mathcal{H}}}{\partial g_{\lambda\xi}}+\tilde{k}^{\beta\xi\mu}\frac{\partial h^{\lambda}}{\partial x^{\beta}}+\tilde{k}^{\lambda\beta\mu}\frac{\partial h^{\xi}}{\partial x^{\beta}}\right) \tag{6.14c}$$

$$\delta\tilde{q}_{\eta}{}^{\lambda\tau\mu}=\epsilon\left(h^{\mu}\frac{\partial\tilde{\mathcal{H}}}{\partial\gamma^{\eta}{}_{\lambda\tau}}-\tilde{q}_{\beta}{}^{\lambda\tau\mu}\frac{\partial h^{\beta}}{\partial x^{\eta}}+\tilde{q}_{\eta}{}^{\beta\tau\mu}\frac{\partial h^{\lambda}}{\partial x^{\beta}}+\tilde{q}_{\eta}{}^{\lambda\beta\mu}\frac{\partial h^{\tau}}{\partial x^{\beta}}\right). \tag{6.14d}$$

6.2.2 *Constructing the Noether Current* $\tilde{j}_{N}^{\mu}$

According to Noether's theorem as formulated in Eq. (6.9), the divergence of the $\tilde{j}_{\mathrm{N}}^{\mu}$ vanishes exactly if $\delta\tilde{\mathcal{H}}-\delta\tilde{\mathcal{H}}\big|_{\mathrm{CT}}=0$. Then, the finite canonical transformation defined by the generating function (6.1) establishes a *symmetry transformation* which leaves the form of the given system invariant under the Diff(M) symmetry group.

In order to construct the current $\tilde{j}_{\mathrm{N}}^{\mu}$ we apply the set of general infinitesimal canonical transformation rules (6.6) to the set of particular infinitesimal transformation rules (6.12) and (6.14) that emerge from the *local* transition map (6.10):

$$-\frac{1}{\epsilon}\delta^{\mu}_{\nu}\,\delta\phi = \frac{\partial\tilde{j}^{\mu}_{\mathrm{N}}}{\partial\tilde{\pi}^{\nu}} \quad \stackrel{!}{=} \delta^{\mu}_{\nu}h^{\beta}\frac{\partial\phi}{\partial x^{\beta}} \tag{6.15a}$$

$$\frac{1}{\epsilon}\delta\tilde{\pi}^{\mu} = \frac{\partial\tilde{j}^{\mu}_{\mathrm{N}}}{\partial\phi} \quad \stackrel{!}{=} h^{\mu}\frac{\partial\tilde{\mathcal{H}}}{\partial\phi} \tag{6.15b}$$

$$-\frac{1}{\epsilon}\delta^{\mu}_{\nu}\,\delta a_{\alpha} = \frac{\partial\tilde{j}^{\mu}_{\mathrm{N}}}{\partial\tilde{p}^{\alpha\nu}} \quad \stackrel{!}{=} \delta^{\mu}_{\nu}\left(h^{\beta}\frac{\partial a_{\alpha}}{\partial x^{\beta}} + a_{\beta}\frac{\partial h^{\beta}}{\partial x^{\alpha}}\right) \tag{6.15c}$$

$$\frac{1}{\epsilon}\delta\tilde{p}^{\alpha\mu} = \frac{\partial\tilde{j}^{\mu}_{\mathrm{N}}}{\partial a_{\alpha}} \quad \stackrel{!}{=} h^{\mu}\frac{\partial\tilde{\mathcal{H}}}{\partial a_{\alpha}} + \tilde{p}^{\beta\mu}\frac{\partial h^{\alpha}}{\partial x^{\beta}} \tag{6.15d}$$

$$-\frac{1}{\epsilon}\delta^{\mu}_{\nu}\,\delta g_{\xi\lambda} = \frac{\partial\tilde{j}^{\mu}_{\mathrm{N}}}{\partial\tilde{k}^{\lambda\xi\nu}} \quad \stackrel{!}{=} \delta^{\mu}_{\nu}\left(h^{\beta}\frac{\partial g_{\xi\lambda}}{\partial x^{\beta}} + g_{\beta\lambda}\frac{\partial h^{\beta}}{\partial x^{\xi}} + g_{\beta\xi}\frac{\partial h^{\beta}}{\partial x^{\lambda}}\right) \tag{6.15e}$$

$$\frac{1}{\epsilon}\delta\tilde{k}^{\lambda\xi\mu} = \frac{\partial\tilde{j}^{\mu}_{\mathrm{N}}}{\partial g_{\lambda\xi}} \quad \stackrel{!}{=} h^{\mu}\frac{\partial\tilde{\mathcal{H}}}{\partial g_{\lambda\xi}} + \tilde{k}^{\beta\xi\mu}\frac{\partial h^{\lambda}}{\partial x^{\beta}} + \tilde{k}^{\lambda\beta\mu}\frac{\partial h^{\xi}}{\partial x^{\beta}} \tag{6.15f}$$

and

$$-\frac{1}{\epsilon}\delta^{\mu}_{\nu}\,\delta\gamma^{\eta}{}_{\lambda\tau} = \frac{\partial\tilde{j}^{\mu}_{\mathrm{N}}}{\partial\tilde{q}_{\eta}{}^{\lambda\tau\nu}} \quad \stackrel{!}{=} \delta^{\mu}_{\nu}\left(h^{\beta}\frac{\partial\gamma^{\eta}{}_{\lambda\tau}}{\partial x^{\beta}} - \gamma^{\beta}{}_{\lambda\tau}\frac{\partial h^{\eta}}{\partial x^{\beta}} + \gamma^{\eta}{}_{\beta\tau}\frac{\partial h^{\beta}}{\partial x^{\lambda}} + \gamma^{\eta}{}_{\lambda\beta}\frac{\partial h^{\beta}}{\partial x^{\tau}} + \frac{\partial^{2}h^{\eta}}{\partial x^{\lambda}\partial x^{\tau}}\right) \tag{6.15g}$$

$$\frac{1}{\epsilon}\delta\tilde{q}_{\eta}{}^{\lambda\tau\mu} = \frac{\partial\tilde{j}^{\mu}_{\mathrm{N}}}{\partial\gamma^{\eta}{}_{\lambda\tau}} \quad \stackrel{!}{=} h^{\mu}\frac{\partial\tilde{\mathcal{H}}}{\partial\gamma^{\eta}{}_{\lambda\tau}} - \tilde{q}_{\beta}{}^{\lambda\tau\mu}\frac{\partial h^{\beta}}{\partial x^{\eta}} + \tilde{q}_{\eta}{}^{\beta\tau\mu}\frac{\partial h^{\lambda}}{\partial x^{\beta}} + \tilde{q}_{\eta}{}^{\lambda\beta\mu}\frac{\partial h^{\tau}}{\partial x^{\beta}}. \tag{6.15h}$$

Integrating Eqs. (6.15) directly yields the $\tilde{j}^{\mu}_{\mathrm{N}}$:

$$\boxed{\tilde{j}^{\mu}_{\mathrm{N}} = h^{\beta}\tilde{B}_{\beta}{}^{\mu} + \frac{\partial h^{\beta}}{\partial x^{\alpha}}\tilde{C}_{\beta}{}^{\alpha\mu} + \frac{\partial^{2}h^{\beta}}{\partial x^{\alpha}\partial x^{\eta}}\,\tilde{q}_{\beta}{}^{\alpha\eta\mu}.} \tag{6.16}$$

$\tilde{B}_{\beta}{}^{\mu}$ abbreviates the sum of all terms emerging from Eqs. (6.15) proportional to h^{β}:

$$\begin{aligned}\tilde{B}_{\beta}{}^{\mu} &= \tilde{\pi}^{\mu}\frac{\partial\phi}{\partial x^{\beta}} + \tilde{p}^{\alpha\mu}\frac{\partial a_{\alpha}}{\partial x^{\beta}} + \tilde{k}^{\alpha\lambda\mu}\frac{\partial g_{\alpha\lambda}}{\partial x^{\beta}} + \tilde{q}_{\eta}{}^{\alpha\lambda\mu}\frac{\partial\gamma^{\eta}{}_{\alpha\lambda}}{\partial x^{\beta}}\\ &\quad - \delta^{\mu}_{\beta}\left(\tilde{\pi}^{\tau}\frac{\partial\phi}{\partial x^{\tau}} + \tilde{p}^{\alpha\tau}\frac{\partial a_{\alpha}}{\partial x^{\tau}} + \tilde{k}^{\alpha\lambda\tau}\frac{\partial g_{\alpha\lambda}}{\partial x^{\tau}} + \tilde{q}_{\eta}{}^{\alpha\lambda\tau}\frac{\partial\gamma^{\eta}{}_{\alpha\lambda}}{\partial x^{\tau}} - \tilde{\mathcal{H}}\right).\end{aligned} \tag{6.17a}$$

$\tilde{C}_{\beta}{}^{\alpha\mu}$ stands for the collection of all terms proportional to $\partial h^{\beta}/\partial x^{\alpha}$:

$$\tilde{C}_{\beta}{}^{\alpha\mu} = \tilde{p}^{\alpha\mu}\,a_{\beta} + \tilde{k}^{\lambda\alpha\mu}g_{\beta\lambda} + \tilde{k}^{\alpha\lambda\mu}g_{\lambda\beta} + \tilde{q}_{\eta}{}^{\lambda\alpha\mu}\gamma^{\eta}{}_{\lambda\beta} + \tilde{q}_{\eta}{}^{\alpha\lambda\mu}\gamma^{\eta}{}_{\beta\lambda} - \tilde{q}_{\beta}{}^{\eta\lambda\mu}\gamma^{\alpha}{}_{\eta\lambda}. \tag{6.17b}$$

$\tilde{B}_{\beta}{}^{\mu}$ is actually the *local* Hamiltonian representation of the canonical energy-momentum tensor density of the total dynamical system consisting of scalar and vector fields in conjunction with a dynamic metric and connection. Moreover, as will be derived in the following section, the terms emerging from $\tilde{C}_{\beta}{}^{\alpha\mu}$ convert the partial derivatives in $\tilde{B}_{\beta}{}^{\mu}$ into covariant derivatives—and hence into the global Hamiltonian representation of the canonical energy-momentum tensor of the total system.

6.2.3 *Conserved Noether Current for Poincaré Transformations*

According to the Hamiltonian form of Noether's theorem, the canonical transformation rules emerging from the generating function $\tilde{\mathcal{F}}_3^{\mu}$, Eq. (6.5), represent a symmetry transformation of the given system exactly if the divergence of the function $\tilde{j}_{\mathrm{N}}^{\mu}(x)$ contained therein vanishes:

$$\frac{\partial \tilde{j}_{\mathrm{N}}^{\mu}}{\partial x^{\mu}} = 0 \qquad \Leftrightarrow \qquad \tilde{j}_{\mathrm{N}}^{\mu} \text{ in } \tilde{\mathcal{F}}_3^{\mu} \text{ defines an (infinitesimal) symmetry transformation.}$$

Then $\tilde{j}_{\mathrm{N}}^{\mu}$ represents the conserved Noether current.

In the coordinate formalism elaborated here the particular symmetry group under consideration can be constructed as a subgroup of the diffeomorphism group. For the particular choice of a linear function $h^{\beta}(x) = h_{0^{\beta}} + h_{1\,\alpha}^{\;\beta}\, x^{\alpha}$, Eq. (6.10) defines infinitesimal Poincaré transformations if $h_{1,\beta\alpha}$ is skew-symmetric:

$$X_{\beta} = x_{\beta} + \epsilon \left(h_{0,\beta} + h_{1,\beta\alpha}\, x^{\alpha} \right), \qquad h_{0,\beta}, h_{1,\beta\alpha} = \text{const.}, \qquad h_{1,(\beta\alpha)} = 0. \quad (6.18)$$

The constant h_0 defines a *translation vector*, while the skew-symmetric (0, 2)-tensor h_1 defines a *Lorentz transformation*. The Noether current (6.16) then reduces to

$$\tilde{j}_{\mathrm{N}}^{\mu} = h_{0,\beta} \tilde{B}^{\beta\mu} + h_{1,[\beta\alpha]} \left(\tilde{B}^{\beta\mu} x^{\alpha} + \tilde{C}^{\beta\alpha\mu} \right),$$

hence the pertaining condition for a conserved Noether current follows as:

$$\frac{\partial \tilde{j}_{\mathrm{N}}^{\mu}}{\partial x^{\mu}} = h_{0,\beta} \frac{\partial \tilde{B}^{\beta\mu}}{\partial x^{\mu}} + h_{1,[\beta\alpha]} \left(\tilde{B}^{\beta\alpha} + \frac{\partial \tilde{B}^{\beta\mu}}{\partial x^{\mu}} x^{\alpha} + \frac{\partial \tilde{C}^{\beta\alpha\mu}}{\partial x^{\mu}} \right) \stackrel{!}{=} 0.$$

Since the coefficients in the expansion (6.18) are arbitrary and independent, this gives two conservation equations [7, 8]:

$$\frac{\partial B^{\beta\mu}}{\partial x^{\mu}} = 0, \qquad B^{[\beta\alpha]} + \frac{\partial C^{[\beta\alpha]\mu}}{\partial x^{\mu}} = 0$$

for both the energy-momentum and the angular momentum, which hold exactly if the system is invariant under Poincaré transformations. The skew-symmetric part of $C^{\beta\alpha\mu}$ defines the of the source vector field

$$C^{[\beta\alpha]\mu} = \tfrac{1}{2}\left(p^{\alpha\mu}a^{\beta} - p^{\beta\mu}a^{\alpha}\right).$$

The Noether current (6.16) generalizes the particular case (6.18) to a general dynamic spacetime geometry if $h^{\beta}(x)$ is no longer restricted to a linear function of x with its 10 constant parameters, but stands for an *arbitrary* differentiable vector function of spacetime. This will be discussed in the following section.

6.2.4 Discussion of the General Conserved Noether Current

For the general case, where $h^{\beta}(x)$ stands for an arbitrary differentiable vector function of spacetime, the divergence of the Noether current (6.16) becomes

$$\begin{aligned}\frac{\partial \tilde{j}_{\mathrm{N}}^{\mu}}{\partial x^{\mu}} &= h^{\beta}\frac{\partial \tilde{B}_{\beta}{}^{\mu}}{\partial x^{\mu}} + \frac{\partial h^{\beta}}{\partial x^{\alpha}}\left(\tilde{B}_{\beta}{}^{\alpha} + \frac{\partial \tilde{C}_{\beta}{}^{\alpha\mu}}{\partial x^{\mu}}\right) \\ &+ \frac{\partial^{2} h^{\beta}}{\partial x^{\alpha}\partial x^{\eta}}\left(\tilde{C}_{\beta}{}^{\alpha\eta} + \frac{\partial \tilde{q}_{\beta}{}^{\alpha\eta\mu}}{\partial x^{\mu}}\right) + \frac{\partial^{3} h^{\beta}}{\partial x^{\alpha}\partial x^{\eta}\partial x^{\mu}}\,\tilde{q}_{\beta}{}^{\alpha\eta\mu} \stackrel{!}{=} 0.\end{aligned} \tag{6.19}$$

As $h^{\beta}(x)$ is supposed to be an arbitrary function of x, Eq. (6.19) has $4 + 16 + 40 + 80 = 140$ independent coefficients and thus defines a finite-dimensional subgroup of the infinite-dimensional diffeomorphism group. We note that in this description the torsion degrees of freedom do not emerge as separate dynamical quantities but are implicitly contained in the Riemann-Cartan tensor, which is constructed on the basis of the 64 non-symmetric connection coefficients $\gamma^{\eta}{}_{\alpha\beta}$. In contrast, the Riemann tensor of Einstein's general relativity emerges from the 40 connection coefficients—referred to as Christoffel symbols—that are symmetric in their lower index pair.

Notice that (6.19) involves a vanishing *partial* derivative of the Noether current $\tilde{j}_{\mathrm{N}}^{\mu}$ and thereby establishes a proper (local) conservation law. In contrast, the field equations emerging from it will turn out to be tensor equations and thus hold in any reference frame.

6.2.4.1 Condition 1: Term Proportional to the Third Partial Derivatives of h^{β}

The only term proportional to the third derivative of h^{β} is the canonical momentum $\tilde{q}_{\beta}{}^{\alpha\eta\mu}$, hence the dual of the partial x^{μ}-derivative of the connection $\gamma^{\beta}{}_{\alpha\eta}$. A necessary and sufficient condition for this term to vanish is that the (generally non-zero) momentum $\tilde{q}_{\beta}^{\ \alpha\eta\mu}$ is skew-symmetric in at least one of the index pairs formed out of α, η, and μ. We choose here the last index pair, namely η and μ, and define

$$\boxed{\tilde{q}_{\beta}^{\ \alpha\eta\mu} = -\tilde{q}_{\beta}^{\ \alpha\mu\eta}.} \tag{6.20}$$

Notice, though, that $\tilde{q}_{\beta}^{\ \alpha\eta\mu}$ need not in addition be skew-symmetric in α and η. Equation (6.19) then simplifies to

$$\frac{\partial \tilde{j}_{\mathrm{N}}^{\mu}}{\partial x^{\mu}} = h^{\beta} \frac{\partial \tilde{B}_{\beta}^{\ \mu}}{\partial x^{\mu}} + \frac{\partial h^{\beta}}{\partial x^{\alpha}} \left(\tilde{B}_{\beta}^{\ \alpha} + \frac{\partial \tilde{C}_{\beta}^{\ \alpha\mu}}{\partial x^{\mu}} \right) + \frac{\partial^2 h^{\beta}}{\partial x^{\alpha} \partial x^{\eta}} \left(\tilde{C}_{\beta}^{\ \alpha\eta} + \frac{\partial \tilde{q}_{\beta}^{\ \alpha\eta\mu}}{\partial x^{\mu}} \right) \stackrel{!}{=} 0. \tag{6.21}$$

6.2.4.2 Condition 2: Terms Proportional to the Second Partial Derivatives of h^{β}

The remaining zero divergence of the Noether current requires in particular that the sum of terms related to the *second* derivatives of $h^{\beta}(x)$ in Eq. (6.21) vanishes. This means with $\tilde{C}_{\beta}^{\ \alpha\eta}$ from Eq. (6.17b) inserted into the last term of Eq. (6.21):

$$\frac{\partial^2 h^{\beta}}{\partial x^{\alpha} \partial x^{\eta}} \left(\frac{\partial \tilde{q}_{\beta}^{\ \alpha\eta\mu}}{\partial x^{\mu}} + \tilde{p}^{\alpha\eta} a_{\beta} + \tilde{k}^{\lambda\alpha\eta} g_{\beta\lambda} + \tilde{k}^{\alpha\lambda\eta} g_{\lambda\beta} + \cancel{\tilde{q}_{\tau}^{\ \lambda\alpha\eta} \gamma^{\tau}_{\ \lambda\beta}} + \tilde{q}_{\tau}^{\ \alpha\lambda\eta} \gamma^{\tau}_{\ \beta\lambda} - \tilde{q}_{\beta}^{\ \tau\lambda\eta} \gamma^{\alpha}_{\ \tau\lambda} \right)$$
$$\stackrel{!}{=} 0.$$

Due to the symmetry of the second partial derivatives of h^{β} in α and η, the term $\tilde{q}_{\tau}^{\ \lambda\alpha\eta} \gamma^{\tau}_{\ \lambda\beta}$ drops out by virtue of the skew-symmetry condition (6.20). As no symmetries in α and η are implied in the remaining terms, the condition

$$\boxed{\frac{\partial \tilde{q}_{\beta}^{\ \alpha\eta\mu}}{\partial x^{\mu}} + \tilde{p}^{\alpha\eta} a_{\beta} + \tilde{k}^{\lambda\alpha\eta} g_{\beta\lambda} + \tilde{k}^{\alpha\lambda\eta} g_{\lambda\beta} - \tilde{q}_{\xi}^{\ \alpha\eta\lambda} \gamma^{\xi}_{\ \beta\lambda} - \tilde{q}_{\beta}^{\ \xi\lambda\eta} \gamma^{\alpha}_{\ \xi\lambda} = 0} \tag{6.22}$$

is sufficient, which can equivalently be expressed as the tensor equation

$$\tilde{q}_{\beta}{}^{\alpha\eta\mu}{}_{;\mu} + \tilde{p}^{\alpha\eta} a_{\beta} + \tilde{k}^{\lambda\alpha\eta} g_{\beta\lambda} + \tilde{k}^{\alpha\lambda\eta} g_{\lambda\beta} - \tilde{q}_{\beta}{}^{\alpha\tau\mu} S^{\eta}{}_{\tau\mu} - 2\tilde{q}_{\beta}{}^{\alpha\eta\mu} S^{\xi}{}_{\mu\xi} = 0. \tag{6.23}$$

Here $S^{\eta}_{\ \tau\mu} \equiv \gamma^{\eta}_{\ [\tau\mu]}$ is obviously the Cartan torsion tensor. It agrees with the corresponding field equation (56) of Ref. [4].

With Eq. (6.22), the condition (6.21) for the divergence of the Noether current now further simplifies to

$$\frac{\partial \tilde{j}_{\mathrm{N}}^{\mu}}{\partial x^{\mu}} = h^{\beta} \frac{\partial \tilde{B}_{\beta}^{\ \mu}}{\partial x^{\mu}} + \frac{\partial h^{\beta}}{\partial x^{\alpha}} \left(\tilde{B}_{\beta}^{\ \alpha} + \frac{\partial \tilde{C}_{\beta}^{\ \alpha\mu}}{\partial x^{\mu}} \right) \stackrel{!}{=} 0. \tag{6.24}$$

6.2.4.3 Condition 3: Terms Proportional to the First Partial Derivatives of h^β

For a generally conserved Noether current, the coefficient proportional to the first derivative of h^β in Eq. (6.24) must vanish as well, hence

$$\tilde{B}_\beta{}^\alpha + \frac{\partial \tilde{C}_\beta{}^{\alpha\mu}}{\partial x^\mu} \stackrel{!}{=} 0. \tag{6.25}$$

Equation (6.25) writes in expanded form with $\tilde{C}_\beta{}^{\alpha\mu}$ from Eq. (6.17b)

$$\tilde{B}_\beta{}^\alpha + \frac{\partial}{\partial x^\mu}\left(\tilde{p}^{\alpha\mu}\, a_\beta + \tilde{k}^{\lambda\alpha\mu} g_{\beta\lambda} + \tilde{k}^{\alpha\lambda\mu} g_{\lambda\beta} + \tilde{q}_\eta{}^{\lambda\alpha\mu}\gamma^\eta{}_{\lambda\beta} + \tilde{q}_\eta{}^{\alpha\lambda\mu}\gamma^\eta{}_{\beta\lambda} - \tilde{q}_\beta{}^{\eta\lambda\mu}\gamma^\alpha{}_{\eta\lambda}\right) = 0,$$

or after inserting Eq. (6.22),

$$\tilde{B}_\beta{}^\alpha + \frac{\partial}{\partial x^\mu}\left(\frac{\partial \tilde{q}_\beta{}^{\alpha\mu\eta}}{\partial x^\eta} + \tilde{q}_\eta{}^{\lambda\alpha\mu}\gamma^\eta{}_{\lambda\beta}\right) = 0.$$

This equation reduces due to the skew-symmetry of $\tilde{q}_\beta{}^{\alpha\mu\eta}$ in its last index pair to

$$\boxed{\tilde{B}_\beta{}^\alpha + \frac{\partial}{\partial x^\mu}\left(\tilde{q}_\eta{}^{\lambda\alpha\mu}\gamma^\eta{}_{\lambda\beta}\right) = 0.} \tag{6.26}$$

As $\tilde{B}_\beta{}^\alpha$—defined by Eq. (6.17a)—is the local representation of the canonical energy-momentum tensor of the *total* system of matter fields and dynamic spacetime, Eq. (6.26) establishes a correlation of this (pseudo-)tensor with the dynamic spacetime. The explicit form of this equation will be discussed in Sect. 6.2.4.5.

6.2.4.4 Condition 4: Term Proportional to h^β

Finally, the remaining term that is proportional to h^β in Eq. (6.24) must separately vanish:

$$\boxed{\frac{\partial \tilde{B}_\beta{}^\mu}{\partial x^\mu} = 0.} \tag{6.27}$$

Equation (6.27) establishes a *local* energy and momentum conservation law of the *total* system of scalar and vector source fields fields on the one hand, and the dynamic spacetime, described by the metric and the connection on the other hand. It turns out to coincide with the divergence of Eq. (6.26) by virtue of the skew-symmetry of $\tilde{q}_\eta{}^{\lambda\alpha\mu}$ in its last index pair:

$$
\begin{aligned}
0 &= \frac{\partial}{\partial x^\mu}\left(\tilde{B}_\beta{}^\mu + \frac{\partial \tilde{q}_\eta{}^{\lambda\mu\alpha}}{\partial x^\alpha}\gamma^\eta{}_{\lambda\beta} + \tilde{q}_\eta{}^{\lambda\mu\alpha}\frac{\partial \gamma^\eta{}_{\lambda\beta}}{\partial x^\alpha}\right) \\
&= \frac{\partial \tilde{B}_\beta{}^\mu}{\partial x^\mu} + \frac{\partial^2 \tilde{q}_\eta{}^{\lambda\mu\alpha}}{\partial x^\mu \partial x^\alpha}\gamma^\eta{}_{\lambda\beta} + \frac{\partial \tilde{q}_\eta{}^{\lambda\mu\alpha}}{\partial x^\alpha}\frac{\partial \gamma^\eta{}_{\lambda\beta}}{\partial x^\mu} + \frac{\partial \tilde{q}_\eta{}^{\lambda\mu\alpha}}{\partial x^\mu}\frac{\partial \gamma^\eta{}_{\lambda\beta}}{\partial x^\alpha} + \tilde{q}_\eta{}^{\lambda\mu\alpha}\frac{\partial^2 \gamma^\eta{}_{\lambda\beta}}{\partial x^\mu \partial x^\alpha} \\
&= \frac{\partial \tilde{B}_\beta{}^\mu}{\partial x^\mu}.
\end{aligned}
$$

Equation (6.27) is thus equivalent to a vanishing divergence of Eq. (6.26) as the divergence of its last term vanishes identically. From Eqs. (6.27) and (6.25), one concludes that

$$
0 = \frac{\partial \tilde{B}_\beta{}^\alpha}{\partial x^\alpha} = -\frac{\partial^2 \tilde{C}_\beta{}^{\alpha\mu}}{\partial x^\alpha \partial x^\mu} = -\frac{\partial^2}{\partial x^\alpha \partial x^\mu}\left(\tilde{C}_\beta{}^{\alpha\mu} + \frac{\partial \tilde{q}_\beta{}^{\alpha\mu\lambda}}{\partial x^\lambda}\right),
$$

which is indeed satisfied owing to Eq. (6.22). This demonstrates the consistency of the set of equations (6.20), (6.22), (6.26), and (6.27), which were obtained from the Noether condition (6.19).

6.2.4.5 Amended Canonical Energy-Momentum Tensor $\tilde{\Theta}_\nu{}^\mu$

We now express Eq. (6.26) in expanded form by inserting the local representation of the canonical energy-momentum tensor $\tilde{B}_\beta{}^\alpha$ from Eq. (6.17a), and the field equation for the partial divergence of $\tilde{q}_\eta{}^{\lambda\alpha\mu}$ from Eq. (6.22):

$$
\begin{aligned}
\tilde{\Theta}_\beta{}^\alpha &\equiv \tilde{B}_\beta{}^\alpha + \frac{\partial}{\partial x^\mu}\left(\tilde{q}_\eta{}^{\lambda\alpha\mu}\gamma^\eta{}_{\lambda\beta}\right) \\
&= \tilde{\pi}^\alpha \frac{\partial \phi}{\partial x^\beta} + \tilde{p}^{\xi\alpha}\frac{\partial a_\xi}{\partial x^\beta} - \tilde{p}^{\xi\alpha}a_\eta\gamma^\eta{}_{\xi\beta} + \tilde{k}^{\xi\lambda\alpha}\frac{\partial g_{\xi\lambda}}{\partial x^\beta} - \tilde{k}^{\lambda\xi\alpha}g_{\eta\lambda}\gamma^\eta{}_{\xi\beta} - \tilde{k}^{\xi\lambda\alpha}g_{\lambda\eta}\gamma^\eta{}_{\xi\beta} \\
&\quad + \tilde{q}_\eta{}^{\xi\lambda\alpha}\left(\frac{\partial \gamma^\eta{}_{\xi\lambda}}{\partial x^\beta} - \frac{\partial \gamma^\eta{}_{\xi\beta}}{\partial x^\lambda} + \gamma^\eta{}_{\tau\beta}\gamma^\tau{}_{\xi\lambda} - \gamma^\eta{}_{\tau\lambda}\gamma^\tau{}_{\xi\beta}\right) \\
&\quad - \delta^\alpha_\beta\left(\tilde{\pi}^\tau \frac{\partial \phi}{\partial x^\tau} + \tilde{p}^{\xi\tau}\frac{\partial a_\xi}{\partial x^\tau} + \tilde{k}^{\xi\lambda\tau}\frac{\partial g_{\xi\lambda}}{\partial x^\tau} + \tilde{q}_\eta{}^{\xi\lambda\tau}\frac{\partial \gamma^\eta{}_{\xi\lambda}}{\partial x^\tau} - \tilde{\mathcal{H}}\right) = 0.
\end{aligned}
\tag{6.28}
$$

The term proportional to $\tilde{q}_\eta{}^{\xi\lambda\alpha}$ is exactly the , defined in the convention of Misner et al. [9] by

$$
R^\eta{}_{\xi\beta\lambda} = \frac{\partial \gamma^\eta{}_{\xi\lambda}}{\partial x^\beta} - \frac{\partial \gamma^\eta{}_{\xi\beta}}{\partial x^\lambda} + \gamma^\eta{}_{\tau\beta}\gamma^\tau{}_{\xi\lambda} - \gamma^\eta{}_{\tau\lambda}\gamma^\tau{}_{\xi\beta}. \tag{6.29}
$$

Hence, this tensor is not put in "by hand" but comes forth from Noether's theorem when a given Poincaré-invariant system is enforced to be form-invariant as well under the general infinitesimal chart transition map (6.10). We remark that the tensor (6.29) is indeed the *Riemann-Cartan tensor*, as it is defined here from a non-symmetric connection, $\gamma^{\eta}{}_{[\xi\lambda]} \not\equiv 0$. Moreover, the torsion—hence the addressed skew-symmetric part of the connection—does *not* emerge as a separate dynamic quantitiy in our description as all terms containing the connection are absorbed into the covariant derivatives and into the Riemann-Cartan tensor.

Equation (6.28) now writes equivalently after merging the partial derivatives with the γ-dependent terms into covariant derivatives:

$$\tilde{\Theta}_{\beta}{}^{\alpha} = \tilde{\pi}^{\alpha}\frac{\partial\phi}{\partial x^{\beta}} + \tilde{p}^{\,\xi\alpha}\, a_{\xi;\beta} + \tilde{k}^{\xi\lambda\alpha}\, g_{\xi\lambda;\beta} - \tilde{q}_{\eta}{}^{\xi\lambda\alpha}\, R^{\eta}{}_{\xi\lambda\beta}$$
$$- \delta^{\alpha}_{\beta}\left(\tilde{\pi}^{\tau}\frac{\partial\phi}{\partial x^{\tau}} + \tilde{p}^{\,\xi\tau}\frac{\partial a_{\xi}}{\partial x^{\tau}} + \tilde{k}^{\xi\lambda\tau}\frac{\partial g_{\xi\lambda}}{\partial x^{\tau}} + \tilde{q}_{\eta}{}^{\xi\lambda\tau}\frac{\partial\gamma^{\eta}{}_{\xi\lambda}}{\partial x^{\tau}} - \tilde{\mathcal{H}}\right) = 0. \quad (6.30)$$

The remaining partial derivatives can similarly be rewritten as tensors if we subtract the corresponding "gauge Hamiltonian" terms from the total Hamiltonian $\tilde{\mathcal{H}}(\tilde{\pi}, \phi, \tilde{p}, a, \tilde{k}, g, \tilde{q}, \gamma)$:

$$\tilde{\mathcal{H}}_0 = \tilde{\mathcal{H}} - \tilde{\mathcal{H}}_{\text{Gauge}}, \quad (6.31)$$

with $\tilde{\mathcal{H}}_{\text{Gauge}}(\tilde{p}, a, \tilde{k}, g, \tilde{q}, \gamma)$ given by

$$\tilde{\mathcal{H}}_{\text{Gauge}} = \tilde{p}^{\,\xi\tau} a_{\eta}\gamma^{\eta}{}_{\xi\tau} + \tilde{k}^{\xi\lambda\tau}\, g_{\eta\xi}\gamma^{\eta}{}_{\lambda\tau} + \tilde{k}^{\lambda\xi\tau}\, g_{\xi\eta}\gamma^{\eta}{}_{\lambda\tau} + \tilde{q}_{\eta}{}^{\xi\lambda\tau}\gamma^{\eta}{}_{\alpha\lambda}\gamma^{\alpha}{}_{\xi\tau}, \quad (6.32)$$

which agrees with the gauge Hamiltonian derived in Eq. (32) of Ref. [4]. The partial derivatives of the fields in Eq. (6.30) are thus converted into covariant derivatives, whereas the partial derivative of the connection reemerges as one-half the Riemann tensor:

$$\tilde{\Theta}_{\nu}{}^{\mu} = \tilde{\pi}^{\mu}\frac{\partial\phi}{\partial x^{\nu}} + \tilde{p}^{\,\alpha\mu} a_{\alpha;\nu} + \tilde{k}^{\alpha\beta\mu} g_{\alpha\beta;\nu} - \tilde{q}_{\eta}{}^{\alpha\beta\mu} R^{\eta}{}_{\alpha\beta\nu}$$
$$- \delta^{\mu}_{\nu}\left(\tilde{\pi}^{\tau}\frac{\partial\phi}{\partial x^{\tau}} + \tilde{p}^{\,\alpha\tau} a_{\alpha;\tau} + \tilde{k}^{\alpha\beta\tau} g_{\alpha\beta;\tau} - \tfrac{1}{2}\tilde{q}_{\eta}{}^{\alpha\beta\tau} R^{\eta}{}_{\alpha\beta\tau} - \tilde{\mathcal{H}}_0\right) = 0. \quad (6.33)$$

$\tilde{\Theta}_{\nu}{}^{\mu}$ obviously stands for the canonical energy-momentum tensor of the closed total system of dynamical fields and spacetime. The non-existing covariant x^{τ}-derivative of the connection $\gamma^{\eta}{}_{\alpha\beta}$ happens to be replaced by $-\frac{1}{2}R^{\eta}{}_{\alpha\beta\tau}$. Moreover, it is not assumed that the covariant derivative of the metric vanishes identically. We thus encounter a full metric-affine theory of gravity including torsion and Non-metricity.

6.3 Field Equations in the Lagrangian Description

The sum in parentheses of Eq. (6.33) is the Legendre transformation of the Hamiltonian density $\tilde{\mathcal{H}}_0$, and thus represents the Lagrangian $\tilde{\mathcal{L}}$ of the total dynamical system consisting of the scalar and vector fields, and the metric and connection:

$$\tilde{\mathcal{L}} = \tilde{\pi}^{\tau} \frac{\partial \phi}{\partial x^{\tau}} + \tilde{p}^{\alpha\tau} a_{\alpha;\tau} + \tilde{k}^{\alpha\beta\tau} g_{\alpha\beta;\tau} - \tfrac{1}{2} \tilde{q}_{\eta}{}^{\alpha\beta\tau} R^{\eta}{}_{\alpha\beta\tau} - \tilde{\mathcal{H}}_0. \tag{6.34}$$

This Lagrangian must be a world scalar density in order for Eq. (6.33) to be a tensor equation, hence to be form-invariant under the Diff(M) symmetry group. Notice that if the covariant Hamiltonian $\tilde{\mathcal{H}}_0$ does not depend on any of the momentum fields, the Lagrangian will automatically be equipped with a Lagrange multiplier setting the corresponding covariant derivative of the field to zero. For example, if the Hamiltonian $\tilde{\mathcal{H}}_0$ does not depend on the field $\tilde{k}^{\alpha\beta\tau}$ then the Euler-Lagrange equation follows as $g_{\alpha\beta;\tau} = 0$ and thus induces . The field equations (6.23) and (6.33) will be rewritten in the following on the basis of this Lagrangian. The canonical momenta are obtained from the Lagrangian (6.34) as

$$\tilde{\pi}^{\mu} = \frac{\partial \tilde{\mathcal{L}}}{\partial\left(\frac{\partial \phi}{\partial x^{\mu}}\right)}, \qquad \tilde{p}^{\alpha\mu} = \frac{\partial \tilde{\mathcal{L}}}{\partial a_{\alpha;\mu}} \tag{6.35a}$$

$$\tilde{k}^{\alpha\beta\mu} = \frac{\partial \tilde{\mathcal{L}}}{\partial g_{\alpha\beta;\mu}}, \qquad -\tfrac{1}{2}\tilde{q}_{\eta}{}^{\alpha\beta\mu} = \frac{\partial \tilde{\mathcal{L}}}{\partial R^{\eta}{}_{\alpha\beta\mu}}. \tag{6.35b}$$

The Noether condition (6.33)—yielding a vanishing canonical energy-momentum tensor $\tilde{\Theta}_{\nu}{}^{\mu} = 0$ of the total system of dynamical fields and spacetime described by $\tilde{\mathcal{L}}$—is then equivalent to:

$$\tilde{\Theta}_{\nu}{}^{\mu} = \frac{\partial \tilde{\mathcal{L}}}{\partial\left(\frac{\partial \phi}{\partial x^{\mu}}\right)} \frac{\partial \phi}{\partial x^{\nu}} + \frac{\partial \tilde{\mathcal{L}}}{\partial a_{\alpha;\mu}} a_{\alpha;\nu} + \frac{\partial \tilde{\mathcal{L}}}{\partial g_{\alpha\beta;\mu}} g_{\alpha\beta;\nu} + 2 \frac{\partial \tilde{\mathcal{L}}}{\partial R^{\eta}{}_{\alpha\beta\mu}} R^{\eta}{}_{\alpha\beta\nu} - \delta^{\mu}_{\nu} \tilde{\mathcal{L}} = 0. \tag{6.36}$$

We may now split the Lagrangian $\tilde{\mathcal{L}}$ of the total system into a Lagrangian $\tilde{\mathcal{L}}_0$ for the dynamics of the base fields ϕ and a_{μ}, and a Lagrangian $\tilde{\mathcal{L}}_{\mathrm{Gr}}$ for the dynamics of the free gravitational field $R^{\eta}{}_{\alpha\beta\nu}$ and the metric $g_{\mu\nu}$ according to

$$\tilde{\mathcal{L}} = \tilde{\mathcal{L}}_0 + \tilde{\mathcal{L}}_R, \qquad \tilde{\mathcal{L}}_0 = \tilde{\mathcal{L}}_0(\phi, \partial\phi, a, \partial a, g, \gamma), \qquad \tilde{\mathcal{L}}_{\mathrm{Gr}} = \tilde{\mathcal{L}}_{\mathrm{Gr}}(\gamma, \partial\gamma, g, \partial g)$$

where each Lagrangian represents separately a world scalar density. As no derivative with respect to the metric appears in Eq. (6.36), we are allowed to divide all terms by $\sqrt{-g}$, whereby the field equation acquires the form of the *generic Einstein-type equation*:

$$2 \frac{\partial \mathcal{L}_{\mathrm{Gr}}}{\partial R^{\eta}{}_{\alpha\beta\mu}} R^{\eta}{}_{\alpha\beta\nu} + \frac{\partial \mathcal{L}_{\mathrm{Gr}}}{\partial g_{\alpha\beta;\mu}} g_{\alpha\beta;\nu} - \delta^{\mu}_{\nu} \mathcal{L}_{\mathrm{Gr}} = - \frac{\partial \mathcal{L}_0}{\partial\left(\frac{\partial \phi}{\partial x^{\mu}}\right)} \frac{\partial \phi}{\partial x^{\nu}} - \frac{\partial \mathcal{L}_0}{\partial a_{\alpha;\mu}} a_{\alpha;\nu} + \delta^{\mu}_{\nu} \mathcal{L}_0. \tag{6.37}$$

The left-hand side of Eq. (6.37), pertaining to $\mathcal{L}_{\mathrm{Gr}}$, can be regarded as the *canonical* energy-momentum tensor T_{Gr} of dynamical spacetime. With the right-hand side the negative canonical energy-momentum tensor T_0 of the system $\mathcal{L}_0$,

$$T_{0\nu}{}^{\mu} \equiv \frac{\partial \mathcal{L}_0}{\partial\left(\frac{\partial \phi}{\partial x^\mu}\right)} \frac{\partial \phi}{\partial x^\nu} + \frac{\partial \mathcal{L}_0}{\partial a_{\alpha;\mu}} a_{\alpha;\nu} - \delta^\mu_\nu \mathcal{L}_0, \tag{6.38}$$

Equation (6.37) thus establishes a *generally covariant energy-momentum balance relation*, with the coupling of spacetime and source fields induced by both, the metric $g_{\alpha\beta}$ and the connection $\gamma^{\alpha}{}_{\xi\tau}$. Hence, the energy-momentum tensor of the closed total system $\mathcal{L} = \mathcal{L}_0 + \mathcal{L}_{\mathrm{Gr}}$ is equal to zero, as already discussed in Sect. 4.5.4.

For the particular case of a covariantly conserved metric, $g_{\alpha\beta;\nu} \equiv 0$, the respective terms in Eq. (6.37) drop out. This yields the for the case of *metric compatibility*:

$$\boxed{2\frac{\partial \mathcal{L}_{\mathrm{Gr}}}{\partial R^{\eta}{}_{\alpha\beta\mu}} R^{\eta}{}_{\alpha\beta\nu} - \delta^\mu_\nu \mathcal{L}_{\mathrm{Gr}} = -T_{0\nu}{}^{\mu}.} \tag{6.39}$$

It applies for all Lagrangians $\mathcal{L}_{\mathrm{Gr}}$ which (i) describe the observed dynamics of the "free" (uncoupled) gravitational field and (ii) entail a consistent field equation with regard to its trace, its symmetries, and its covariant derivatives. Obviously, the expression

$$T_{\mathrm{Gr}}{}^{\mu}_{\nu} \equiv 2\frac{\partial \mathcal{L}_{\mathrm{Gr}}}{\partial R^{\eta_{\alpha\beta\mu}}} R^{\eta_{\alpha\beta\nu}} - \delta^\mu_\nu \mathcal{L}_{\mathrm{Gr}}$$

can be interpreted as the energy-momentum tensor pertaining to the Lagrangian $\mathcal{L}_{\mathrm{Gr}}$ of the free (uncoupled) gravitational field. This recovers the Zero-Energy Universe conjecture discussed in Sect. 4.5.4 as Eq. (6.37) reduces to $\Theta^{\nu^\mu} \equiv T_{\mathrm{Gr}^{\nu\mu}} + T_{0^{\nu\mu}} = 0$. In general, the covariant and contravariant representations of Eq. (6.39) are not necessarily symmetric in ν and μ and thus include a possible spin of the source field and the then emerging torsion of spacetime. We thus obtain the following relation of spin and torsion:

$$\boxed{\frac{\partial \mathcal{L}_{\mathrm{Gr}}}{\partial R^{\eta_{\alpha\beta\nu}}} R^{\eta^{\mu}_{\alpha\beta}} - \frac{\partial \mathcal{L}_{\mathrm{Gr}}}{\partial R^{\eta_{\alpha\beta\mu}}} R^{\eta^{\nu}_{\alpha\beta}} = T_{0^{[\nu\mu]}}.} \tag{6.40}$$

For the Hilbert Lagrangian—and even for Lagrangians with additional quadratic terms in $R^{\eta}_{\alpha\beta^\mu}$—the left-hand side of this equation simplifies to the skew-symmetric part of the Ricci tensor

$$R^{[\nu\mu]} = 8\pi G\, T_0{}^{[\nu\mu]}. \tag{6.41}$$

This is equivalent to (4.158) when we realize that $M_p^2 \equiv 1/8\pi G$ and $g_3 = 0$ here.

6.4 Summary

The coordinate formulation of theories of gravity, where vierbeins can be absorbed in the metric tensor and Lorentz indices avoided, can be established once fields with half-integer spin are excluded. The differences to the formulation in Chap. 4 arise not only due to the missing fermion field, though. Since the variation proceeds with respect to metric and the affine connection, rather than by the vierbein and spin connection fields, we encounter a modified definition of the pertinent momentum fields. Applying this framework to the Hamiltonian formulation of the Noether theorem had a threefold purpose. Firstly, to formulate the link between the vierbein and the coordinate description of spacetime used in earlier work [3–5], secondly to show the differences in the interpretation of fields, and thirdly to demonstrate how the absence of fermions changes the description of the dynamics of spacetime.

References

1. J. Struckmeier, D. Vasak, J. Kirsch, Generic theory of geometrodynamics from Noether's theorem for the Diff(M) symmetry group, in *Discoveries at the Frontiers*, ed. by J. Kirsch et al. (Springer Nature, Switzerland AG, 2020), pp. 143–181. https://doi.org/10.1007/978-3-030-34234-0_12. arXiv:1807.03000 [gr-qc]
2. E. Noether, Invariante Variationsprobleme. Nachrichten Königl. Ges. Wiss. Göttingen, Math. **57**, 235 (1918)
3. J. Struckmeier, A. Redelbach, Covariant Hamiltonian field theory. Int. J. Mod. Phys. E **17**, 435–491 (2008). https://doi.org/10.1142/s0218301308009458. arXiv:0811.0508
4. J. Struckmeier et al., Canonical transformation path to gauge theories of gravity. Phys. Rev. D **95**, 124048 (2017). https://doi.org/10.1103/PhysRevD.95.124048. arXiv:1704.07246
5. J. Struckmeier et al., Canonical transformation path to gauge theories of gravity II — Spacetime coupling of spin-0 and spin-1 particle fields. Int. J. Mod. Phys. E **28**(1), 1950007 (2019). https://doi.org/10.1142/S0218301319500071
6. J.A. Schouten, *Ricci-Calculus* (Springer, Berlin, 1954). 978-3-642- 05692-5. https://doi.org/10.1007/978-3-662-12927-2
7. F.W. Hehl, On the energy tensor of spinning massive matter in classical field theory and general relativity. Rep. Math. Phys. **9**(3), 55–82 (1976)
8. F.W. Hehl, On energy-momentum and spin/helicity of quark and gluon fields, in *Proceedings, 15th Workshop on High Energy Spin Physics (DSPIN-13): Dubna, Russia*, Oct 8–12 (2014). arXiv:1402.0261 [gr-qc]
9. Ch.W. Misner, K.S. Thorne, J.A. Wheeler, *Gravitation* (W. H. Freeman and Company, New York, 1973)

Chapter 7
A Note on Birkhoff's Theorem

Seven years after Schwarzschild published a vacuum solution [1] of Einstein's field equation [2] for a spherically symmetric spacetime, Birkhoff was able to show that any spherically symmetric solution of the vacuum field equation must be static and given by the Schwarzschild metric. This result is called Birkhoff's theorem [3]. Ramaswamy and Yasskin proved in [4] that Birkhoff's theorem is valid even for $R + R^2$ gravity.

Here we shall show first that in pure quadratic ansatz for the free gravitational Hamiltonian, the existence of time-dependent solutions cannot be excluded. For the CCGG theory with the quadratic-linear ansatz (4.107) we then observe how the linear term restricts that solution variety.

The object our analysis is based on is an uncharged spherical stellar object, and we look only for *analytic* solutions in the outer region. For simplicity and to align with Birkhoff's approach we assume metricity and zero torsion. Then the CCGG Eq. (4.146) simplifies to

$$Q_{\nu}{}^{\mu} - \frac{1}{8\pi G}\left(R_{\nu}{}^{\mu} - \tfrac{1}{2}\delta_{\nu}^{\mu}\, R + \Lambda\delta_{\nu}^{\mu}\right) = 0 \tag{7.1}$$

with some yet unspecified Λ, and

$$Q_{\nu}{}^{\mu} := g_1\left(R_{\alpha\beta\gamma\nu}\, R^{\alpha\beta\gamma\mu} - \tfrac{1}{4}\delta_{\nu}^{\mu}\, R_{\alpha\beta\gamma\delta}\, R^{\alpha\beta\gamma\delta}\right). \tag{7.2}$$

As discussed in [5], we shall analyze first to what extent Birkhoff's theorem holds for solutions of the purely quadratic part, i.e. for the equation

$$Q_{\nu}{}^{\mu} = 0\,, \tag{7.3}$$

D. Vasak et al., *Covariant Canonical Gauge Gravity*, FIAS Interdisciplinary Science Series, https://doi.org/10.1007/978-3-031-43717-5_7

and check whether also Einstein's equation

$$R_\nu{}^\mu - \tfrac{1}{2}\delta^\mu_\nu R + \Lambda\delta^\mu_\nu = 0 \tag{7.4}$$

and thus Eq. (7.1) are satisfied.

7.1 Static Schwarzschild-de Sitter Solutions

We consider first the static case and start by setting up the Schwarzschild ansatz for the metric

$$g_{\mu\nu} = \begin{pmatrix} e^{2\kappa(r)} & 0 & 0 & 0 \\ 0 & -e^{2\lambda(r)} & 0 & 0 \\ 0 & 0 & -r^2 & 0 \\ 0 & 0 & 0 & -r^2\sin\theta \end{pmatrix} \tag{7.5}$$

The components $R^\eta{}_{\alpha\xi\beta}$ of the Riemann-Cartan curvature tensor are

$$R^\eta{}_{\alpha\xi\beta} = \frac{\partial\gamma^\eta{}_{\alpha\beta}}{\partial x^\xi} - \frac{\partial\gamma^\eta{}_{\alpha\xi}}{\partial x^\beta} + \gamma^\tau{}_{\alpha\beta}\gamma^\eta{}_{\tau\xi} - \gamma^\tau{}_{\alpha\xi}\gamma^\eta{}_{\tau\beta}. \tag{7.6}$$

Assuming now both metricity and zero torsion

$$g_{\mu\nu\,;\,\alpha} = 0, \qquad \gamma^\kappa{}_{\alpha\beta} = \gamma^\kappa{}_{\beta\alpha}, \tag{7.7}$$

the affine connection reduces to the Christoffel symbol (2.55) (Levi-Civita connection)

$$\gamma^\kappa_{\alpha\beta} = \left\{ \begin{matrix} \kappa \\ \alpha\beta \end{matrix} \right\} = \frac{1}{2} g^{\kappa\lambda}\left(\frac{\partial g_{\alpha\lambda}}{\partial x^\beta} + \frac{\partial g_{\beta\lambda}}{\partial x^\alpha} - \frac{\partial g_{\alpha\beta}}{\partial x^\lambda} \right).$$

For the particular ansatz (7.5) and the conditions (7.7), merely six components of the Riemann tensor (7.6) are independent [6]. Explicitly, these components are

$$\begin{aligned} R_{0101} &= e^{2\kappa}\left(\kappa'' - \kappa'\lambda' + \kappa'^2\right) & R_{0202} &= r\,\kappa'\,e^{2\kappa-2\lambda} \\ R_{0303} &= r\,\kappa'\,e^{2\kappa-2\lambda}\,\sin^2\theta & R_{1212} &= r\,\lambda' \\ R_{1313} &= r\,\lambda'\,\sin^2\theta & R_{2323} &= r^2(e^{-2\lambda} - 1)\sin^2\theta \end{aligned} \tag{7.8}$$

where the primes (′) denote the derivative with respect to r. The Riemann tensor with the components (7.8) is now inserted into the quadratic equation (7.3). All non-diagonal elements are identical zero. The main diagonal terms $Q_0{}^0, \ldots, Q_3{}^3$ are not identically zero:

$$\begin{aligned}
Q_0{}^0 &= \left\{\left[\kappa'' + \kappa'\left(\kappa' - \lambda'\right)\right]^2 + 2r^{-2}\left(\kappa' - \lambda'\right)\left(\kappa' + \lambda'\right)\right\} e^{-4\lambda} - r^{-4}\left(e^{-2\lambda} - 1\right)^2 \\
Q_1{}^1 &= \left\{\left[\kappa'' + \kappa'\left(\kappa' - \lambda'\right)\right]^2 - 2r^{-2}\left(\kappa' - \lambda'\right)\left(\kappa' + \lambda'\right)\right\} e^{-4\lambda} - r^{-4}\left(e^{-2\lambda} - 1\right)^2 \\
Q_2{}^2 &= -\left\{\left[\kappa'' + \kappa'\left(\kappa' - \lambda'\right)\right]^2\right\} e^{-4\lambda} + r^{-4}\left(e^{-2\lambda} - 1\right)^2 \\
Q_3{}^3 &= Q_2{}^2.
\end{aligned}$$

The tensor elements $Q_0{}^0$ and $Q_1{}^1$ differ only in the sign of the terms proportional to r^{-2}, hence, with Eq. (7.3):

$$Q_0{}^0 - Q_1{}^1 = 4r^{-2}\left(\kappa' + \lambda'\right)\left(\kappa' - \lambda'\right) e^{-4\lambda} = 0.$$

The condition $Q_0{}^0 - Q_1{}^1 = 0$ is thus globally satisfied for all functions $\lambda(r)$ if

$$\kappa'(r) = -\lambda'(r) \qquad \text{or} \qquad \kappa'(r) = \lambda'(r) \tag{7.9}$$

holds. In either case, the main diagonal terms coincide, up to the sign, $Q_2{}^2 = Q_3{}^3 = -Q_1{}^1 = -Q_0{}^0$. We now distinguish two cases.

Case 1:$\kappa' = -\lambda'$: The tensor element $Q_2{}^2$ simplifies to

$$Q_2{}^2 = -\left[\left(-\lambda'' + 2\lambda'^2\right) e^{-2\lambda}\right]^2 + \left[r^{-2}\left(e^{-2\lambda} - 1\right)\right]^2.$$

Thus, the conditions $Q_0{}^0, \ldots, Q_3{}^3 = 0$ equally state

$$\left(-\lambda'' + 2\lambda'^2\right)^2 = \left[r^{-2}\left(1 - e^{2\lambda}\right)\right]^2,$$

which is equivalent to the pair of nonlinear second order equations

$$\lambda'' - 2\lambda'^2 \pm \frac{1 - e^{2\lambda}}{r^2} = 0. \tag{7.10}$$

The positive sign in Eq. (7.10) entails the following analytic solution for $\lambda(r)$

$$\lambda(r) = -\tfrac{1}{2} \ln\left(1 - \frac{c_1}{r} + c_2 r^2\right),$$

hence, by virtue of $\kappa'(r) = -\lambda'(r)$

$$\kappa(r) = \tfrac{1}{2} \ln c_3\left(1 - \frac{c_1}{r} + c_2 r^2\right),$$

which yields the Schwarzschild-de Sitter metric [7, 8]

$$g_{00} = 1 - \frac{2m}{r} + \tfrac{1}{3}\Lambda r^2 \tag{7.11a}$$

$$g_{11} = -\left(1 - \frac{2m}{r} + \tfrac{1}{3}\Lambda r^2\right)^{-1} \tag{7.11b}$$

with $c_1 = 2m$, $c_2 = \frac{1}{3}\Lambda$, and $c_3 = 1$ the integration constants. This metric *simultaneously* satisfies the Einstein equation (7.4) with cosmological constant Λ. We point out that the case with the minus sign in Eq. (7.10) is not analytically solvable and, as mentioned, does not fall within our scope of considering only analytic solutions.

Case 2:$\kappa' = +\lambda'$: The equation ${Q_2}^2 = 0$ reduces to

$${Q_2}^2 = -\left[\lambda'' e^{-2\lambda}\right]^2 + \left[r^{-2}\left(e^{-2\lambda} - 1\right)\right]^2 = 0$$

which yields a second pair of nonlinear second order equations

$$\lambda'' \pm \frac{1 - e^{2\lambda}}{r^2} = 0. \tag{7.12}$$

Again we find only for the positive sign an analytic solution for $\lambda(r)$

$$\lambda(r) = \tfrac{1}{2}\ln\left[\frac{r^2}{c_1^2}\left(1 + \tan^2\left(\frac{c_2 - r}{c_1}\right)\right)\right].$$

Hence, by virtue of $\kappa'(r) = \lambda'(r)$, it yields the metric

$$g_{00} = \frac{r^2}{c_1^2}\cos^{-2}\left(\frac{r - c_2}{c_1}\right) \tag{7.13}$$

$$g_{11} = -\frac{r^2}{c_1^2}\cos^{-2}\left(\frac{r - c_2}{c_1}\right), \tag{7.14}$$

which does not satisfy the Einstein equation (7.4), and can be ruled out as unphysical, see Fig. 7.1.

As result we state that with the above mentioned restrictions also for the extended Einstein equation Eq. (7.1) of the CCGG theory only the Schwarzschild-de Sitter solution exists, and thus confirm the second statement of Birkhoff's theorem.

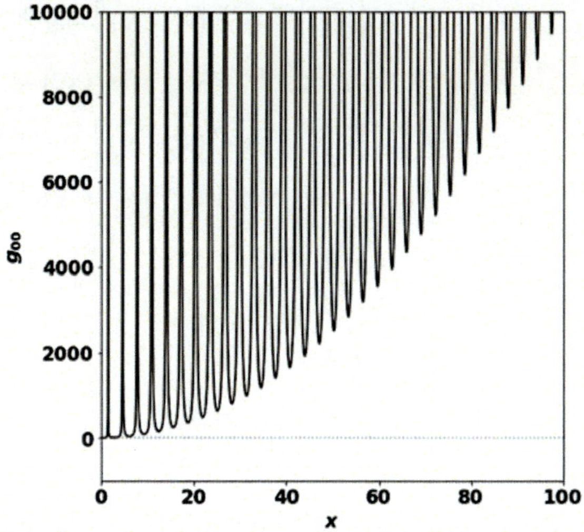

Fig. 7.1 The $g_{00} = (x/\cos x)^2$-component of the metric as analytic solution for the positive sign of Eq. (7.12) versus the scaled radius $x = r/c_1$ with $c_2 = 0$

7.2 Time Dependent Solutions for the Quadratic Term

We now allow a time dependence of the metric and start with the more general ansatz

$$g_{\mu\nu} = \begin{pmatrix} e^{2\kappa(r,t)} & 0 & 0 & 0 \\ 0 & -e^{2\lambda(r,t)} & 0 & 0 \\ 0 & 0 & -r^2 & 0 \\ 0 & 0 & 0 & -r^2 \sin\theta \end{pmatrix}. \tag{7.15}$$

With this ansatz and the conditions (7.7), eight components of the Riemann tensor (7.6) do not vanish. Explicitly, the modified components as compared to Eq. (7.6) are

$$\begin{aligned} R_{0101} &= e^{2\kappa}\left(\kappa'' - \kappa'\,\lambda' + \kappa'^2\right) - e^{2\lambda}\left(\ddot{\lambda} - \dot{\lambda}\,\dot{\kappa} + \dot{\lambda}^2\right) \\ R_{0212} &= r\,\dot{\lambda} \\ R_{0313} &= r\,\dot{\lambda}\sin^2\theta, \end{aligned} \tag{7.16}$$

where the dots ($\dot{}$) denote the derivative with respect to time t. Inserting the Riemann tensor into the quadratic equation (7.3) yields also non-vanishing non-diagonal terms. In total, the non-zero components of $Q_\xi{}^\alpha$ are

$$\begin{aligned}
Q_0{}^0 &= \left\{e^{-2\lambda}\Big[\kappa'' + \kappa'\left(\kappa' - \lambda'\right)\Big] - e^{-2\kappa}\left[\ddot{\lambda} + \dot{\lambda}\left(\dot{\lambda} - \dot{\kappa}\right)\right]\right\}^2 \\
&\quad + r^{-2}\, 2e^{-4\lambda}\left(\kappa' - \lambda'\right)\left(\kappa' + \lambda'\right) - r^{-4}\left(e^{-2\lambda} - 1\right)^2 \\
Q_1{}^1 &= \left\{e^{-2\lambda}\Big[\kappa'' + \kappa'\left(\kappa' - \lambda'\right)\Big] - e^{-2\kappa}\left[\ddot{\lambda} + \dot{\lambda}\left(\dot{\lambda} - \dot{\kappa}\right)\right]\right\}^2 \\
&\quad - r^{-2}\, 2e^{-4\lambda}\left(\kappa' - \lambda'\right)\left(\kappa' + \lambda'\right) - r^{-4}\left(e^{-2\lambda} - 1\right)^2 \\
Q_2{}^2 &= -\left\{e^{-2\lambda}\Big[\kappa'' + \kappa'\left(\kappa' - \lambda'\right)\Big] - e^{-2\kappa}\left[\ddot{\lambda} + \dot{\lambda}\left(\dot{\lambda} - \dot{\kappa}\right)\right]\right\}^2 \\
&\quad + r^{-4}\left(e^{-2\lambda} - 1\right)^2 \\
Q_3{}^3 &= Q_2{}^2 \\
Q_0{}^1 &= -4r^{-2}\,\dot{\lambda}\left(\kappa' - \lambda'\right)e^{-4\lambda} \qquad (7.17) \\
Q_1{}^0 &= 4r^{-2}\,\dot{\lambda}\left(\kappa' - \lambda'\right)e^{-2(\lambda+\kappa)}. \qquad (7.18)
\end{aligned}$$

The tensor elements $Q_0{}^0$ and $Q_1{}^1$ differ only in the sign of the terms proportional to r^{-2}, hence

$$Q_0{}^0 - Q_1{}^1 = 4r^{-2}\left(\kappa' - \lambda'\right)\left(\kappa' + \lambda'\right)e^{-4\lambda} \overset{!}{=} 0.$$

In analogy to the static case the condition $Q_0{}^0 - Q_1{}^1 = 0$ is thus globally satisfied for all functions $\lambda(r, t)$ if

$$\kappa' = \pm\lambda'.$$

The conditions for the non-diagonal terms to vanish are obtained from Eqs. (7.17) and (7.18)

$$Q_0{}^1 = Q_1{}^0 = 0 \quad \Leftrightarrow \quad \dot{\lambda} = 0 \ \text{ or } \ \lambda' = \kappa'. \qquad (7.19)$$

For $\dot{\lambda} = 0$, the analysis returns to the previously discussed case $\kappa' = -\lambda'$ in Sect. 7.1. A time-dependent solution for the metric thus exists only for $\kappa'(r, t) = \lambda'(r, t)$. The non-zero components of $Q_\nu{}^\mu$ for this case are

$$\begin{aligned}
Q_0{}^0 &= \left\{e^{-2\lambda}\lambda'' - e^{-2\kappa}\left[\ddot{\lambda} + \dot{\lambda}\left(\dot{\lambda} - \dot{\kappa}\right)\right]\right\}^2 - \left(\frac{e^{-2\lambda} - 1}{r^2}\right)^2 \\
Q_1{}^1 &= Q_0{}^0 = -Q_2{}^2 = -Q_3{}^3.
\end{aligned}$$

The equations $Q_\nu{}^\mu = 0$ are thus satisfied for

$$\left\{\lambda'' - e^{2(\lambda-\kappa)}\left[\ddot{\lambda} + \dot{\lambda}\left(\dot{\lambda} - \dot{\kappa}\right)\right]\right\}^2 = \left(\frac{1 - e^{2\lambda}}{r^2}\right)^2,$$

which yields the pair of nonlinear second order partial differential equations

$$\lambda'' - e^{2(\lambda-\kappa)}\left[\ddot{\lambda} + \dot{\lambda}\left(\dot{\lambda} - \dot{\kappa}\right)\right] \pm \frac{1-e^{2\lambda}}{r^2} = 0. \tag{7.20}$$

For the time-independent case $\dot{\lambda} \equiv 0$, Eq. (7.20) reduces to the the previously derived Eq. (7.12), as required. Defining $\kappa(r,t) \equiv \lambda(r,t)$—which is compatible with the condition $\lambda' = \kappa'$ from Eq. (7.19)—yields

$$\lambda'' - \ddot{\lambda} \pm \frac{1-e^{2\lambda}}{r^2} = 0. \tag{7.21}$$

For the positive sign in Eq. (7.21), the equation is solved by

$$\lambda(r,t) \equiv \kappa(r,t) = \frac{1}{2}\left[\ln\left(\frac{r^2}{t^2}\right) + i\pi\right] \tag{7.22}$$

Hence, the metric is given by

$$g_{00} = -g_{11} = -\frac{r^2}{t^2} \tag{7.23}$$

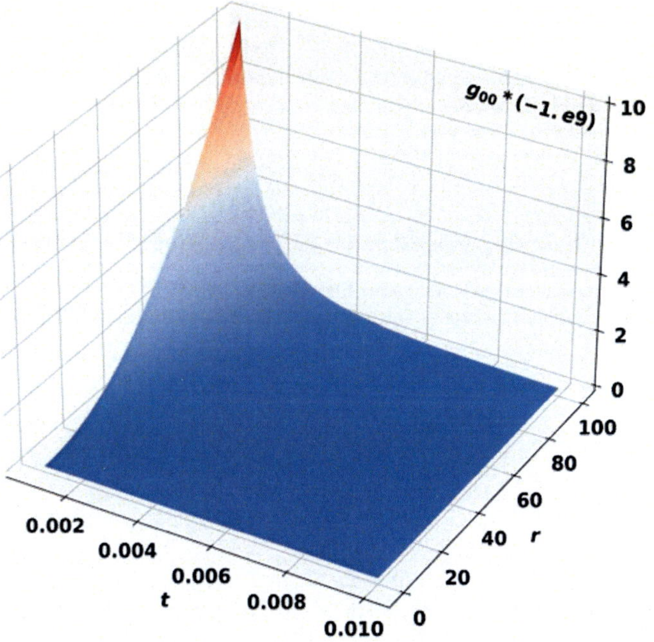

Fig. 7.2 The g_{00}-component of the metric versus the radial distance r and time t as the analytic solution for the positive sign of Eq. (7.21)

and represented in Fig. 7.2. This result can never be a solution of the total Eq. (7.1), regardless of its plausibility, as it does not solve the Einstein equation according to Birkhoff's theorem.

The above analysis thus leaves only the Schwarzschild metric to simultaneously solve both, the pure quadratic and the linear Einstein-Hilbert gravity, Eqs. (7.3) and (7.4). Whether the case

$$0 \neq Q_\nu{}^\mu = R_\nu{}^\mu - \tfrac{1}{2}\delta^\mu_\nu R + \Lambda\delta^\mu_\nu \tag{7.24}$$

leads to further analytical solutions remains to be investigated. Moreover, at this point even the existence of time-dependent (numerical) solutions of the Eq. (7.1) cannot be excluded. How the solution variety might be enriched by the presence of torsion is another interesting topic for future research.

References

1. K. Schwarzschild, Über das Gravitationsfeld eines Massenpunktes nach der Einsteinschen Theorie. Sitzungsberichte der Königlich-Preussischen Akademie der Wissenschaften **1916**, 189–196 (1916). https://doi.org/10.1007/978-3-642-58084-0_21
2. A. Einstein, Die Feldgleichungen der Gravitation. Sitzungsberichte der Königlich-Preussischen **1915**, 844–847 (1915)
3. G.D. Birkhoff, *Relativity and Modern Physics* (Havard University Press, 1927)
4. S. Ramaswamy, P.B. Yasskin, Birkhoff theorem for an $R + R^2$ theory of gravity with torsion. Phys. Rev. D **19**, 2264–2267 (1979). https://doi.org/10.1103/PhysRevD.19.2264. https://link.aps.org/doi/10.1103/PhysRevD.19.2264
5. D. Kehm et al., Violation of Birkhoff's theorem for pure quadratic gravity action. Astron. Nachr./AN 338.9-10, 1015–1018. https://doi.org/10.1002/asna.201713421. https://doi.org/10.1002/asna.201713421
6. G. Stephenson, Quadratic lagrangians and general relativity. Nuovo Cimento **9.2**, 263 (1958). https://doi.org/10.1007/bf02724929
7. W. De Sitter, Mon. Notes R Astronom. Soc. **76**, 699 (1916)
8. A. Thomas, *Moore* (University Science Books, A General Relativity Workbook, 2012). 978-1891389825

Chapter 8
Implications to Cosmology

In this chapter the question is addressed whether the CCGG formulation of the gauge theory of gravity has the potential to solve problems in cosmology that have not yet been solved within the framework of the ΛCDM model.

Attempts to explain both "dark" concepts by modifying the standard model of particles has not been validated in any of the many and extremely costly experiments carried out in the past, and little hope is manifest for changing this in the near future. An alternative and attractive research avenue for explaining the gaps between theory and observations relies on modifications of General Relativity. The numerous, so called extended theories of gravity, have delivered a variety of models linking dark energy to advanced geometrical features of spacetime, be it additional gravitational (scalar) fields, effects of torsion, or non-metricity [1–9]. The large majority of those extended theories relies, though, on ad-hoc model assumptions, often inconsistent with other observations, lacking physical justification or even being mathematically questionable.

We approach this question on the basis of the "Zero-Energy Universe" conjecture discussed in Sect. 4.5.4. It, in the first place, sheds new light on the so called "Cosmological constant problem", which is often called the largest discrepancy between theory and experiment. With the vacuum energy of spacetime cancelling the vacuum energy of matter a residual torsional cosmological field is shown to substitute dark energy, i.e. to drive the late-time expansion of the Universe.

8.1 On the Cosmological Constant

The cosmological constant Λ, which appears in Einstein's equations, is interpreted as the energy density of the vacuum [10], and is associated with the accelerated expansion of the Universe. However, the observed value of Λ is about 120 orders of magnitude smaller than the value of vacuum energy derived from quantum theoretical

D. Vasak et al., *Covariant Canonical Gauge Gravity*, FIAS Interdisciplinary Science Series, https://doi.org/10.1007/978-3-031-43717-5_8

considerations, a mismatch that the ΛCDM model cannot explain, haunting the community as the "cosmological constant problem".

Here the cosmological constant issue is approached starting from Eqs. (4.151) to (4.152), following Ref. [11]:

$$g_1 \left(R_{\alpha\beta\gamma}{}^{\nu}\, R^{\alpha\beta\gamma\mu} - \tfrac{1}{4} g^{\mu\nu} R_{\alpha\beta\gamma\delta}\, R^{\alpha\beta\gamma\delta}\right) - \frac{1}{8\pi G}\left(R^{(\mu\nu)} - \tfrac{1}{2} g^{\mu\nu}\, R + \lambda g^{\mu\nu}\right)$$
$$- g_3 \left(S_{\alpha\beta}{}^{\mu}\, S^{\alpha\beta\nu} - \tfrac{1}{2} g^{\mu\nu}\, S_{\alpha\beta\gamma}\, S^{\alpha\beta\gamma}\right) = T_0^{(\mu\nu)} . \tag{8.1}$$

As shown in Sect. 4.5.4, λ is inverse proportional to the coupling constant g_1, which controls the degree of "deformation" [12] of the Einstein equation versus General Relativity:

$$\lambda \equiv \frac{3M_{\mathrm{p}}^2}{2g_1} . \tag{8.2}$$

M_{p} is the reduced Planck mass that, for physical reasons, is related to the coupling constants g_1 and g_2 via

$$M_{\mathrm{p}}^2 \equiv \frac{1}{8\pi G} = 2g_1\, g_2 . \tag{8.3}$$

It is important to realize that the relation of λ to M_{p} emerges from the postulated quadratic-linear ansatz for semi-classical gauge gravity, and is thus a solely classical geometrical contribution to the cosmological constant.

As discussed in Sect. 4.5.4 the left-hand side of Eq. (8.1) is the negative strain-energy tensor Θ_{Gr} as defined in Eq. (4.152), giving

$$\Theta_{\mathrm{Gr}}^{\mu\nu} + T_0^{\mu\nu} = 0 . \tag{8.4}$$

Based on this three scenarios are analyzed now.

8.1.1 Spacetime Void of any Matter

In absence of matter with even the vacuum energy of matter vanishing, we have

$$T_0^{\mu\nu} = 0 .$$

We expect then the spacetime to sit in its *static* ground state where the momentum fields of gravity vanish. For vanishing momenta of spacetime the canonical equations yield

$$q_i{}^{j\alpha\beta} = 0 \implies R^i{}_{j\alpha\beta} = \bar{R}^i{}_{j\alpha\beta} ; \tag{8.5a}$$
$$k_i{}^{\alpha\beta} = 0 \implies S_i{}^{\alpha\beta} = 0 , \tag{8.5b}$$

where $\bar{R}_i{}^{j\alpha\beta}$ is defined in Eq. (4.111). By substituting $\bar{R}_i{}^{j\alpha\beta}$ for $R_i{}^{j\alpha\beta}$ on the left-hand side of Eq. (8.1), we find

$$g_1 \left(\bar{R}_{\alpha\beta\gamma}{}^{\nu}\, \bar{R}^{\alpha\beta\gamma\mu} - \tfrac{1}{4} g^{\mu\nu} \bar{R}_{\alpha\beta\gamma\delta}\, \bar{R}^{\alpha\beta\gamma\delta}\right) - 2g_1 g_2 \left(\bar{R}^{(\mu\nu)} - \tfrac{1}{2} g^{\mu\nu}\, \bar{R}\right) = 6 g_1 g_2^2\, g^{\mu\nu}.$$

Hence $\Theta^{\mu\nu}_{\mathrm{Gr}} = 6g_1 g_2^2\, g^{\mu\nu} - 6g_1 g_2^2\, g^{\mu\nu} = 0$. The strain-energy of spacetime thus vanishes, and the relation (8.4) is satisfied. This holds for both, positive and negative values of g_2, i.e. for the dS and AdS spacetimes, respectively.

8.1.2 Spacetime with Matter in Vacuum State

If, on the other hand, matter exists but is globally in its (quantum) vacuum state (or infinitely far away from any real matter distribution), the stress-energy tensor reduces to [10]

$$T_0^{\mu\nu} = g^{\mu\nu}\, \theta_{\mathrm{vac}}\,.$$

Physical wisdom backed up by observations tells us that spacetime is flat and torsion-free (Minkowski), and static in such a configuration, i.e. $R_i{}^{j\alpha\beta} = 0$ and $S_i{}^{\alpha\beta} = 0$. This leads via Eq. (8.1) to

$$-\,\Theta^{\mu\nu}_{\mathrm{Gr,flat}} = -\frac{\lambda}{8\pi G}\, g^{\mu\nu} = g^{\mu\nu}\, \theta_{\mathrm{vac}}\,, \tag{8.6}$$

and gives with Eq. (8.2)

$$-\frac{3M_{\mathrm{p}}^2}{2g_1} = \frac{\theta_{\mathrm{vac}}}{M_{\mathrm{p}}^2}. \tag{8.7}$$

In general, we infer that observing the physical geometry of spacetime to be Minkowskian in absence of real matter allows to fix the physical value of the constant g_1 by the given value of the vacuum energy of matter [13, 14]. Taking for example Weinberg's "naive" estimate $\theta_{\mathrm{vac}} \approx M_{\mathrm{p}}^4$ [10] of the quantum zero-point energy of matter based on a Planck-scale ultra-violet cutoff, the value of the coupling constant is calculated to $g_1 \approx -3/2$. Since g_1 and g_2 must be constants, and by Eq. (8.3) have the same sign, the ground state of the Universe is the AdS geometry in this case.

8.1.3 Spacetime with Real Matter

In the following we consider the general situation of a system of real matter embedded in a dynamical spacetime with torsion. The total stress-energy tensor on the right-hand side of Eq. (8.1) consists now of real particular matter and radiation and of the related bulk vacuum energy:

$$T_0^{\mu\nu} = T_{0,\mathrm{real}}^{\mu\nu} + g^{\mu\nu}\,\theta_{\mathrm{vac}}\,. \tag{8.8}$$

Then with the relations (8.2) and (8.3) the trace of Eq. (8.1) reduces to

$$R + 8\pi G\, g_3\, S^2 - 4\lambda = 8\pi G\, \theta_0\,, \tag{8.9}$$

as the quadratic gravity is trace-free, $\theta_0 := g_{\mu\nu}\, T_0^{\mu\nu}$, and $S^2 := S_{\alpha\beta\gamma}\, S^{\alpha\beta\gamma}$.

Now in this general case we also have to take into account effects of the more complex dynamical geometry and of graviton vacuum fluctuation, giving

$$R+8\pi G\, g_3\, S^2-4\lambda \equiv R_{\mathrm{LC}}+R_{\mathrm{geom}}+R_{\mathrm{quant}}+8\pi G\, g_3\, S^2-4\lambda=8\pi G\,(4\theta_{\mathrm{vac}}+\theta_{\mathrm{real}})\,. \tag{8.10}$$

Here R_{LC} is the "classical", Levi-Civita version of the Ricci tensor as known from General Relativity, R_{geom} denotes contribution from dynamical spacetime geometry, in the first place from torsion dynamics which is determined by the canonical equations with the full gravity Hamiltonian including the last term in Eq. (4.107). R_{quant} stands for some yet unspecified graviton vacuum fluctuations. This compares directly with the trace of Einstein's field equation with the observed cosmological constant Λ and the "normalized" stress-energy tensor $T_{\mathrm{real}}^{\mu\nu} = T_0^{\mu\nu} - g^{\mu\nu}\,\theta_{\mathrm{vac}}$ that is void of vacuum energy:

$$R_{\mathrm{LC}} - 4\Lambda = 8\pi G\, \theta_{\mathrm{real}}. \tag{8.11}$$

Combining now Eqs. (8.7), (8.10) and (8.11) implies

$$\tfrac{1}{4}\left(R_{\mathrm{geom}} + R_{\mathrm{quant}} + 8\pi G\, g_3\, S^2\right) \equiv -\Lambda\,. \tag{8.12}$$

The concluding statement is thus that in the linear-quadratic gravity ansatz the vacuum energy of matter is eliminated, due to the law of Zero-Energy Universe, by the spacetime vacuum energy λ constructed from the fundamental coupling constants of the theory. The observed cosmological constant Λ is merely the current value of a residual correction of geometrical origin based on the dynamics of torsion[1] with possible contributions from quantum fluctuations of gravitons. It is a dynamical entity that might be called the *dark-energy function*.

The above reasoning shows that the Cosmological constant problem can be explained by an elimination mechanism of the bulk vacuum energies of spacetime and matter. At the first sight that appears as another "fine tuning" exercise for g_1 of $\mathcal{O}(10^{-120})$. At a second sight it is quite natural, though, to assign values to fundamental constants in order to reproduce empirical findings. Here fixing g_1 versus θ_{vac} goes back to alignment with the weak gravity limit, and to ensuring that static spacetime in absence of, or in infinite distance from, any real matter is inertial.

[1] The impact of the torsion-related corrections of the Einstein equation on cosmology has been discussed in [1, 4, 14–18].

However, uncertainties about the actual value of the constant g_1 remain due to the yet unknown value of the vacuum energy of matter. Albeit the estimate of $\theta_{\text{vac}} \approx M_{\text{p}}^4$ is at the heart of the cosmological constant problem, it is not supported by all field- or string-theoretical calculations, and can vary in value and sign almost arbitrarily. Assuming for example the extreme case supported by the string theory, $\theta_{\text{vac}} \to 0$, then according to Eq. (8.7) we find [13] $g_1 \to \infty$.

A vacuum energy density of matter in the meV range, a value discussed previously for example in Ref. [19] that corresponds to the order of magnitude of the cosmological constant of the ΛCDM model, leads to $g_1 \sim -10^{119}$.

The facts collected here stand to reason that the combination of linear (Einstein-Cartan) and (trace-free) quadratic gravity in the classical gauge-field theory of gravity, and the empirical knowledge of Newton's constant and astronomical observations, can explain both the existence *and* to a significant extent also the magnitude of the cosmological constant. In fact, the above reasoning is valid regardless of the actual value of θ_{vac}, for a compensating value of g_1 can always be found as the solution of Eq. (8.7).

8.2 Review of the Standard ΛCDM Model

One of the most important hypotheses of modern cosmology, the "Cosmological Principle", states that our position in the Universe is not preferred in any way, but rather typical, i.e., the Universe smoothed by averaging over sufficiently large scales is assumed to be isotropic (and hence spatially homogeneous) at any of its points. To do justice to it, we are led to the Friedman-Lemaître-Robertson-Walker (FLRW) metric, where the FLRW line element in spherical co-moving coordinates (t, r, θ, ϕ) is

$$ds^2 = dt^2 - a^2(t)\left[\frac{dr^2}{1 - K_0 r^2} + r^2(d\theta^2 + \sin^2\theta d\phi^2)\right] \tag{8.13}$$

The parameter K_0 fixes the type of the underlying spatial geometry: $K_0 = 0$ flat, $K_0 > 0$ spherical, $K_0 < 0$ hyperbolic. The dimensionless scale factor $a(t)$ characterizes the relative size of space-like hypersurfaces at different times and is the only dynamical freedom left.[2] If t_0 is the current age of the Universe, $a(t_0) = 1$ is set for today. According to a further important hypothesis of cosmology, derived from the observation of receding distant galaxies, is that the Universe is subject to an accelerating expansion implying that $a(t)$ increases with time. The evolution of $a(t)$ depends on the content of matter of the Universe and on the model of gravity.

[2] In the FLRW model we have the choice to either define the scale parameter a *or* the spatial curvature parameter K as dimensionless, but not both at the same time. This is often ignored in the literature. Here a is dimensionless and K has the dimension L^{-2}, and $R_K := K^{1/2}$ is for positive K (closed Universe) the radius of the 3D space.

In the ΛCDM cosmology, the latter is General Relativity, i.e. Einstein's field equation

$$\bar{R}^{\mu}{}_{\nu} - \frac{1}{2}\delta^{\mu}_{\nu}\bar{R} + \Lambda_0\delta^{\mu}_{\nu} = -8\pi G\,\bar{T}^{\mu}{}_{\nu}\ . \tag{8.14}$$

It can be derived from Eq. (8.1), if $g_1 = 0$ is set and a torsion-free geometry is assumed (c.f. Eq. (4.150)). The bars indicate that the quantities are then calculated with Christoffel symbols, a convention used already in earlier chapters. Notice that here and in the following the stress-energy tensor is denoted simply by $\bar{T}^{\mu}{}_{\nu}$, without the subscript 0.

The material content is given by the symmetric stress-energy tensor $\bar{T}^{\mu}{}_{\nu}$. In the Friedman Universe matter is assumed to consist of perfect fluids made of classical particles and radiation. In the co-moving frame, i.e. in a frame that moves along with the common flow of the fluid constituents,

$$\bar{T}^{\mu}{}_{\nu} = \sum_{i=r,m} \mathrm{diag}(\rho_i, -p_i, -p_i, -p_i). \tag{8.15}$$

ρ_i and p_i are functions of the global time t only, and the index i tallies just two basic types of matter, namely particles ("dust", $i = m$) and radiation ($i = r$). The particle matter is itself a sum over all standard-model particles, and in the cold dark matter model (CDM) it also includes hypothetical dark particles. Radiation, on the other hand, includes not only genuine photon energy density but also contribution from neutrinos or other highly relativistic particles where mass is negligible compared to their kinetic energy.

The equation of state (EOS) for a perfect fluid is assumed [20] to have the linear (barotropic) form

$$p_i = \omega_i\,\rho_i, \tag{8.16}$$

where $\omega_m = 0$ and $\omega_r = \frac{1}{3}$.

Inserting Eq. (8.13) in Eq. (8.14) yields the well known Friedman-Lemaître equations for $a(t)$:

$$\left(\frac{\dot{a}}{a}\right)^2 + \frac{K_0}{a^2} - \frac{1}{3}\Lambda = \frac{8\pi G}{3}\sum_{i=m,r}\rho_i \tag{8.17a}$$

$$\frac{\ddot{a}}{a} + \frac{\dot{a}^2}{a^2} + \frac{K_0}{a^2} - \frac{2}{3}\Lambda = -\frac{4\pi G}{3}\sum_{i=m,r}(\rho_i + 3p_i)\,. \tag{8.17b}$$

Here the dot denotes time derivatives with respect to the universal time t, e.g. $\dot{a} \equiv da/dt$. Under the assumptions, the fluids are inert and non-interacting, the combination of Eqs. (8.17a) and (8.17b) fixes the dynamics of each individual fluid:

$$\dot{\rho}_i = -3\frac{\dot{a}}{a}(\rho_i + p_i) = -3\frac{\dot{a}}{a}\rho_i\,(1 + \omega_i)\,. \tag{8.18}$$

This is consistent with the covariant energy-momentum conservation $\bar{\nabla}_\mu \bar{T}^{\nu\mu} = 0$ and implies the scaling law [20, 21]:

$$\rho_i(t) \sim a^{-n_i} \qquad \Rightarrow \qquad \rho_i\, a^{n_i} = \text{const}, \tag{8.19}$$

with the definition $n_i \equiv 3(\omega_i + 1)$, i.e. $n_r = 4$ and $n_m = 3$.

The cosmological parameters Ω_i, $i = m, r, \Lambda, K_0$ entering the density contributions to the total energy density ρ are introduced according to the conventions of the ΛCDM model by:

$$\rho_m := \rho_{crit}\, \Omega_m\, a^{-3} \tag{8.20a}$$
$$\rho_r := \rho_{crit}\, \Omega_r\, a^{-4} \tag{8.20b}$$
$$\rho_\Lambda := \rho_{crit}\, \Omega_\Lambda \tag{8.20c}$$
$$\rho_K := \rho_{crit}\, \Omega_{K_0}\, a^{-2} \tag{8.20d}$$

with ρ_K describing the energy content assigned to the spatial curvature, and

$$\rho_{crit} := \frac{3H_0^2}{8\pi G} \tag{8.21a}$$
$$\Omega_\Lambda := \frac{1}{3}\frac{\Lambda}{H_0^2} \tag{8.21b}$$
$$\Omega_{K_0} := -\frac{K_0}{H_0^2} \tag{8.21c}$$

Note that all Ωs are dimensionless. The parameter $H_0 = H(a_0)$ denotes the present-day value of the Hubble function

$$H(a) := \frac{\dot{a}}{a}\,. \tag{8.22}$$

Utilizing now the dimensionless time $\tau = t H_0$ and denoting henceforth derivatives $d/d\tau$ by an overdot, Eqs. (8.17a) and (8.17b) are recast to

$$\dot{a}^2 + V_0(a) = \Omega_{K_0} \tag{8.23a}$$
$$\ddot{a}a + \dot{a}^2 - 2M(a)\, a^2 - \Omega_{K_0} = 0, \tag{8.23b}$$

where

$$V_0(a) := -\Omega_m a^{-1} - \Omega_r a^{-2} - \Omega_\Lambda a^2 \tag{8.24a}$$
$$M(a) := \frac{1}{4}\Omega_m a^{-3} + \Omega_\Lambda\,. \tag{8.24b}$$

Obviously (8.23a) has the form of a relation in which the dynamics of $a(\tau)$ is described as that of a particle with "mass" 2 and "energy" Ω_{K_0} at position $a(\tau)$ moving in a potential V_0. Note that, due to

$$\frac{\mathrm{d}a}{\mathrm{d}t} = \frac{\mathrm{d}a}{\mathrm{d}\tau}\frac{\mathrm{d}\tau}{\mathrm{d}t} = \dot{a}\, H_0\,,$$

Equation (8.23a) can be re-written in terms of the normalized Hubble function as

$$E^2(a) := \frac{H^2(a)}{H_0^2} = \frac{1}{a^2}\left[-V_0(a) + \Omega_{K_0}\right] = \Omega_m a^{-3} + \Omega_r a^{-4} + \Omega_\Lambda + \Omega_{K_0} a^{-2}\,. \tag{8.25}$$

For $a(\tau_0) = a_0 = 1$ this gives $E(1) = 1$, and thus

$$\Omega_m + \Omega_r + \Omega_\Lambda + \Omega_{K_0} = 1\,. \tag{8.26}$$

The so called *Concordance model* based on current observational status and Einstein-Hilbert gravity shows that the cosmological constant ($\Omega_\Lambda \approx 0.7$) dominates the densities of matter ($\Omega_m \approx 0.3$) and radiation ($\Omega_r \approx 5 \times 10^{-5}$). The exact value of $H_0 = h * H_{100}$, where $H_{100} = 100\,\mathrm{km/(s\,Mpc)}$, is subject to uncertainty ("Hubble tension"), though, and varies roughly between $h = 0.6$ and $h = 0.8$, depending on which observations are analyzed (early vs. late Universe measurements) [22].

8.3 The Torsion Model

While in earlier approximate calculations torsion has been identified as a good candidate for a dynamical dark energy [14], a detailed model of the torsion field was missing. We show that for classical matter this can be achieved since the torsion tensor is totally anti-symmetric as shown in Sect. 2.6. Consider now

$$\nabla_\mu \bar{T}^{(\nu\mu)} = \bar{\nabla}_\mu \bar{T}^{(\nu\mu)} + K^\nu{}_{\alpha\mu}\, \bar{T}^{(\alpha\mu)} + K^\mu{}_{\alpha\mu}\, \bar{T}^{(\nu\alpha)} = 0. \tag{8.27}$$

with the contortion tensor defined in Eq. (2.57). However, with Eq. (4.157) we identify $K^\nu{}_{\alpha\mu} \equiv S^\nu{}_{\alpha\mu}$, and it immediately follows that

$$\nabla_\nu T^{(\mu\nu)} \equiv \nabla_\nu \bar{T}^{(\mu\nu)} = \bar{\nabla}_\nu \bar{T}^{(\mu\nu)} = 0 \tag{8.28}$$

holds for the covariant conservation of the stress-energy tensor. This allows to align the matter model with the assumption common in standard ΛCDM cosmology, and apply the scaling laws Eq. (8.19) for (conserved) matter and radiation.

It is important to stress that with the validity of the covariant derivative of stress-energy tensor, Eq. (8.28), a similar condition must be imposed on the strain-energy tensor:

$$\nabla_\mu \bar{T}^{(\nu\mu)} \stackrel{!}{=} -\nabla_\mu \Theta^{(\nu\mu)}_{\text{Gr}} = \nabla_\mu \left(g_3 \, S^{\nu\alpha\beta} \, S^\mu{}_{\alpha\beta} - T^{(\nu\mu)}_{\text{Gr}} \right), \tag{8.29}$$

which is a constraint for the spacetime dynamics. However, once Eq. (8.4) is solved for all x, this constraint is trivially satisfied.

It is convenient to express the totally anti-symmetric torsion tensor as

$$S_{\alpha\mu\nu} = \frac{1}{\sqrt{3!}} \epsilon_{\alpha\mu\nu\sigma} \, s^\sigma, \tag{8.30}$$

with the totally anti-symmetric covariant Levi-Civita tensor density $\epsilon_{\alpha\mu\nu\sigma}$. This relation can be reversed to express the axial vector density s^σ using the contravariant tensor density $\epsilon^{\alpha\mu\nu\sigma}$:

$$s^\sigma = \frac{1}{\sqrt{3!}} \epsilon^{\sigma\alpha\mu\nu} \, S_{\alpha\mu\nu}. \tag{8.31}$$

In order to preserve the Cosmological Principle, we pursue a similar ansatz as done in [4], and set for s^σ a time-like vector density

$$s^\sigma = (s_0, 0, 0, 0), \tag{8.32}$$

where s_0 is a scalar function depending only on time. Thus, the g_3 proportional term in Eq. (8.1) can be evaluated into

$$S^{\xi\alpha\mu} S_{\xi\alpha\nu} - \frac{1}{2} \delta^\mu_\nu \, S^{\xi\alpha\beta} S_{\xi\alpha\beta} = -\left(\frac{s_0}{\sqrt{3!}} \right)^2 \begin{pmatrix} 3 & 0 & 0 & 0 \\ 0 & 1 & 0 & 0 \\ 0 & 0 & 1 & 0 \\ 0 & 0 & 0 & 1 \end{pmatrix}. \tag{8.33}$$

This concept provides a complete description of a Riemann-Cartan metric-compatible geometry with total anti-symmetric torsion.

8.4 The Extended Friedman Equations

Applying the FLRW metric to Eq. (8.1) with a totally anti-symmetric torsion, and inserting the stress-tensor (8.15) with the scaling laws (8.20), leads to the extended Friedman equation

$$- 8\pi G g_1 \left[\left(\frac{(\frac{da}{dt})^2 + K_0}{a} \right)^2 - \left(\frac{d^2 a}{dt^2} \right)^2 \right] + \left(\frac{da}{dt} \right)^2 + K_0 - \frac{1}{3}\Lambda a^2 + B(a, s_0)$$

$$= \frac{8\pi G}{3} a^2 \sum_{i=m,r} \rho_i \tag{8.34a}$$

$$a \frac{d^2 a}{dt^2} + \left(\frac{da}{dt} \right)^2 + K_0 - \frac{2}{3}\Lambda a^2 + (8\pi G g_3 - 1)\frac{s_0^2}{6} a^2 = \frac{4\pi G}{3} a^2 \rho_m \tag{8.34b}$$

where

$$B(a, s_0) := 8\pi G g_1 \left[\frac{s_0^2}{6} \left(-\frac{s_0^2}{6} a^2 + 5 \left(\frac{da}{dt} \right)^2 + 2K_0 \right) - \frac{1}{6} \left(\left(\frac{ds_0}{dt} \right)^2 a^2 + 2 \frac{ds_0}{dt} s_0 \frac{da}{dt} a \right) \right]$$
$$+ (8\pi G g_3 - 1) \frac{s_0^2}{6} a^2 .$$

The constant Λ is here a free parameter that will be analyzed in the later following calculations. For $s_0 = 0$ and thus $B = 0$ these equations reduce to the familiar Friedman equations (8.23). If we use the assumptions of the ΛCDM model for the densities Eq. (8.20), and deploy the dimensionless time variable $\tau = t\, H_0$, we obtain the transformed set of dimensionless equations generalizing Eqs. (8.23):

$$\dot{a}^2 + V(a, s) = \Omega_K \tag{8.35a}$$

$$\ddot{a}\, a + \dot{a}^2 - 2M(a)\, a^2 - \Omega_K + (\Omega_s - 1)s^2 a^2 = 0 \tag{8.35b}$$

with the definitions

$$M(a) := \frac{1}{4}\Omega_m a^{-3} + \Omega_\Lambda \tag{8.36a}$$

$$\Omega_g := \frac{1}{32\pi G\, H_0^2\, g_1} \tag{8.36b}$$

$$\Omega_s := 8\pi G g_3 \tag{8.36c}$$

$$s := \frac{s_0}{\sqrt{3!} H_0} \tag{8.36d}$$

$$V(a, s) := V_0(a) + V_{geo}(a) + V_{tor}(a, s) \tag{8.36e}$$

$$V_0(a) := -\Omega_m a^{-1} - \Omega_r a^{-2} - \Omega_\Lambda a^2 \tag{8.36f}$$

$$V_{geo}(a) := -\frac{M}{\Omega_g - M} \left(\frac{3}{4}\Omega_m a^{-1} + \Omega_r a^{-2} \right) \tag{8.36g}$$

$$V_{tor}(a, s) := -\frac{1}{\Omega_g - M} \left[\frac{1}{4}(\dot{s}a + s\dot{a})^2 - s^2 \dot{a}^2 \right.$$
$$\left. - \frac{\Omega_s}{2} s^2 a^2 \left(\frac{\dot{a}^2}{a^2} + (\frac{\Omega_s}{2} - 1)s^2 - \Omega_K a^{-2} \right) \right] + (\Omega_s - 1)s^2 a^2 . \tag{8.36h}$$

These equations are complemented by the conservation law Eq. (8.29) where the initially occurring 3rd derivative of a was replaced with the help of Eq. (8.35b):

$$\begin{aligned}
&-\frac{3}{2}\Omega_{\mathrm{m}}a^{-1}\dot{a}\ddot{a}+a^{2}\ddot{a}s\,[\,5\dot{a}s-(2\Omega_{\mathrm{s}}-1)a\dot{s}\,]\\
&-(a\dot{s}+\dot{a}s)\left[\,a^{3}\ddot{s}+2a^{3}s^{3}-5a\dot{a}^{2}s+3a^{2}\dot{a}\dot{s}-4\Omega_{\mathrm{g}}(\Omega_{\mathrm{s}}-1)a^{3}s+2\Omega_{\mathrm{K}}as\,\right]\\
&=0\,.
\end{aligned} \tag{8.37}$$

It should be emphasized that the Eqs. (8.35) are not solvable in torsion-free geometry where $s=0$. This can be seen by taking the time derivative of the first equation and thus eliminating $\ddot{a}$ in the 2nd equation of (8.35). In this way one obtains [23] for $s\equiv 0$:

$$\frac{1}{2}\frac{\mathrm{d}V}{\mathrm{d}a}+\frac{V}{a}+2Ma=0\,.$$

Inserting the potential $V=V_0+V_{\mathrm{geo}}$ gives the obviously wrong relation $0.75\,\Omega_{\mathrm{m}}a+\Omega_{\mathrm{r}}=0$. We therefore conclude that *torsion is necessary for a linear-quadratic gravity Hamiltonian as defined in Eq. (4.107).*

For the complete system with torsion we notice a seemingly possible inconsistency, since we have obtained three equations for just two functions, $a(\tau)$ and $s(\tau)$. An analogous problem exists for the Einstein-Friedman equations, too, but there the conservation law is automatically satisfied, since the left-hand side vanishes due to the Bianchi identities for torsion free geometry, and the right-hand side vanishes with the selected model for matter and radiation. Here $\nabla^{\mu}Q_{\mu\nu}=0$ resp. Equation (8.37) is satisfied, if $\nabla^{\mu}T_{\mu\nu}=0$ *and* if a and s are solutions of Eq. (8.4) resp. the Friedman equations (8.35). However, Eq. (8.37) can be used to validate the numerically determined solutions [24].

A recent study [25] indeed suggests that there is no analytical solution across all three cosmological epochs. However, we will see in the next sections that, for selected parameter sets, numerical solutions exist (which also satisfy the conservation law).

8.5 Interpreting Friedman Equations as Energy Balance

In order to connect this formulation with the Zero-Energy condition, we replace the potential V in the first Friedman equation with the energy densities $\hat{\rho}_i$ defined relative to the critical density (8.21a):

$$H^2=\frac{\dot{a}^2}{a^2}=\frac{1}{a^2}(-V+\Omega_{\mathrm{K}})=\hat{\rho}_{\mathrm{m}}+\hat{\rho}_{\mathrm{r}}+\hat{\rho}_{\Lambda}+\hat{\rho}_{\mathrm{K}}+\hat{\rho}_{\mathrm{geo}}+\hat{\rho}_{\mathrm{tor}} \tag{8.38}$$

with

$$
\begin{aligned}
\hat{\rho}_{\rm m} &= \Omega_{\rm m}\, a^{-3} \\
\hat{\rho}_{\rm r} &= \Omega_{\rm r}\, a^{-4} \\
\hat{\rho}_{\Lambda} &= \Omega_{\Lambda} \\
\hat{\rho}_{\rm K} &= \Omega_{\rm K}\, a^{-2} \\
\hat{\rho}_{\rm geo} &= \Omega_{\rm geo}(a) := -V_{\rm geo}\, a^{-2} \\
\hat{\rho}_{\rm tor} &= \Omega_{\rm tor}(a, s) := -V_{\rm tor}\, a^{-2} \, .
\end{aligned}
\tag{8.39}
$$

Thus we recover the Zero-Energy Universe condition of Eq. (8.4) in the form

$$
\hat{\rho}_{\rm st} + \hat{\rho}_{\rm matter} = 0\,,
$$

with

$$
\hat{\rho}_{\rm matter} := \hat{\rho}_{\rm r} + \hat{\rho}_{\rm m} \tag{8.40}
$$

$$
\hat{\rho}_{\rm st} := -H^2(\dot{a}, a) + \hat{\rho}_{\Lambda} + \hat{\rho}_{\rm K} + \hat{\rho}_{\rm geo} + \hat{\rho}_{\rm tor}\,. \tag{8.41}
$$

Obviously $\hat{\rho}_{\rm st} \leq 0$ for $\hat{\rho}_{\rm matter} \geq 0$. Since the latter condition is assumed to be always satisfied, $\hat{\rho}_{\rm st}$ must contain a "phantom term" that absorbs the energy density of matter. Hereby is the Hubble parameter, that depends on the expansion velocity of the Universe, naturally identified with the kinetic energy of spacetime, while the other geometry and matter related terms play the role of potential densities.

8.6 Numerical Study

For a numerical analysis of a system of differential equations it is reasonable to put the equations under investigation into the form $y' = f(x, y)$, since numerous proven solution methods are available for this purpose. To achieve this in the present case, the second derivative in the Friedman equation (8.35b) has to be removed by introducing a new variable. It seems natural to choose the (dimensionless) Hubble function $H(a) = \dot{a}/a$. After some algebra we then get

$$
\dot{a} = a\, H \tag{8.42a}
$$

$$
\dot{H} = -2H^2 + 2M + \Omega_K a^{-2} - (\Omega_{\rm s} - 1)s^2 \tag{8.42b}
$$

$$
\begin{aligned}
\dot{s} = -Hs + 2\Bigg[& H^2 s^2 - M\left(\frac{3}{4}\Omega_{\rm m} a^{-3} + \Omega_{\rm r} a^{-4}\right) \\
& + \frac{\Omega_{\rm s}}{2} s^2 \left(H^2 + (\frac{\Omega_{\rm s}}{2} - 1)s^2 - \Omega_K a^{-2}\right) \\
& + (\Omega_{\rm g} - M)\left(H^2 + V_0 a^{-2} - \Omega_K a^{-2} + (\Omega_{\rm s} - 1)s^2\right)\Bigg]^{1/2} \,.
\end{aligned}
\tag{8.42c}
$$

It should be mentioned that, in principle, both signs can appear in front of the root. The negative sign, however, led in test calculations either to numerically instable or physically implausible results, e.g. to a growing scaling factor for $\tau \to 0$, and is therefore excluded from further analysis here. The conservation law (8.37), expressed with the variables a, H, s, now reads

$$\left[\frac{3}{2}\Omega_{\mathrm{m}} a^{-3} H - s\left(5Hs - (2\Omega_{\mathrm{s}} - 1)\dot{s}\right)\right]\left[H^2 - \Omega_{\mathrm{K}} a^{-2} - 2M + (\Omega_{\mathrm{s}} - 1)s^2\right]$$
$$- (Hs + \dot{s})\left[\ddot{s} + 2s^3 - 5H^2 s + 3H\dot{s} - 4\Omega_{\mathrm{g}}(\Omega_{\mathrm{s}} - 1)s + 2\Omega_{\mathrm{K}} a^{-2} s\right] = 0\,. \tag{8.43}$$

These equations have to be solved with suitable boundary conditions. Without loss of generality we set the variable $a(\tau_1) = 1$ for $\tau_1 = 1$, the present time. Since the Hubble function H has already been "normalized" to H_0, its present value, $H(1) = 1$ is automatically valid. The unit of the universal time t is then the Hubble time, $t_0 = 1/H_0$, usually referred to as the age of the Universe. However, there is no obvious choice for the initial value of the torsion parameter $s_1 \equiv s(1)$. Even the relation for the cosmological parameters derived from the first Friedman equation does not provide any remedy.

We mark today's values of the relative densities (8.39) with index 1 and obtain for the last two functions

$$\begin{aligned}
\Omega_{\mathrm{geo},1} &= \frac{(\frac{1}{4}\Omega_{\mathrm{m}} + \Omega_\Lambda)\,(\frac{3}{4}\Omega_{\mathrm{m}} + \Omega_{\mathrm{r}})}{\Omega_{\mathrm{g}} - \frac{1}{4}\Omega_{\mathrm{m}} - \Omega_\Lambda} \\
\Omega_{\mathrm{tor},1} &= \frac{\frac{1}{4}(\dot{s}_1 + s_1)^2 - s_1^2 - \frac{\Omega_{\mathrm{s}}}{2}\, s_1^2\,(1 + (\frac{\Omega_{\mathrm{s}}}{2} - 1)s_1^2 - \Omega_{\mathrm{K}})}{\Omega_{\mathrm{g}} - \frac{1}{4}\Omega_{\mathrm{m}} - \Omega_\Lambda} - (\Omega_{\mathrm{s}} - 1)\, s_1^2\,,
\end{aligned} \tag{8.44}$$

Equation (8.26) is then generalized to a relation containing besides s_1 also its unknown first derivative $\dot{s}_1$:

$$1 = \Omega_{\mathrm{m}} + \Omega_{\mathrm{r}} + \Omega_\Lambda + \Omega_{\mathrm{K}} + \Omega_{\mathrm{geo},1} + \Omega_{\mathrm{tor},1}\,. \tag{8.45}$$

Hence compared with the standard parameter set of the Concordance Model with just Ω_{m}, Ω_{r}, Ω_Λ and Ω_{K}, there are three new independent parameters in this theory, namely g_1 in Ω_{g}, g_3 in Ω_{s}, and the initial value s_1, for which no specific observational data are available and whose range of values cannot be limited a priori—except that Ω_{g} must be non-zero, otherwise g_1 becomes infinite, and the root in Eq. (8.42c) must be real.

8.7 Results and Verification

We start the numerical calculation applying a 4th order Runge-Kutta method with step size adjustment, and set arbitrarily $s_1 = 0.5$. To achieve comparability of the torsion terms in Eq. (8.1) with the Einstein tensor, we set $2g_3 = 1/8\pi G$ from which $\Omega_s = 0.5$ follows according to the definition (8.36c). Furthermore we use for $\Omega_m h^2 = (\Omega_b + \Omega_c)h^2$, Ω_Λ and h the cosmological parameters from *Planck* TT,TE,EE+lowE+lensing, [22], set $\Omega_r h^2 = 2.47 \cdot 10^{-5}$ corresponding to the temperature $T = 2.725$ K of CMB, and $\Omega_K = 0$ assuming a flat Universe, see Table 8.1.

In a first calculation the value of Ω_Λ is set following the standard ΛCDM cosmology, according to which $\Omega_\Lambda = 1 - \Omega_m - \Omega_r$ holds.

An example depicted in Fig. 8.1 shows the result for $\Omega_g = -1$. We observe that the system of Eqs. (8.42) is completely solvable (left panel) and consistent in the given domain which was tested by substituting the solutions for a, H, and s into Eq. (8.43). From a physical point of view, it is noteworthy that the scaling factor a is significantly different from zero at $t = 0$, and even for negative times. Times less than zero are no

Table 8.1 The ΛCDM parameter set h, Ω_i from Planck Collaboration [22], Table 2, column 5

h	Ω_m	Ω_r	Ω_Λ	Ω_K
0.674	0.315	$5.44 \cdot 10^{-5}$	0.685	0.000

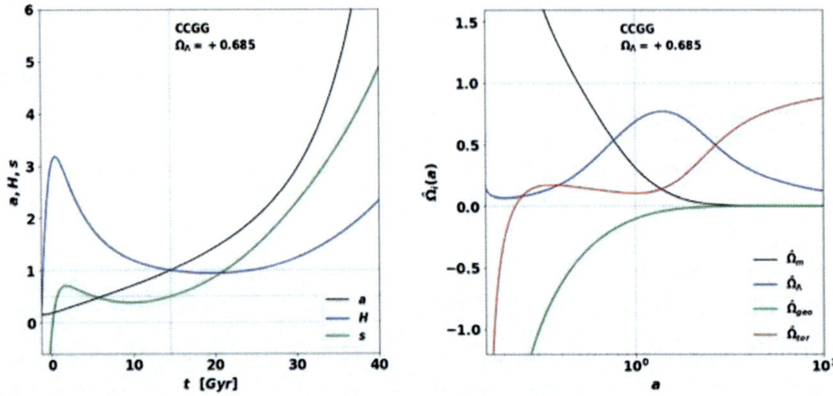

Fig. 8.1 Left panel: a, H, and s as a function of the cosmological time t, where the scaling factor a reaches its minimum for a time less than zero and increases strictly monotonically thereafter. Right panel: Evolution of the fractional density parameters $\hat{\Omega}_i(a) = \hat{\rho}_i(a)/H^2(a)$ as function of a. For the sake of clarity, the parameters $\hat{\Omega}_r$ and $\hat{\Omega}_K$ have been neglected as they do not contribute in the given domain. Note that the representation of $\hat{\Omega}_{tor}(a(t), s(t))$ as function of a is only feasible because $a(t)$ grows strictly monotonically with time. The vertical dashed lines mark present day

objection against our theory, though, since the present time was arbitrarily identified with the Hubble time $t = 1/H_0$.[3]

An important physical aspect is revealed in Fig. 8.1 (right panel) showing the evolution of the fractional density parameters defined by $\hat{\Omega}_i(a) := \hat{\rho}_i(a)/H^2(a)$ for $i = \mathrm{m}, \Lambda, \mathrm{geo}, \mathrm{tor}$. For large scale factors a, the torsion term $\hat{\Omega}_{\mathrm{tor}}$ dominates all others, in particular $\hat{\Omega}_\Lambda$ is diminishing.

According to the discussion in Sect. 8.1, only a small residue remains of the cosmological constant Λ.

We therefore test in a second calculation whether torsion can explain the observed phenomena attributed to dark energy, and set $\Lambda = 0$ and $\Omega_\Lambda = 0$, respectively. With this assumption it is easy to show that the Eqs. (8.42) can be solved exactly in the asymptotic domain $\tau \to \infty$. From

$$a(\tau \to \infty) = e^{H_\infty \tau}, \tag{8.46}$$

with an asymptotically constant Hubble function H_∞, we have $\dot{H} = 0$ and via Eq. (8.42b)

$$s_\infty^2 = \frac{2H_\infty^2}{1 - \Omega_{\mathrm{s}}}. \tag{8.47}$$

Equation (8.42c) then gives

$$H_\infty^2 = 2\,\Omega_{\mathrm{g}} \frac{(1 - \Omega_{\mathrm{s}})^2}{3 - 5\,\Omega_{\mathrm{s}}}. \tag{8.48}$$

Now the yet free parameter Ω_{g} can be chosen such that the Hubble function has the same asymptotic behavior as in the Concordance model, $H_\infty = H_{\mathrm{GR},\infty} = \sqrt{\Omega_{\mathrm{GR},\Lambda}}$, with $\Omega_{\mathrm{GR},\Lambda} = 0.685$ from Table 8.1:

$$\Omega_{\mathrm{g}} = \frac{1}{2}\, H_{\mathrm{GR},\infty}^2 \, \frac{3 - 5\,\Omega_{\mathrm{s}}}{(1 - \Omega_{\mathrm{s}})^2}. \tag{8.49}$$

In addition, we assume $s_1 = s_\infty$ which means

$$s_1 = H_{\mathrm{GR},\infty} \left(\frac{2}{1 - \Omega_{\mathrm{s}}} \right)^{1/2}. \tag{8.50}$$

The only remaining free parameter Ω_{s} must be less than 1 by Eq. (8.47), and greater than 0.6 by Eq. (8.49) to ensure $\Omega_{\mathrm{g}} < 1$ respectively $g_1 < 0$, and to maintain AdS geometry as discussed in Sect. 8.1. Thus we set for the following example calculation $\Omega_{\mathrm{s}} = 0.92$. As we can see in Fig. 8.2, left panel, there is an exponential progression of

[3] The Hubble time agrees with the true age of the Universe only when a uniform expansion is assumed. We do not use a zero-time adjustment here, since this leads to different time axes for different calculations. In particular, the comparability with the standard model would be lost.

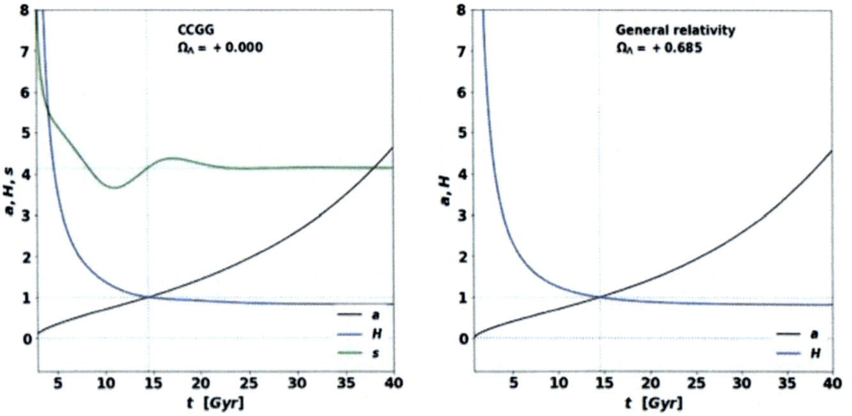

Fig. 8.2 The results for the CCGG theory, calculated with $\Omega_\Lambda = 0$, and the General Relativity, calculated with $\Omega_\Lambda = 0.685$, in comparison. Obviously, both $a(t)$ and $H(t)$ show an almost identical course

$a(t)$, which agrees excellently with the result for the standard ΛCDM Universe (right panel) calculated with the same parameters h, $\Omega_{\rm m}$, and $\Omega_{\rm r}$, but with $\Omega_{\rm GR,\Lambda} = 0.685$.

We therefore conclude: The presented results suggest that torsion is well suited to play the role of dark energy. A calculation with an increased $h = 0.740$, aligned to the local observations of the Cepheids in the Large Magellanic Cloud by Riess et al. [26], shows that the exponential growth of $a(t)$ is steeper just as expected for larger h, and the Hubble time is correspondingly shorter. But the qualitative course does not change [24].

Figure 8.3 illustrates that the density fraction $\hat{\Omega}_{\rm tor,1}$ dominates in the CCGG model for large times in a similar way as $\hat{\Omega}_\Lambda$ in the standard model. Even the relation $\hat{\Omega}_\Lambda/\hat{\Omega}_{\rm m} \sim 0.7/0.3$, as occurs in the standard model, is reproduced by $\hat{\Omega}_{\rm tor,1}/\hat{\Omega}_{\rm m}$ in the CCGG model with a vanishing cosmological constant. Thus the so-called "coincidence problem", i.e. the observation that the densities of dark matter and dark energy are currently of the same order of magnitude, is obsolete or requires at least a new interpretation.

A further indication, that the cosmic dynamics could be caused solely by torsion, is given by Fig. 8.4: As already mentioned in Sect. 8.2, one can interpret Eq. (8.35a) as equation of motion of a mass point of position $a(\tau)$ in the potential V. The two total potentials, $V_{tot}(a, s, \Omega_\Lambda = 0)$ of the CCGG model and $V_{tot}(a, \Omega_\Lambda = 0.685)$ of general relativity (black curves), take almost the same temporal course and therefore provide the same dynamics of $a(\tau)$.

Another interesting insight is found in Ref. [25]. Resolving Eq. (8.49) for $\Omega_{\rm s}$,

$$\Omega_{\rm s,\pm} = -\frac{5}{4}\frac{{H_{\rm GR,\infty}}^2}{\Omega_{\rm g}} + 1 \pm \left[\left(\frac{5}{4}\frac{{H_{\rm GR,\infty}}^2}{\Omega_{\rm g}}\right)^2 - \frac{H^2_{\rm GR,\infty}}{\Omega_{\rm g}}\right]^{1/2}, \tag{8.51}$$

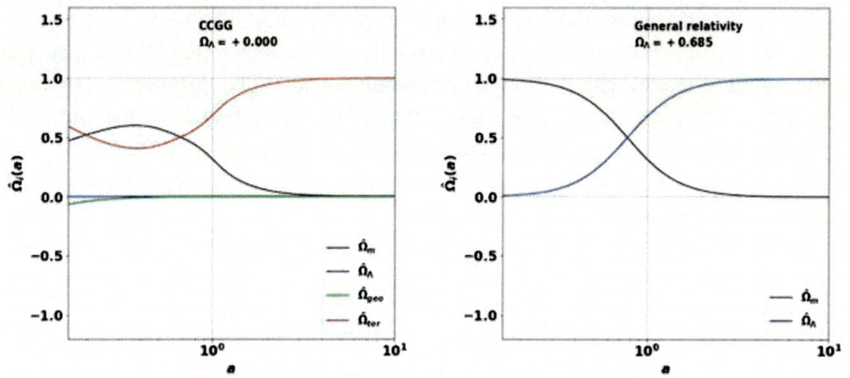

Fig. 8.3 Comparison of the fractional densities: $\hat{\Omega}_{\text{tor}}(a)$, red curve on the left, in close agreement with $\hat{\Omega}_\Lambda(a)$, blue curve on the right, for $a \geq 1$. According to our preposition there is no contribution of the cosmological constant (blue curve on the left). As expected, the fractional matter densities $\hat{\Omega}_{\text{m}}(a)$ fade away in an expanding Universe (black curves). The same is true for the fractional density $\hat{\Omega}_{\text{geo}}(a)$ of a deformed spacetime, which flattens out with increasing expansion (green curve on the left)

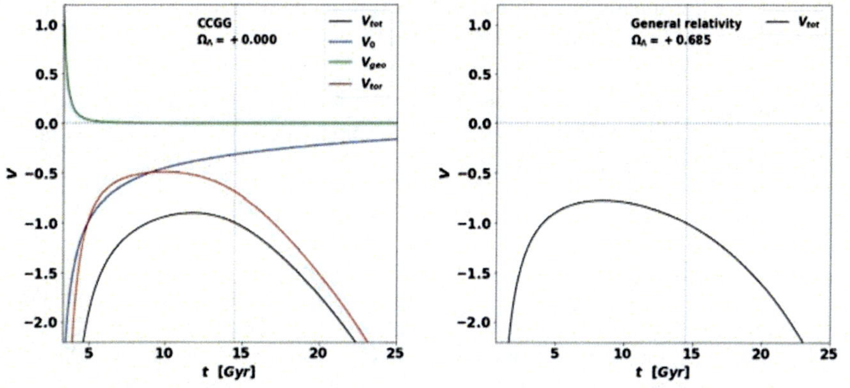

Fig. 8.4 Juxtaposition of the total potentials of CCGG and GR. The left panel additionally shows the individual contributions of V_0, V_{geo} and V_{tor}

shows that Ω_{s} has a real value only if

$$\Omega_{\text{g}} \leq \frac{25}{16} H^2_{\text{GR},\infty} \,. \tag{8.52}$$

Replacing now Ω_{g} by its definition (8.36b), and using Eq. (8.7) relating the deformation parameter g_1 and the vacuum energy of matter θ_{vac} for $\Omega_\Lambda = 0$ gives

$$\theta_{\text{vac}} \geq -\frac{75}{8} M_p^2 \, H_0^2 \, H^2_{\text{GR},\infty} \approx -10^{-120} \, M_p^4 \,. \tag{8.53}$$

This essentially means that the vacuum energy of matter must be positive.

We conclude the analysis by comparing the predictions of the CCGG theory and the standard model with the low-z cosmological data.[4] The distance measurements of the standard candles, i.e. the light curves of type Ia supernovae, are particularly suitable for this purpose, since their absolute magnitude M is known with very good accuracy: $M \approx 19.25 \pm 0.2$ [28]. Distance estimates are then derived from the luminosity distance [29]

$$d_L = \sqrt{\frac{L_{int}}{4\pi\mathcal{F}}} = (1+z)\int_0^z \frac{dz'}{H(z')} \tag{8.54}$$

where L_{int} is the intrinsic luminosity and $\mathcal{F}$ is the measured flux from the observed astronomical object. The distance modulus μ is related to the logarithm of the luminosity distance via

$$\mu(z) = m - M = 5\,\log_{10}\left(\frac{d_L}{Mpc}\right) + 25 \tag{8.55}$$

where m is the apparent magnitude of the explored light source. For our calculations we used the reduced Pantheon data set from [30] provided on github.[5]

As shown in Fig. 8.5, the agreement of the CCGG values with the observed values is slightly better than the GR prediction as z increases. However, more extensive analyses across the full cosmological data set are needed, using a sampling approach such as MCMC methods, to confirm this finding.

8.8 Summary

The starting point of this chapter was the question, what contribution the CCGG theory can make to clarifying currently unsolved cosmological problems. We summarize:

- The deformation of spacetime compensates the vacuum energy predicted by quantum theory, where the deformation in the CCGG model was described by a quadratic approach of the free Hamiltonian. This follows from the Zero-Energy Universe conjecture and, according to Eq. (8.7), holds regardless of the model on which the vacuum energy calculation is based. The deformation parameter g_1 introduced in this approach is dimensionless and should be interpreted as a natural constant, the value of which depends on the vacuum energy density of matter.
- In cosmology based on the CCGG model of gravity, the features of dark energy are derived as purely geometric properties of spacetime, without the use of any

[4] See also [27] where torsion has been included by substituting the cosmological constant by a scale depending function $\Lambda(a)$.

[5] https://github.com/dscolnic/Pantheon/blob/master/Binned_data/lcparam_DS17f.txt.

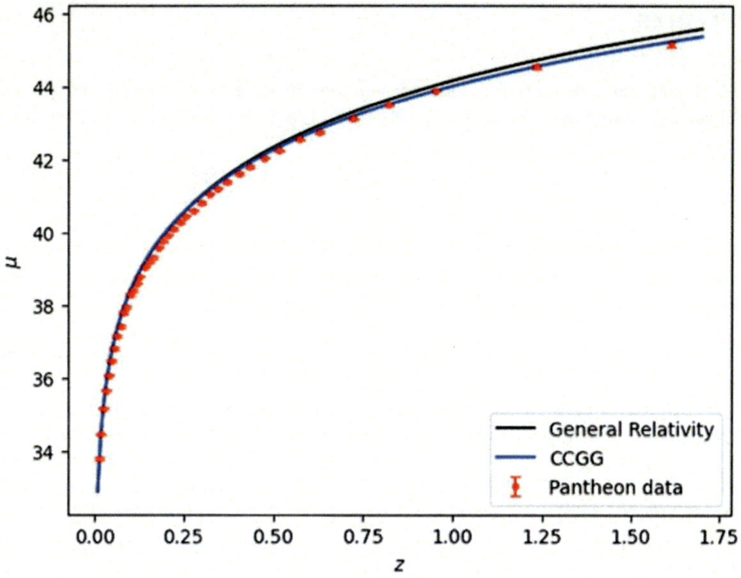

Fig. 8.5 μ values derived from the observed SNeIa light curves compared to the model prediction of CCGG and GR

auxiliary fields or of quantity like Λ. In addition to the fundamental principles of the CCGG theory, we assumed the Cosmological Principle, the conservation of the stress-energy tensor, and the scaling laws for the energy densities. This extended the Friedman-Lemaître equations of the ΛCDM model with torsion modelled by a time-like homogeneous axial vector $(s_0(t), 0, 0, 0)$. The numerical results for the scaling factor and the Hubble function show a very good agreement with the Concordance model, i.e. results of the standard model with a non-vanishing Λ. Moreover the observed data of the distance modulus were slightly better reproduced by the CCGG model.

- As mentioned above, the coincidence problem in its current formulation is not applicable in the context of CCGG cosmology.[6]

Although the above results do not represent a definite confirmation, they nevertheless offer a reasonable indication that torsion can resolve both mysteries, the magnitude and the coincidence problems, ascribed to the cosmological constant, a quantity which is the subject of much speculation in modern physics. Whether the two additional parameters Ω_g and Ω_s can combine the geometrical and torsional contributions to relieve the Hubble tension, perhaps as a model of Early Dark Energy (a review of such models can be found in [31]), is an open question for further research.

[6] To what extent it is justified at all to speak of a remarkable coincidence due to similar values of the energy densities of torsion and mass in the present, could be deepened in a discussion of anthropic principles, but is not pursued further here.

References

1. M. Tsamparlis, Methods for deriving solutions in generalized theories of gravitation: the Einstein-Cartan theory. Phys. Rev. D **24**, 1451 (1981). https://doi.org/10.1103/PhysRevD.24.1451
2. Y. Yoon, Conformally coupled induced gravity with gradient torsion. Phys. Rev. D **59**, 127501 (1999). https://doi.org/10.1103/PhysRevD.59.127501. https://link.aps.org/doi/10.1103/PhysRevD.59.127501
3. S.B. Medina, M. Nowakowski, D. Batic, D.: Einstein-Cartan cosmologies. Ann. Phys. **400** (2018). https://doi.org/10.1016/j.aop.2018.11.002
4. D. Kranas et al., Friedmann-like universes with torsion. European Phys. J. C **79**, 341 (2019). https://doi.org/10.1140/epjc/s10052-019-6822-4
5. G. Unger, N. Popławski, Big bounce and closed universe from spin and torsion. Astrophys. J. **870.2**, 78 (2019). https://doi.org/10.3847/1538-4357/aaf169
6. G.W. Milton, A possible explanation of dark matter and dark energy involving a vector torsion field, in *Universe 8.6*, p. 298 (2022). https://doi.org/10.3390/universe8060298. arXiv: 2003.11587 [gr-qc]
7. D. Benisty et al., Inflation from fermions with curvature-dependent mass. Phys. Rev. D D **100.4**, 043523 (2019). https://doi.org/10.1103/PhysRevD.100.043523. arXiv: 1905.03731 [gr-qc]
8. S. Bahamonde et al., Teleparallel gravity: from theory to cosmology. Rep. Prog. **86.2**, 026901 (2023). https://doi.org/10.1088/1361-6633/ac9cef
9. D. Iosifidis, L. Ravera, The cosmology of quadratic torsionful gravity. European Phys. J. C **81.8** (2021). https://doi.org/10.1140/epjc/s10052-021-09532-8
10. S. Weinberg, The cosmological constant problem. Rev. Mod. Phys. **61**, 1 (1989)
11. D. Vasak et al., On the cosmological constant in the deformed Einstein-Cartan gauge gravity in De Donder-Weyl Hamiltonian formulation. Astron. Nachr. (2022). https://doi.org/10.1002/asna.20220069,2022. arXiv: 2209.00501 [gr-qc]
12. F.J. Herranz, M. Santander, (Anti)de Sitter/Poincare symmetries and representations from Poincare/Galilei through a classical deformation approach. J. Phys. **41**, 015204 (2008). https://doi.org/10.1088/1751-8113/41/1/015204. arXiv: math-ph/0612059
13. D. Vasak et al., Covariant canonical gauge gravitation and cosmology. J. Phys. Conf. Ser. **1194**, 012108 (2019). arXiv:1812.00578
14. D. Vasak, J. Kirsch, J. Struckmeier, Dark energy and inflation invoked in Covariant Canonical Gauge Theory of Gravity (CCGG) by locally contorted space-time. Eur. Phys. J. Plus **135**, 404 (2020). https://doi.org/10.1140/epjp/s13360-020-00415-7
15. P. Minkowski, On the cosmological equations in the presence of a spatially homogeneous torsion field. Phys. Lett. **173**, 247 (1986)
16. S. Capozziello, G. Lambiase, C. Stornaiolo, Geometric classification of the torsion tensor in space-time. Annalen Phys. **10**, 713–727 (2001). https://doi.org/10.1002/1521-3889(200108)10:8<713::AID-ANDP713>3.0.CO;2-2. arXiv: gr-qc/0101038
17. S. Capozziello, C. Stornaiolo, Torsion tensor and its geometric interpretation. Ann. Fond. Broglie **32.2-3**, 196–214 (2007)
18. N.J. Popławski, Cosmology with torsion: An alternative to cosmic inflation. Phys. Lett. B **694**, 181–185 (2010). [Erratum: Phys. Lett. B **701**, 672–672 (2011)]
19. J. Prat et al., Vacuum energy density measured from cosmological data. JCAP **06.06**, 015 (2022)
20. Weinberg, S.: *Gravitation and Cosmology: Principles and Applications of the General Theory of New York* (Wiley 1972). ISBN: 978-81-265-1755-8
21. D.D. Reid et al., The picture of our universe: a view from modern cosmology (2002). arXiv:astro-ph/0209504
22. N. Aghanim et al., Planck 2018 results. VI. Cosmological parameters. Astron. Astrophys. **641**, A6 (2020). [Erratum: Astron. Astrophys. **652**, C4 (2021)]. https://doi.org/10.1051/0004-6361/201833910. arXiv: 1807.06209 [astro-ph.CO]

23. D. Vasak, J. Kirsch, J. Struckmeier, Rigorous derivation of dark energy and inflation as geometry effects in Covariant Canonical Gauge Gravity. Astron. Nachr. **342.1-2** (2021). ed. by C.A. Zen Vasconcellos et al., pp. 81–88. https://doi.org/10.1002/asna.202113885. arXiv: 2101.04379 [gr-qc]
24. J. Kirsch et al., Torsion driving cosmic expansion. European Phys. J. C (2023). https://doi.org/10.1140/epjc/s10052-023-11571-2. https://link.springer.com/content/pdf/10.1140/epjc/s10052-023-11571-2.pdf
25. A. van de Venn et al., Torsional dark energy in quadratic gauge gravity. European Phys. J. (2023). https://doi.org/10.1140/epjc/s10052-023-11397-y. arXiv: 2211.11868 [gr-qc]
26. A.G. Riess et al., *Large Magellanic Cloud Cepheid Standards Provide a 1% Foundation for the Determination of the Hubble Constant and Stronger Evidence for Physics beyond* Λ *CDM* (2019). arXiv:1903.07603 [astro-ph.CO]. https://doi.org/10.3847
27. D. Benisty et al., Low-Redshift constraints on covariant canonical gauge theory of gravity. Eur. Phys. J. C **81**, 125 (2021)
28. D. Richardson et al., Absolute-magnitude distributions of supernovae. Astronom. J. **147.5** (2014). https://doi.org/10.1088/0004-6256/147/5/118
29. E. Rebhan, Theoretische Physik: Relativitatstheorie und Kosmologie. ger. Springer eBook Collection. Description based upon print version of record (Spektrum Akademischer Verlag, Heidelberg, 2012). Online. Ressource (XVIII, 608 S, digital). ISBN: 978-3-8274-2315-3. https://doi.org/10.1007/978-3-8274-2315-3. http://dx.doi.org/10.1007/978-3-8274-2315-3
30. D.M. Scolnic et al., The complete light-curve sample of spectroscopically confirmed SNe Ia from Pan-STARRS1 and cosmological constraints from the combined pantheon sample. Astrophys. J. **859.2**, 101 (2018). https://doi.org/10.3847/1538-4357/aab9bb
31. V. Poulin, T.L. Smith, T. Karwal, The Ups and Downs of Early Dark Energy solutions to the Hubble tension: a review of models, hints and constraints circa 2023 (2023). arXiv: 2302.09032 [astro-ph.CO]. https://doi.org/10.48550/arXiv.2302.09032

Index

D. Vasak et al., *Covariant Canonical Gauge Gravity*, FIAS Interdisciplinary Science Series, https://doi.org/10.1007/978-3-031-43717-5